ND | Springer Series in **Nonlinear Dynamics**

W0107177

Springer Series in **Nonlinear Dynamics**

Series Editors: F. Calogero, B. Fuchssteiner, G. Rowlands, H. Segur, M. Wadati and V. E. Zakharov

V.E. Zakharov (Ed.)

What Is Integrability?

With Contributions by
F. Calogero   N. Ercolani   H. Flaschka
V.A. Marchenko   A.V. Mikhailov
A.C. Newell   E.I. Schulman   A.B. Shabat
E.D. Siggia   V.V. Sokolov   M. Tabor
A.P. Veselov   V.E. Zakharov

Springer-Verlag

Berlin Heidelberg New York London
Paris Tokyo Hong Kong Barcelona

Professor Dr. Vladimir E. Zakharov

Landau Institute for Theoretical Physics, USSR Academy of Sciences,
ul. Kosygina 2, SU-117334 Moscow, USSR

Series Editor:

Professor Francesco Calogero

Dipartimento di Fisica, Università di Roma "La Sapienza",
Piazzale Aldo Moro 2, I-00185 Roma, Italy

ISBN 978-3-642-88705-5   ISBN 978-3-642-88703-1 (eBook)
DOI 10.1007/978-3-642-88703-1

Library of Congress Cataloging-in-Publication Data. What is integrability? / V. E. Zakharov (ed.) ; with contributions by F. Calogero . . . [et al.]. p. cm.– (Springer series in nonlinear dynamics) Includes bibliographical references. ISBN 978-3-642-88705-5 1. Differential equations, Partial–Numerical solutions. 2. Hamiltonian systems. 3. Nonlinear theories. 4. Mathematical physics. I. Zakharov, Vladimir Evgen'evich. II. Calogero, F. III. Series. QC20.7.D5W43 1990 530.1'55353–dc20 90-38242

© Springer-Verlag Berlin Heidelberg 1991
Softcover reprint of the hardcover 1st edition 1991

2157/3140-543210 – Printed on acid-free paper

Foreword

The idea of devoting a complete book to this topic was born at one of the Workshops on Nonlinear and Turbulent Processes in Physics taking place regularly in Kiev. With the exception of E. D. Siggia and N. Ercolani, all authors of this volume were participants at the third of these workshops. All of them were acquainted with each other and with each other's work. Yet it seemed to be somewhat of a discovery that all of them were and are trying to understand the same problem – the problem of integrability of dynamical systems, primarily Hamiltonian ones with an infinite number of degrees of freedom. No doubt that they (or to be more exact, we) were led to this by the logical process of scientific evolution which often leads to independent, almost simultaneous discoveries.

Integrable, or, more accurately, exactly solvable equations are essential to theoretical and mathematical physics. One could say that they constitute the "mathematical nucleus" of theoretical physics whose goal is to describe real classical or quantum systems. For example, the kinetic gas theory may be considered to be a theory of a system which is trivially integrable: the system of classical noninteracting particles. One of the main tasks of quantum electrodynamics is the development of a theory of an integrable perturbed quantum system, namely, noninteracting electromagnetic and electron-positron fields. Another well-known example is that in solid-state physics where linear equations describe a system of free oscillators representing atoms connected to each other by linear elastic forces. On the other hand, nonlinear forces yield nonlinear equations for this system.

Nonlinear integrable systems were discovered as early as the 18th century. At that time only a few were known and with no real understanding of their characteristics and solutions. Now, however, it is correct to state that it is impossible to overestimate their importance in the development of all areas of science.

Among their applications is the integrable problem arising for the motion of a particle in a central field, associated with atomic and nuclear physics. The problem of a particle moving in the fields of two Coulomb centers is fundamental to celestial mechanics and molecular physics. Also in molecular and nuclear physics the integrability of the Euler problem for the motion of a heavy rigid body is used. The development of the theory of gyroscopes would have been impossible without the Lagrange solution of a symmetric top in a gravitational field. Only one of the classical nonlinear integrable systems, namely, the Kovalewsky top, has not yet found direct physical applications. But within mathematics this problem is of great importance and led to the discovery of a method later developed by Painlevé, which is one of the basic subjects of this volume.

During most of the 19th century mathematicians tried to find new nonlinear integrable systems, but mostly in vain. The works of A. Poincaré put an end to these efforts: he showed that among dynamical systems, integrable ones are the exception. Already a small perturbation usually prevents integrability. Poincaré's results, obtained in the 1880s, dramatically reduced interest in the search for new integrable systems. This remained the case for more than the first half of this century.

The situation changed dramatically in the last two decades. 1987 marked the 20th anniversary of the publication of the well-known paper by C. S. Gardner, J. M. Green, M. D. Kruskal, and R. M. Miura, introducing a new, very powerful method, the inverse scattering transform (IST). In this paper the IST was applied to the famous Korteveg–de Vries equation (KdV). Before long it was shown that the inverse scattering method is applicable to a number of important nonlinear partial differential equations which had been known for a long time. The first of these are the nonlinear Schrödinger and sine-Gordon equations. All these equations were recognized to be Hamiltonian with an infinite number of conservation laws. In the course of developing the IST it could be shown that all of them are, at least in the rapidly decreasing or periodic cases, completely integrable systems in the classical sense. This led to the foundation of "soliton factories" at the beginning of the seventies.

The three equations mentioned above were the first examples of nonlinear integrable field-theoretical systems with an infinite number of degrees of freedom. At present more than several dozen such equations are known and their number is rising continuously. Among them there are the particularly important Weil equations describing an axially symmetric stationary gravitational field and the duality equations of the Yang–Mills theory. The evolution of the theory of quantum nonlinear systems with a large number of degrees of freedom is intimately connected with that of the classical inverse scattering method.

The tremendous significance of such equations and of the associated concepts to physics and mathematics led to the need to comprehend the notion of integrability at a more precise level and to understand the role of integrable systems in mathematical physics. This book attempts to make a step in this direction.

The book begins with a paper by F. Calogero which is, loosely speaking, devoted to "the origin" of integrable systems. A multiscale expansion method is described for a rather wide class of essentially nonlinear partial differential equations. As a result, a number of models with a high degree of universality arises, based on equations that are, as a rule, integrable. Their integration may be accomplished in different ways: either by a simple change of variables (C-integrability) or via the IST technique (S-integrability). But the conclusion is the same, i.e., the integrability of a dynamical system is related to its universality. This conclusion is undoubtedly of heuristic value. It is worth emphasizing here that not only partial differential equations but also more general pseudo-differential ones could be used as a starting point for analysis. After applying the multiscale expansion method, the results, as derived for chosen models, are found to be true on a more general level. This can be shown easily if the original system possesses a Hamiltonian structure.

Several examples of this type may be found in the paper by V. E. Zakharov and E. I. Schulman which is primarily devoted to a quite different question: how can we determine whether a given system is integrable or not? This problem has recently become more and more urgent, and is therefore thoroughly addressed in this volume. There are essentially three approaches to solving it, all discussed here. They originate from classical work initiated in the previous century. The approach used in the paper by E. D. Siggia and N. Ercolani and in the contribution by H. Flaschka, A. C. Newell and M. Tabor is essentially based on the classic paper by S. Kovalewskaya discussing the integrability of a top in a gravitational field.

Kovalewskaya observed that the majority of known integrable systems is integrated in terms of elliptic and, consequently, meromorphic functions and thus cannot have any movable critical points. This particular condition of the nonexistence of movable critical points led subsequently to the integrable equation for the Kovalewsky top. Kovalewskaya's idea was pursued further by Painlevé. This method of verifying the integrability of equations through an analysis of the arrangement of critical points of their solutions in the complex plane is called the Painlevé test. In the contribution by Flaschka, Newell, and Tabor the Painlevé test is used on partial differential equations and is proved to be a powerful tool. It allows not only to verify the integrability of systems but also, in the case of a positive answer, it helps to find their Lax representation as a compatibility condition (imposed on an overdetermined linear system), symmetries, and Hirota's bilinear form.

One of the highlights of the third workshop in Kiev was the demonstration (by A. C. Newell) of the power of the Painlevé test as applied to the integrable system found by A. V. Mikhailov and A. B. Shabat. It is worth noting that in spite of all the advances of the Painlevé test there is no reliable assurance for systems not satisfying this test to be definitely nonintegrable. It should also be added that further research is required to provide an even more solid mathematical foundation for this quite useful and successful method.

The next paper in the volume is from A. V. Mikhailov, V. V. Sokolov, and A. B. Shabat. They develop a symmetry approach originating from the famous Sophus Lie. The question posed is under which conditions does a class of partial differential equations admit a nontrivial group of local symmetry transformations (depending on a finite number of derivatives). In the cases under consideration the authors succeed in constructing a complete classification of systems possessing symmetries. They also prove that when a few symmetries exist it follows that there are actually an infinite number of them. It should be noted that in this paper not only Hamiltonian but also dissipative systems are considered which cannot be integrable in the classical sense but may be C-integrable, i.e., they may be reduced to linear systems by changing variables.

The paper by V. E. Zakharov and E. I. Schulman is based on Poincaré's works. Rather than choosing some differential equations and transforming them to their Fourier representation where differential and pseudo-differential operators differ only in coefficient functions, a Hamiltonian translationally invariant system is taken as the starting point. The question posed is whether at least one additional invariant motion for this system exists. It is shown that the existence of such an integral implies rather important conclusions, discussed thoroughly in the paper. They are formulated as restrictions on the perturbation series in the vicinity of linearized (and trivially integrable) systems. In particular, the existence of an additional invariant of motion implies the existence of an infinite number of invariants. This result agrees with the paper by Mikhailov, Sokolov, and Shabat. An extremely important result of this report is to make clear that the existence of an infinite set of invariants of motion does not always mean integrability in Liouville's sense. The set of integrals may be incomplete. Effective criteria for identifying such cases are presented.

The contribution by A. P. Veselov is devoted to systems with discrete time and thus has significant applications in physics. In this paper the particular concept of integrability of systems of this type is defined. The contribution by V. A. Marchenko devoted to the solution of the Cauchy problem of the KdV equation (with nondecaying boundary conditions at infinity) lies to some degree outside the general scope of this volume. It has been incorporated here, however, because it seems to me that the inclusion of a classic paper of modern mathematical physics can only increase the value and beauty of any presentation of associated problems.

Moscow, August 1990   *V. E. Zakharov*

Contents

Contributors

Calogero, Francesco
  Dipartimento di Fisica, Universitá di Roma "La Sapienza",
  Istituto Nazionale di Fisica Nucleare, Sezione di Roma,
  I-00185 Roma, Italy

Ercolani, Nicholas
  Department of Mathematics, University of Arizona,
  Tucson, AZ 85721, USA

Flaschka, Hermann
  Department of Mathematics, University of Arizona,
  Tucson, AZ 85721, USA

Marchenko, Vladimir A.
  Institute for Low Temperature Physics and Engineering,
  Ukranian SSR Academy of Sciences, 47 Lenin Avenue,
  310164 Kharkov, USSR

Mikhailov, Aleksandr V.
  Landau Institute for Theoretical Physics, USSR Academy of Sciences,
  ul. Kosygina 2, 117334 Moscow, USSR

Newell, Alan C.
  Department of Mathematics, University of Arizona,
  Tucson, AZ 85721, USA

Schulman, Evgenii I.
  P.P. Shirshov Institute for Oceanology, USSR Academy of Sciences
  Krasikova Street 23, 117218 Moscow, USSR

Shabat, Aleksei B.
  Landau Institute for Theoretical Physics, USSR Academy of Sciences,
  ul. Kosygina 2, 117334 Moscow, USSR

Siggia, Eric D.
  Laboratory of Atomic & Solid State Physics, Clark Hall, Cornell University,
  Ithaca, NY 14853-2501, USA

Sokolov, Vladimir V.
  Institute of Mathematics, Tukaeva 50, Ufa, USSR

Tabor, Michael
Department of Applied Physics, Columbia University,
New York, NY 10012, USA

Veselov, Aleksandr P.
Department of Mathematics and Mechanics, Moscow State University,
119899 Moscow, USSR

Zakharov, Vladimir E.
Landau Institute for Theoretical Physics, USSR Academy of Sciences,
ul. Kosygina 2, 117334 Moscow, USSR

Why Are Certain Nonlinear PDEs Both Widely Applicable and Integrable?

F. Calogero

Summary

Certain "universal" nonlinear evolution PDEs can be obtained, by a limiting procedure involving rescalings and an asymptotic expansion, from very large classes of nonlinear evolution equations. Because this limiting procedure is the correct one to evince weakly nonlinear effects, these universal model equations show up in many applicative contexts. Because this limiting procedure generally preserves integrability, these universal model equations are likely to be integrable, since for this to happen it is sufficient that the very large class from which they are obtainable contain just one integrable equation. The relevance and usefulness of this approach, to understand the integrability of known equations, to test the integrability of new equations and to obtain novel integrable equations likely to be applicable, is tersely discussed. In this context, the heuristic distinction is mentioned among "C-integrable" and "S-integrable" nonlinear PDEs, namely, equations that are, linearizable by an appropriate *Change of variables*, and equations that are integrable via the *Spectral transform* technique; and several interesting C-integrable equations are reported.

Introduction

It is moot whether the question put by the title of this paper may be justifiably asked, and appropriately answered, in a scientific context. Maybe the only appropriate response to it is to reiterate the statement beautifully formulated, more than three centuries ago, by *Galileo*: "Questo grandissimo libro che continuamente ci sta aperto innanzi agli occhi (io dico l'universo) ... è scritto in lingua matematica" ["this great book that stands always open before our eyes (I mean the universe)... is written in mathematical language"] [1]. Yet in this paper, that is largely based on joint work with *Eckhaus* [2–4], we try and suggest a less metaphysical explanation. As it has been known for some time and as we show below, certain "universal" nonlinear evolution PDEs can be obtained, by a limiting procedure involving rescalings and an asymptotic expansion, from very large classes of nonlinear evolution equations; for instance, from the class of autonomous nonlinear evolution equations whose linear part is dispersive but otherwise arbitrary, and whose nonlinear part depends in an analytic but otherwise arbitrary manner on the dependent variable and its derivatives. Because this limiting procedure is the correct one to evince nonlinear effects, the universal model equations obtained in this manner (of which the nonlinear Schrödinger equation in $1+1$ dimensions is the prototype) show up in many disparate, applicative contexts: they are widely applicable. Because this limiting procedure generally preserves integrability, these universal model equations are likely to be integrable, since in order for this to happen it is sufficient that the very large class from which they are obtainable contain just one integrable equation; indeed, while the fact that an arbitrarily given equation turns out to be integrable may be seen as an exceptionable event, the fact that a very large class of equations contain at least one integrable equation may be considered normal, i.e., not exceptional; hence a universal model equation that is obtainable by a limiting procedure from (all!) the equations of a large class is likely to be integrable, provided the limiting procedure preserves integrability. And note that this argument may also be run backwards; if a universal model equation, obtainable via a limiting procedure that preserves integrability from all the equations of a large class, turns out *not* to be integrable, then *none* of the equations contained in the large class is integrable; hence this approach also yields *necessary conditions for integrability* of wide applicability, as we explain in more detail below [4].

As indicated in the following section in a specific context, the argument outlined above is based in part on solid results and in part on heuristic considerations. In particular, the unfolding of the limiting procedure relevant to describe the dynamics in the regime of weak nonlinearity, and the identification of the corresponding model equations, is based on a precise and reliable algorithm; these results are mathematically correct, although more work than has been done until now would be required to back them with rigorous estimates and to turn them into rigorous statements relating solutions of the original equations (belonging to the large class) to solutions of the model equations (obtained via the limiting procedure from the large class and being relevant to describe the behavior in the regime of weak nonlinearity). The assertion about the preservation of the property of integrability through the limiting procedure is instead based on plausible arguments, which appear quite convincing and are supported by many examples, but could not be characterized as rigorous theorems; indeed, they could not be formulated as such (let alone proved), as long as no precise definition of "integrability" is available for nonlinear evolution PDEs. Indeed, at this stage we must be satisfied with the *heuristic* notions of "C-integrability" and "S-integrability"; the former corresponds to the possibility of linearization via an appropriate *Change of variables* (whose precise nature is left vague at this stage; we generally have in mind an explicit redefinition of the dependent variable; although in some case the independent variables might also be transformed, see below); the latter denotes solvability via the *Spectral transform* technique (or, equivalently, the *inverse Scattering method*; see, for instance, [5,6]). As a rule of thumb, the property of C-integrability is more stringent than S-integrability; in some sense, C-integrability implies S-integrability, but not vice versa (however, if the definition of S-integrability is made precise by requiring the existence of an infinity of *local* conservation laws, then there are C-integrable equations, linearizable by a nonlocal change of dependent variable, that are not S-integrable; see below for an interesting example).

The approach outlined above and described in more detail below provides an answer, which we deem convincing if lacking in mathematical rigor, to the question posed in the title of this paper; thereby explaining "what had hitherto appeared to us a puzzling miracle, namely, the fact that certain nonlinear PDEs appear in many applications *and* are integrable" [2]. It provides moreover a powerful, if heuristic, methodology, to understand the integrability of known equations, to test the integrability of new equations, and to obtain novel integrable equations likely to be applicable.

Surely the ideas outlined above and described in more detail below are not quite new; for instance, the basic fact that, in $1 + 1$ dimensions, the nonlinear Schrödinger equation is generally the appropriate tool to describe any situation characterized by dispersion and weak nonlinearity has been known for decades. Perhaps the first to formulate this kind of result in general form has been *Taniuti* (see, for instance, the papers by him and by his collaborators and colleagues in [7]). More recently *Eckhaus* [8] has substantially clarified and streamlined the derivation of model equations appropriate to describe the asymptotic regime of weak nonlinearity, by showing how they can be obtained just by introducing via appropriate rescaling "slow" independent variables, together with an asymptotic expansion of the dependent variable (see below). Moreover, the synergism of this approach with the question of integrability has been already exploited by *Zakharov* and *Kuznetsov* [9] to relate several known integrable nonlinear PDEs; these authors also mentioned the possibility to use this approach in order to obtain novel integrable equations, but they did not push the methodology far enough to actually produce such results.

This paper is organized as follows. Section 1 contains an outline of the main ideas, in a specific context suitable to their presentation; it is based on joint work with *Eckhaus* (see, in particular, [3, 4]). Section 2 displays several model equations, in $1 + 1$ and $2 + 1$ dimensions, that have emerged from this kind of approach; it is based on joint work with *Eckhaus* [2, 3] and *Maccari* [10]. Section 3 reports several C-integrable equations, that are in some way related to some of the equations of Sect. 2. Section 4 contains some concluding remarks. Throughout the paper the emphasis is on the presentation of results, rather than on their derivation, for which the interested reader is referred in each case to the relevant literature.

1. The Main Ideas in an Illustrative Context

Consider the class of nonlinear evolution PDEs characterized by the following (rather general) structure:

$$Du(x,t) = F[u, u_x, u_t, u_{xx}, u_{xt}, u_{tt}, \ldots] . \tag{1.1}$$

Here $u(x,t)$ is the dependent variable; we restrict attention to real variables and real equations.

The left-hand side of (1.1) is the linear part of this equation; we assume the linear differential operator D to have the form

$$D = \frac{\partial^2}{\partial t^2} + \sum_{j=0}^{J} (-1)^j b_j \frac{\partial^{2j}}{\partial x^{2j}} . \tag{1.2}$$

The quantities b_j are real constants. We assume moreover that, at least for some value of the real constant k (on which our attention will be focused), the quantity

$$\omega^2(k) = \sum_{j=0}^{J} b_j k^{2j} \tag{1.3}$$

is *positive*, so that its square root, $\omega(k)$, is real (the determination of the sign of $\omega(k)$ is optional). We moreover assume that $\omega(k)$ is not linear in k, namely that the *group velocity*

$$v(k) = \frac{d\omega(k)}{dk} = \sum_{j=1}^{J} j b_j \frac{k^{2j-1}}{\omega(k)} \tag{1.4}$$

is not constant (k-independent), i.e., we exclude the case $b_j = b_1 \delta_{j1}$.

These conditions are sufficient to guarantee that the linear part of (1.1) will be *dispersive*, so that the equation

$$Du(x,t) = 0 \tag{1.5}$$

admits as a (real) solution the (real) traveling wave

$$u(x,t) = A \exp\{i[kx - \omega(k)t]\} + \text{c.c.} , \tag{1.6a}$$

$$u(x,t) = 2|A| \cos[kx - \omega(k)t + \alpha] , \tag{1.6b}$$

and, more generally, the "wave packet" solution,

$$u(x,t) = \int dk' A(k') \exp\{i[k'x - \omega(k')t]\} + \text{c.c.} . \tag{1.7}$$

Let us recall that the solution (1.7), if it is initially localized both in x-space and in k-space (namely, if both $u(x,0)$ is concentrated around a position x_0 and its Fourier transform $\hat{u}(k',0) = A(k')$ is concentrated around a value k), generally evolves, at least at large time, as a wave packet whose envelope travels with the constant *group velocity* $v(k)$, see (1.4), and gets slowly dispersed (i.e., delocalized; its peak amplitude decreases proportionally to $t^{-1/2}$).

The form (1.2) is not the only one that gives rise to dispersive waves; for instance another possibility is the "odd" operator

$$D = \frac{\partial}{\partial t} + \sum_{j=0}^{J} (-1)^j a_j \frac{\partial^{2j+1}}{\partial x^{2j+1}} , \tag{1.8}$$

with the a_j being real constants. We restrict our consideration in this section to the "even" operator (1.2), following closely the treatment of [3]. The case of a *linear* operator of type (1.8) has been treated in [2]; the results are analogous, albeit somewhat more complicated, than those discussed in this section, and they are included among those reported in the following section.

Note that the amplitude A of the dispersive wave solution (1.6) of the linear equation (1.5) is constant. The question we are going to discuss below is what happens if the time evolution is determined by the *nonlinear* equation (1.1) rather than the linear equation (1.5), under the assumption of "weak nonlinearity" (see below); in which case a solution such as (1.6) may still (approximately) hold, but with the amplitude A being a slowly varying function of space and time. Indeed, our interest will be focused just on the equations governing that kind of wave modulation. Such equations are of course relevant only on a "slow" time scale and over a large, hence "coarse-grained", space scale, whence their "universal" character, namely their structural independence from the specific features of the nonlinear evolution equation (1.1), be they the parameters that characterize the linear operator D (namely, the values of J and of the constants b_j; see (1.2)), or the detailed nature of the right-hand side of (1.1), that constitutes the nonlinear part of this equation and to whose description we now turn.

We assume that F is an analytic nonlinear real function of the dependent variable u and of its derivatives, namely we assume that it admits, for small ε, the expansion

$$F[\varepsilon u, \varepsilon u_x, \varepsilon u_t, \varepsilon u_{xx}, \ldots] = \sum_{m=2}^{M} \varepsilon^m F^{(m)}[u, u_x, u_t, u_{xx}, \ldots] + o(\varepsilon^M) , \tag{1.9}$$

for any $M \geq 2$ (actually, for the results given below it is sufficient that this expansion hold up to some small value of M, say $M = 3$ or $M = 5$, as the case may be; see below). Here the functions $F^{(m)}$ are homogeneous polynomials of degree m in u and its derivatives:

$$F^{(m)}[u, \ldots] = \sum_{\mu=0}^{m} (u_t)^\mu \sum_{j_1=0}^{\infty} \sum_{j_2=j_1}^{\infty} \cdots \sum_{j_{m-\mu}=j_{m-\mu-1}}^{\infty}$$
$$\times c_{j_1 j_2 \cdots j_{m-\mu}}^{(m)(\mu)} u^{(j_1)} u^{(j_2)} \ldots u^{(j_{m-\mu})} . \tag{1.10}$$

Here and below we use the synthetic notation

$$u^{(j)} \equiv \frac{\partial^j u}{\partial x^j} \, . \tag{1.11}$$

In writing the expression (1.10), we have assumed that no higher *time*-derivatives than the first appear in the rhs of (1.1). This assumption is introduced here merely to simplify the notation (which is already sufficiently complicated as it is; see below). Note that (1.10) identifies uniquely the *real* constants $c_{j_1 j_2 \cdots j_{m-\mu}}^{(m)(\mu)}$; of course, in any actual application, only a subset of these constants does not vanish. To elucidate this notation, let us display in explicit detail the structure of (1.10) for $m = 2$ and $m = 3$:

$$
\begin{aligned}
F^{(2)}[u, \ldots] = {} & c^{(2)(2)} u_t^2 \\
& + c_0^{(2)(1)} u_t u + c_1^{(2)(1)} u_t u_x + c_2^{(2)(1)} u_t u_{xx} + \ldots \\
& + c_{00}^{(2)(0)} u^2 + c_{01}^{(2)(0)} u u_x + c_{02}^{(2)(0)} u u_{xx} + \ldots \\
& + c_{11}^{(2)(0)} u_x^2 + c_{12}^{(2)(0)} u_x u_{xx} + \ldots \\
& + c_{22}^{(2)(0)} u_{xx}^2 + \ldots \, ,
\end{aligned}
\tag{1.12a}
$$

$$
\begin{aligned}
F^{(3)}[u, \ldots] = {} & c^{(3)(3)} u_t^3 \\
& + c_0^{(3)(2)} u_t^2 u + c_1^{(3)(2)} u_t^2 u_x + c_2^{(3)(2)} u_t^2 u_{xx} + \ldots \\
& + c_{00}^{(3)(1)} u_t u^2 + c_{01}^{(3)(1)} u_t u u_x + c_{02}^{(3)(1)} u_t u u_{xx} + \ldots \\
& + c_{11}^{(3)(1)} u_t u_x^2 + c_{12}^{(3)(1)} u_t u_x u_{xx} + \ldots \\
& + c_{000}^{(3)(0)} u^3 + c_{001}^{(3)(0)} u^2 u_x + c_{002}^{(3)(0)} u^2 u_{xx} + c_{003}^{(3)(0)} u^2 u_{xxx} + \ldots \\
& + c_{011}^{(3)(0)} u u_x^2 + c_{012}^{(3)(0)} u u_x u_{xx} + c_{013}^{(3)(0)} u u_x u_{xxx} + \ldots \\
& + c_{111}^{(3)(0)} u_x^3 + c_{112}^{(3)(0)} u_x^2 u_{xx} + c_{113}^{(3)(0)} u_x^2 u_{xxx} + \ldots \\
& + c_{122}^{(3)(0)} u_x u_{xx}^2 + c_{123}^{(3)(0)} u_x u_{xx} u_{xxx} + \ldots \\
& + c_{222}^{(3)(0)} u_{xx}^3 + c_{223}^{(3)(0)} u_{xx}^2 u_{xxx} + \ldots \, .
\end{aligned}
\tag{1.12b}
$$

For instance, the sine-Gordon equation,

$$u_{tt} - u_{xx} + c \sin(\gamma u) = 0 \, , \tag{1.13}$$

corresponds to (1.1, 9, 10) with

$$
b_0 = c\gamma \, , \quad b_1 = 1 \, , \quad c_{00 \ldots 00}^{(2m+1)(0)} = (-1)^{m-1} \frac{c\gamma^{2m+1}}{(2m+1)!} \, ,
\tag{1.14}
$$
$$m = 1, 2, 3, \ldots$$

and all other b and c constants vanishing.

Note that the sum in the right-hand side of (1.9) starts from $m = 2$; this corresponds to the separation of (1.1) into a *linear* left-hand side and a *nonlinear* right-hand side. To investigate the regime of *weak* nonlinearity, it is expedient to replace in (1.1) the dependent variable u by εu, and to treat ε as a small parameter, thereby replacing (1.1), via (1.9), by

$$Du = \sum_{m=2}^{M} \varepsilon^{m-1} F^{(m)}[u, \dots], \tag{1.15}$$

with M chosen sufficiently large to guarantee that all relevant contributions are taken into account (see below).

Of course, for $\varepsilon = 0$, this nonlinear evolution equation goes over into the linear equation (1.5), that admits as a solution the traveling wave (1.6). As already indicated above, our main interest is to investigate solutions of (1.1) that are close, for small ε, to this traveling wave. To this end it is convenient to introduce the formal Fourier/asymptotic expansion

$$u(x, t) = \sum_{n=-\infty}^{+\infty} \exp\{in[kx - \omega(k)t]\} \varepsilon^{\gamma_n} \psi_n(\xi, \tau) \tag{1.16}$$

into (1.15), and to determine the evolution of the coefficients $\psi_n(\xi, \tau)$ of the various Fourier modes, in particular the evolution of the amplitude

$$\psi_1(\xi, \tau) \equiv \psi(\xi, \tau) \tag{1.17}$$

describing the modulation of the dominant mode [that coincides of course with the traveling wave (1.6); compare (1.16) with (1.6), and note that $\gamma_1 = \gamma_{-1} = 0$, and $\gamma_n \geq 0$ for $n \neq 1, -1$; see below]. But let us first pause to justify and explain the ansatz (1.16).

First of all we record the conditions

$$\gamma_n = \gamma_{-n}, \quad \psi_n(\xi, \tau) = \psi^*_{-n}(\xi, \tau) \tag{1.18}$$

that correspond to the reality of $u(x, t)$.

Next, to motivate the introduction of this ansatz, we suggest to imagine for a moment to solve (1.15) by iteration, starting from the traveling wave (1.6). The iteration would clearly produce higher harmonics of the basic dispersive wave (1.6), due to the nonlinear character of the rhs of (1.15); as well as the zeroth order harmonic, due to the interference of the basic dispersive wave $\exp[i(kx - \omega t)]$ with its complex conjugate. These are precisely the terms that appear in the ansatz (1.16).

The exponents γ_n account for the fact that whenever nonlinear effects come into play, they carry an element of smallness, associated with positive powers of ε. It turns out (see below) that in most cases a consistent choice for these exponents is

$$\gamma_n = n - 1 \quad \text{for} \quad n = 1, 2, 3, \dots . \tag{1.19}$$

Note that for the moment we keep open the determination of the exponent

$$\gamma_0 \equiv r \tag{1.20}$$

whose value shall depend mainly on the structure of the linear operator D (see below).

Finally, some comments on the amplitudes $\psi_n(\xi, \tau)$ of the various modes. First of all let us emphasize that, in writing the expansion (1.16), we do not imply that these functions, $\psi_n(\xi, \tau)$, are independent of ε, but merely that they remain finite in the limit $\varepsilon \to 0$ (this will be accounted for by a proper choice of the exponents γ_n, see above). Second, and most importantly, we define the "slow" variables ξ and τ as

$$\xi = \varepsilon^p(x - vt) ,\tag{1.21}$$

$$\tau = \varepsilon^q t .\tag{1.22}$$

Here the *positive* parameters p and q set the scale, in space and time, over which the nonlinear effects become relevant. Note that, in principle, the values of p and q could be set arbitrarily, as one can always rescale the independent variables at one's whim. But the choice of too large values for p or q would yield equations that contain divergences in the $\varepsilon \to 0$ limit; while the choice of too small values for p or q would yield uninteresting results in the $\varepsilon \to 0$ limit, corresponding essentially to the linear equation (1.5), and signifying that such scales are not slow enough to allow for the nonlinear effects to build up and become relevant. Note moreover that the definition of ξ, see (1.21), implies looking at the system in a reference frame that moves with the group velocity, see (1.4), appropriate to the specific traveling wave that constitutes the basic approximation; again, one might make a different choice, but only the choice (1.21) leads to interesting results, for reasons analogous to those we have just mentioned (by following the carrier wave with its group velocity one can evince the effects produce by a weak nonlinearity; namely the effects which remain significant even in the $\varepsilon \to 0$ limit).

Note that the ansatz (1.16) with (1.17–20) implies the asymptotic relation

$$u(x, t) = \psi(\xi, \tau) \exp[i(kx - \omega\tau)] + \text{c.c.} + \varepsilon^r \psi_0(\xi, \tau)$$
$$+ \varepsilon\{\psi_2(\xi, \tau) \exp[2i(kx - \omega\tau)] + \text{c.c.}\} + O(\varepsilon^2) .\tag{1.23}$$

As already mentioned, our main interest is the determination of the evolution equation satisfied, in the $\varepsilon \to 0$ limit, by the function $\psi(\xi, \tau)$, accounting for the modulation of the amplitude of the dominant mode $\exp[i(kx - \omega t)]$ (the carrier wave), due to (weak) nonlinear effects.

Let us now outline the procedure to obtain this result; except for the choice of the three parameters r, p, and q, this computation is purely algorithmic, so that it could be performed by computer, using an algebraic manipulation program.

Firstly, let us insert the ansatz (1.16) in the left-hand side of (1.15) (namely, in the linear part of this equation). This is easy:

$$Du = \sum_{n=-\infty}^{+\infty} \exp[in(kx - \omega t)]\varepsilon^{\gamma_n} D_n \psi_n(\xi, \tau) ,\tag{1.24}$$

$$D_n = \left(in\omega - \varepsilon^p v \frac{\partial}{\partial\xi} + \varepsilon^q \frac{\partial}{\partial\tau}\right)^2 + \sum_{j=0}^{J}(-1)^j b_j \left(ink + \varepsilon^p \frac{\partial}{\partial\xi}\right)^{2j} ,\tag{1.25}$$

$$D_n = A_n + iw_n \varepsilon^p \frac{\partial}{\partial \xi} + B_n \varepsilon^{2p} \frac{\partial^2}{\partial \xi^2} - 2inw\varepsilon^q \frac{\partial}{\partial \tau}$$

$$+ \varepsilon^{2q} \frac{\partial^2}{\partial \tau^2} - 2v\varepsilon^{p+q} \frac{\partial}{\partial \xi \partial \tau} + O(\varepsilon^{2p}) + O(\varepsilon^{2q}), \tag{1.26}$$

$$A_0 = b_0 ; \quad A_1 = 0 ;$$

$$A_n = \sum_{j=0}^{J} b_j k^{2j}(n^{2j} - n^2) = \omega^2(nk) - n^2\omega^2(k), \quad n \geq 2, \tag{1.27}$$

$$B_n = v^2 - b_1 - \sum_{j=2}^{J} j(2j - 1)b_j(kn)^{2(j-1)}, \tag{1.28}$$

$$w_0 = w_1 = 0 ;$$

$$w_n = 2n \sum_{j=2}^{J} jb_j k^{2j-1}(1 - n^{2(j-1)}) = 2[n\omega(k)v(k) - \omega(nk)v(nk)]. \tag{1.29}$$

Note that, for the basic mode $n = 1$, both A_1 and w_1 vanish, hence the dominant contributions, see (1.26), are of order ε^{2p} (assuming B_1 does not vanish) and ε^q. Hence a preferred choice that shall often be the appropriate one sets

$$q = 2p. \tag{1.30}$$

With this choice we get:

$$D_0 = b_0 + B_0 \varepsilon^{2p} \frac{\partial^2}{\partial \xi^2} + O(\varepsilon^{2p}), \tag{1.31a}$$

$$D_1 = \varepsilon^{2p} \left(B_1 \frac{\partial^2}{\partial \xi^2} - 2i\omega \frac{\partial}{\partial \tau} \right) + O(\varepsilon^{2p}), \tag{1.31b}$$

$$D_n = A_n + iw_n \varepsilon^p \frac{\partial}{\partial \xi} + O(\varepsilon^p), \quad n \geq 2. \tag{1.31c}$$

Note that the above results correspond to the "rule of thumb" substitution of the space derivative $\partial/\partial x$ by $ink + \varepsilon^p \partial/\partial \xi$, and of the time derivative $\partial/\partial t$ by $-in\omega - v\varepsilon^p \partial/\partial \xi + \varepsilon^{2p} \partial/\partial \tau$. Hence, in the $\varepsilon \to 0$ limit, derivatives get replaced by constants.

Let us then turn to the discussion of the right-hand side of (1.15). Clearly, by inserting the ansatz (1.16) into the polynomials $F^{(m)}[u, \ldots]$ and by rearranging the Fourier series, one obtains the expressions

$$F^{(m)}[u, \ldots] = \sum_{n=-\infty}^{+\infty} \exp\{in[kx - \omega(k)t]\} f_n^{(m)}. \tag{1.32}$$

Hence, via (2.24), the evolution equation (1.15) yields the sequence of equations

$$\varepsilon^{\gamma_n} D_n \psi_n(\xi, \tau) = f_n, \quad n = 0, 1, 2, \ldots, \tag{1.33}$$

where of course,

$$f_n = \sum_{m=2}^{M} \varepsilon^{m-1} f_n^{(m)} \,. \tag{1.34}$$

Note that the evaluation of the expressions $f_n^{(m)}$ is purely algorithmic (although it may turn out to be rather cumbersome, especially for larger values of m and n; but in general only the quantities $f_n^{(m)}$ with small values of m and n play a role, see below). It is easily seen that the first few $f_n^{(m)}$ have the following explicit form [3];

$$f_0^{(2)} = g(1,-1)|\psi_1|^2 + \varepsilon^p \left[g(1,-1')\psi_1 \frac{\partial \psi_{-1}}{\partial \xi} + g(1',-1)\psi_{-1} \frac{\partial \psi_1}{\partial \xi} \right]$$
$$+ \varepsilon^{2r} g(0,0)\psi_0^2 + \varepsilon^2 g(2,-2)|\psi_2|^2 + \dots \,, \tag{1.35a}$$

$$f_0^{(3)} = 2\varepsilon \text{Re}\{g(1,1,-2)\psi_1^2 \psi_{-2}\} + \varepsilon^r g(1,0,-1)|\psi_1|^2 \psi_0$$
$$+ \varepsilon^{3r} g(0,0,0)\psi_0^3 + \varepsilon^{2+r} g(2,0,-2)|\psi_2|^2 \psi_0$$
$$+ 2\varepsilon^3 \text{Re}\{g(2,1,-3)\psi_1 \psi_2 \psi_{-3}\}$$
$$+ 2\varepsilon^{1+p} \text{Re}\left\{ g(1,1',-2)\psi_1 \psi_{-2} \frac{\partial \psi_1}{\partial \xi} + g(1,1,-2')\psi_1^2 \frac{\partial \psi_{-2}}{\partial \xi} \right\} + \dots \,, \tag{1.35b}$$

$$f_0^{(4)} = g(1,1,-1,-1)|\psi_1|^4 + \varepsilon^2 g(1,0,0,-1)|\psi_1|^2 \psi_0^2$$
$$+ \varepsilon^2 \{ g(2,1,-1,-2)|\psi_1|^2 |\psi_2|^2 + 2\text{Re}[g(1,1,1,-3)\psi_1^3 \psi_{-3}]\}$$
$$+ 2\varepsilon^{1+r} \text{Re}\{g(1,1,0,-2)\psi_1^2 \psi_0 \psi_{-2}\}$$
$$+ 2\varepsilon^p \text{Re}\left\{ g(1,1',-1,-1)|\psi_1|^2 \psi_{-1} \frac{\partial \psi_1}{\partial \xi} \right\} + \dots \,, \tag{1.35c}$$

$$f_0^{(5)} = O(\varepsilon, \varepsilon^r) \,; \tag{1.35d}$$

$$f_1^{(2)} = \varepsilon^r g(1,0)\psi_1 \psi_0 + \varepsilon g(2,-1)\psi_{-1}\psi_2 + \varepsilon^3 g(3,-2)\psi_{-2}\psi_3$$
$$+ \varepsilon^{p+r} \left[g(1,0')\psi_1 \frac{\partial \psi_0}{\partial \xi} + g(1',0)\psi_0 \frac{\partial \psi_1}{\partial \xi} \right]$$
$$+ \varepsilon^{p+1} \left[g(2',-1)\psi_{-1} \frac{\partial \psi_2}{\partial \xi} + g(2,-1')\psi_2 \frac{\partial \psi_{-1}}{\partial \xi} \right] + \dots \,, \tag{1.36a}$$

$$f_1^{(3)} = g(1,1,-1)|\psi_1|^2 \psi_1 + \varepsilon^{2r} g(1,0,0)\psi_1 \psi_0^2$$
$$+ \varepsilon^{1+r} g(2,0,-1)\psi_0 \psi_{-1}\psi_2 + \varepsilon^2 [g(2,1,-2)\psi_1 |\psi_2|^2$$
$$+ g(3,-1,-1)\psi_{-1}^2 \psi_3]$$
$$+ \varepsilon^p \left[g(1,1',-1)|\psi_1|^2 \frac{\partial \psi_1}{\partial \xi} + g(1,1,-1')\psi_1^2 \frac{\partial \psi_{-1}}{\partial \xi} \right] + \dots \,, \tag{1.36b}$$

$$f_1^{(4)} = \varepsilon [g(1,1,1,-2)\psi_1^3 \psi_{-2} + g(2,1,-1,-1)|\psi_1|^2 \psi_{-1}\psi_2]$$
$$+ \varepsilon^r g(1,1,0,-1)|\psi_1|^2 \psi_1 \psi_0 + \dots \,, \tag{1.36c}$$

$$f_1^{(5)} = g(1,1,1,-1,-1)|\psi_1|^4 \psi_1 + \dots \,; \tag{1.36d}$$

$$f_2^{(2)} = g(1,1)\psi_1^2 + \varepsilon^{1+r} g(2,0)\psi_0\psi_2 + \varepsilon^2 g(3,-1)\psi_{-1}\psi_3$$

$$+ \varepsilon^p g(1,1')\psi_1 \frac{\partial\psi_1}{\partial\xi} + \dots , \tag{1.37a}$$

$$f_2^{(3)} = \varepsilon g(2,1,-1)|\psi_1|^2\psi_2 + \varepsilon^r g(1,1,0)\psi_1^2\psi_0 + \dots , \tag{1.37b}$$

$$f_2^{(4)} = g(1,1,1,-1)|\psi_1|^2\psi_1^2 + \dots , \tag{1.37c}$$

$$f_2^{(5)} = O(\varepsilon,\varepsilon^r) ; \tag{1.37d}$$

$$f_3^{(2)} = \varepsilon g(2,1)\psi_1\psi_2 + \dots , \tag{1.38a}$$

$$f_3^{(3)} = g(1,1,1)\psi_1^3 + \dots , \tag{1.38b}$$

$$f_3^{(4)} = O(\varepsilon,\varepsilon^r) , \tag{1.38c}$$

$$f_3^{(5)} = O(1) . \tag{1.38d}$$

The terms omitted (denoted by the dots) are, in each formula, of lower order than those displayed. The coefficients g are constants; their determination in terms of the constants c that characterize the right-hand side of (1.15) (see (1.10)) is an algorithmic task that is described in some detail in the Appendix A of [3]. We report here the explicit expressions of some of these constants:

$$g(0,0) = c_{00}^{(2)(0)} , \tag{1.39a}$$

$$g(1,0) = -\omega c_0^{(2)(1)} + 2c_{00}^{(2)(0)} + \sum_{j=1}^{\infty}(ik)^j c_{0j}^{(2)(0)} , \tag{1.39b}$$

$$g(1,1) = -\omega^2 c^{(2)(2)} - i\omega \sum_{j=0}^{\infty}(ik)^j c_j^{(2)(1)}$$

$$+ \sum_{j_1=0}^{\infty}\sum_{j_2=j_1}^{\infty}(ik)^{j_1+j_2} c_{j_1 j_2}(2)(0) , \tag{1.39c}$$

$$g(1,-1) = 2\omega^2 c^{(2)(2)} - 2\omega \sum_{j=0}^{\infty}(-1)^j k^{2j+1} c_{2j+1}^{(2)(1)}$$

$$+ \sum_{j_1=0}^{\infty}\sum_{j_2=j_1}^{\infty}(ik)^{j_1+j_2} c_{j_1 j_2}^{(2)(0)}[(-1)^{j_1} + (-1)^{j_2}] , \tag{1.39d}$$

$$g(2,-1) = 4\omega^2 c^{(2)(2)} - i\omega \sum_{j=0}^{\infty}(ik)^j c_j^{(2)(1)}[-2^j + 2(-1)^j]$$

$$+ \sum_{j_1=0}^{\infty}\sum_{j_2=j_1}^{\infty}(ik)^{j_1+j_2} c_{j_1 j_2}^{(2)(0)}[2^{j_1}(-1)^{j_2} + 2^{j_1}(-1)^{j_2}] , \tag{1.39e}$$

$$g(1,1,-1) = -3i\omega^3 c^{(3)(3)}$$

$$+ \omega^2 \sum_{j=0}^{\infty} (ik)^j c_j^{(3)(2)} [2 - (-1)^j]$$

$$+ i\omega \sum_{j_1=0}^{\infty} \sum_{j_2=j_1}^{\infty} (ik)^{j_1+j_2} c_{j_1 j_2}^{(3)(1)} [1 - (-1)^{j_1} - (-1)^{j_2}]$$

$$+ \sum_{j_1=0}^{\infty} \sum_{j_2=j_1}^{\infty} \sum_{j_3=j_2}^{\infty} (ik)^{j_1+j_2+j+3} c_{j_1 j_2 j_3}^{(3)(0)} [(-1)^{j_1} + (-1)^{j_2} + (-1)^{j_3}] .$$

$$(1.39f)$$

Note that the values of these coefficients depend generally on k [except for $g(0,0)$; see (1.39a)], and that at least some of them are generally complex (for a more detailed discussion of the reality properties of these coefficients see Appendix A of [3]).

We are now in the position to proceed with the derivation of the equation satisfied, in the $\varepsilon \to 0$ limit, by $\psi(\xi, \tau)$ (see (1.17)).

For the reasons outlined above, we assume that the relation $q = 2p$ holds (see (1.30)). We can then use the formulae (1.31). Inserting them into (1.33), with $n = 0, 1, 2$, and using (1.34, 35a, b, 36a, b, 37a, b, 18–20), we get:

$$\varepsilon^r \left[b_0 + \varepsilon^{2p} B_0 \frac{\partial^2}{\partial \xi^2} + O(\varepsilon^{2p}) \right] \psi_0 = \varepsilon g(1, -1)|\psi|^2 + O(\varepsilon) , \tag{1.40}$$

$$\varepsilon^{2p}[B_1 \psi_{\xi\xi} - 2i\omega \psi_\tau] + O(\varepsilon^{2p})$$
$$= \varepsilon^{1+r} g(1,0)\psi\psi_0 + \varepsilon^2 g(2,-1)\psi^*\psi_2 + \varepsilon^2 g(1,1,-1)|\psi|^2\psi + O(\varepsilon^2, \varepsilon^{1+r}) , \tag{1.41}$$

$$\varepsilon \left[A_2 + i\varepsilon^p w_2 \frac{\partial}{\partial \xi} + O(\varepsilon^p) \right] \psi_2 = \varepsilon g(1,1)\psi^2 + O(\varepsilon) . \tag{1.42}$$

If $A_2 \neq 0$ and $g(1,1) \neq 0$, in the limit $\varepsilon \to 0$ the last of these equations yields

$$\psi_2 = \frac{g(1,1)}{A_2} \psi^2 , \tag{1.43}$$

since p is positive.

If $b_0 = A_0 \neq 0$ and $g(1,-1) \neq 0$, (1.40) suggests setting

$$r = 1 , \tag{1.44}$$

and in the $\varepsilon \to 0$ limit it yields

$$\psi_0 = \frac{g(1,-1)}{b_0} |\psi|^2 . \tag{1.45}$$

We may now insert these determinations of ψ_2, r, and ψ_0 into (1.41). It is then clear that the appropriate determination of p is

$$p = 1 \tag{1.46}$$

and in the $\varepsilon \to 0$ limit one gets for ψ the *nonlinear Schrödinger equation* (NLS)

$$-2i\omega\psi_\tau + B_1\psi_{\xi\xi} = \eta|\psi|^2\psi \tag{1.47}$$

with

$$\eta = \frac{g(1,-1)g(1,0)}{b_0} + \frac{g(2,-1))g(1,1)}{A_2} + g(1,1,-1) . \tag{1.48}$$

Note that the coefficient η is generally a function of the wave number k, since the coefficients g (see (1.39) and (1.3)) as well as A_2 (see (1.27)) depend on k. Also note that η need not be real. Indeed, while the coefficients b_0 and A_2 are real, the coefficients g may be complex. But there are subclasses of nonlinear evolution equations for which η is automatically real. For instance, the "even" class of evolution equations, whose right-hand side contains only terms having overall an even number of differentiations, namely the class such that the coefficients $c_{j_1 j_2 \ldots j_{m-\mu}}^{(m)(\mu)}$ vanish if the integer $(\mu + \sum_1^{m-\mu} j_s)$ is odd (see (1.1,9,10,15)), yields coefficients $g(1,0)$, $g(1,1)$, $g(2,-1)$ and $g(1,1,-1)$ that are all real (note that the coefficient $g(1,-1)$ in (1.39) is always real). Of course, this "even" class may also be identified by its invariance under the transformation $x \to -x$, $t \to -t$.

Note incidentally that if η is real and none of the three real constants ω, B_1, and η vanishes, then by appropriate "cosmetic" rescalings of the independent and dependent variables the NLS equation can be recast into the canonical form

$$i\psi + \psi_{\xi\xi} + s|\psi|^2\psi = 0 , \tag{1.49}$$

with

$$s = -\text{sign}\frac{\eta}{B_1} . \tag{1.50}$$

Let us emphasize the *universal character* of (1.47): This NLS equation determines, in the regime of weak nonlinearity, *for any equation of the class* (1.1), the (slow) modulation of the amplitude of the dominant mode for any solution of (1.1) "close" to a traveling wave solution of the linear part of (1.1). The universal character of this result is underscored by the vastness of the class (1.1) of nonlinear evolution equations. But in fact this result is even more general, since the NLS equation would also emerge from other classes of nonlinear evolution equations (including integrodifferential equations and finite-difference equations); roughly speaking, NLS emerges from all evolution equations characterized by a linear part which is "dispersive" and a nonlinear part which is "analytic". Indeed, this fact has been known for a long time, although the degree of precision with which this result has been formulated has improved over time (and there certainly remains room for additional improvement, as mentioned above).

It should also be noted, however, that the emergence of the NLS equation has required some conditions, in particular the validity of certain *inequalities*, such as $\omega \neq 0$, $B_1 \neq 0$, $b_0 \neq 0$ and $A_2 \neq 0$ (see above); and it is moreover clear that, whenever η vanishes, one is left with the linear Schrödinger equation rather than the nonlinear Schrödinger equation. Hence the cases when the derivation of the NLS equation breaks down are, in some sense, not generic; they emerge only if some quantity vanishes. Yet, as we shall see, their study is quite important; indeed the identification of their relevance has been a major recent development [2–4].

But for the moment let us still elaborate on the present finding.

The *universal character* of the NLS equation that has just been emphasized accounts, of course, for its *wide applicability*; this is a consequence of the fact that in many, disparate, applicative contexts, the governing equations are indeed characterized by a linear part that is dispersive and a nonlinear part that is analytic, and moreover, the situation of applicative interest corresponds to a regime of weak nonlinearity with a dispersive wave of given wave number k playing the dominant role; then the interest is naturally focused on the modulation of the amplitude of the carrier wave, due to the (weak) nonlinear effects which, as we have just seen, is generally governed by the NLS equation.

Of course, in these applicative contexts the validity of the approximation that yields the NLS equation (when ε is small but not quite vanishing) may also be at issue, as well as questions of stability; in this context a relevant role is generally played by those special values (if any) of the wave number k at which the approximation breaks down because one of the conditions under which the NLS equation has been derived ceases to hold (for instance, B_1, A_2, or η, vanish).

Our main interest in the present context is instead focused on the interplay of these findings with the issue of integrability. To illustrate this point, imagine taking, as a starting point of the analysis, an equation belonging to the class (1.1) that is itself "integrable" (without specifying for the moment the precise significance of this term). Then the model equation that is produced by the analysis (namely, in the present context, the NLS equation; but the argument is valid more generally, see below) must also be "integrable" (in the same sense), since its solution may be obtained, by taking an appropriate asymptotic limit, from solutions (appropriately chosen, so that they represent a weakly perturbed traveling wave) of the original equation.

This argument, of course, lacks precision, if not cogency. But in fact, in every specific case we have investigated (see below), it can be turned into a fully reliable proof of integrability, by analyzing the procedure that underlies the "integrability" of the original equation, by inserting into it the ansatz (1.16), by investigating the asymptotic limit $\varepsilon \to 0$, and by thereby evincing an analogous technique, applicable to the model equation, that demonstrates explicitly its integrability and indeed provides generally a constructive technique to solve it.

For instance, by taking as a starting point of the analysis the sine-Gordon equation (1.13), that is of course known to be S-integrable (i.e., integrable by the Spectral transform technique [5, 6]), one immediately obtains the NLS equation (1.47), with

$$\omega = (c\gamma + k^2)^{1/2} , \qquad (1.51a)$$

$$B_1 = -\left(\frac{1+k^2}{c\gamma}\right)^{-1} = \frac{-c\gamma}{\omega^2} \, , \tag{1.51b}$$

$$\eta = \tfrac{1}{2}c\gamma^3 \, , \tag{1.51c}$$

[since (1.14) implies $g(1,0) = g(1,1) = g(1,-1) = g(2,-1) = 0$, $g(1,1,-1) = c\gamma^3/2$]. Hence, it may be concluded that the NLS equation (1.47) (with ω, B_1 and η real constants) must itself be S-integrable; which is of course well-known to be the case [5, 6].

Note incidentally that this well-known conclusion applies to (1.49) with either determination of s; since (1.50, 51a, b), together with the assumed positivity of ω^2, imply $s = \text{sign}(\gamma^2)$; this may be positive or negative, depending whether γ is real or imaginary, these being precisely the two cases in which the sine-Gordon equation (1.13) is itself S-integrable; provided, of course, that c is correspondingly real or imaginary so as to preserve the reality of the equation (to be sure, if γ and c are imaginary, sine-Gordon is replaced by sinh-Gordon).

Moreover, it is actually possible, by following the technique mentioned above (first introduced by *Zakharov* and *Kuznetsov* [9]), to derive, from the Lax pair that underlines the S-integrability of the sine-Gordon equation (1.13), the Lax pair that demonstrates the S-integrability of the NLS equation (1.47) or (1.49).

Let us now proceed and discuss an apparent paradox that has sparked much recent research. Imagine applying this method taking as a starting point of the analysis an equation that is C-integrable, namely linearizable by an appropriate change of variables; for instance, the nonlinear evolution PDE [11]

$$u_t - u_{xxx} = 3\varepsilon^2(u_{xx}u^2 + 3u_x^2 u) + 3\varepsilon^4 u_x u^4 \tag{1.52}$$

that is linearized by the following change of dependent variable [12]

$$v(x,t) = u(x,t)\exp\left\{\varepsilon^2 \int^x dx'[u(x',t)]^2\right\} \, , \tag{1.53}$$

yielding

$$v_t - v_{xxx} = 0 \, . \tag{1.54}$$

(To be sure, the nonlinear evolution PDE (1.52) does not belong to the class (1.1, 2), since the differential operator that characterizes its linear part is odd rather than even; but it does belong to the general class of real nonlinear evolution PDEs whose linear part is dispersive and whose nonlinear part is analytic, to which the technique described above is generally applicable. The applicability of such an approach to a class of nonlinear evolution PDEs that includes (1.52) has been indeed demonstrated in [2]. Our motivation for focusing here on this example is its "historical" relevance [2, 11]. Several other instances of C-integrable nonlinear evolution PDEs are exhibited below.)

Since the NLS equation is obtained generally as final outcome of the application of this method, one should expect that it would also emerge from the

C-integrable equation (1.52). But then (for reasons that have been outlined above, and that will be further illustrated below) one should conclude that the NLS equation itself is C-integrable, rather than being merely S-integrable; namely, there should exist an appropriate change of dependent variable, perhaps analogous to (1.53) (and indeed obtainable by introducing in (1.53) the appropriate asymptotic *ansatzen* for $u(x, t)$ and $v(x, t)$; see (1.16)), that would linearize the NLS equation. Yet there are strong reasons to believe that this is not the case, namely, that the NLS equation, although S-integrable, is not C-integrable.

The way out of this paradox is provided by the following circumstance: by applying to (1.52) the limiting procedure "one seems to get the nonlinear Schrödinger equation, but with a vanishing numerical coefficient in front of the nonlinear term!" [11].

This finding provided an escape from the paradox: when the nonlinear term in the NLS equation is missing, this equation becomes just the (linear!) Schrödinger equation; hence the expectation that in this case it should be C-integrable (having being obtained from a C-integrable equation) is indeed realized (of course the linear Schrödinger equation, being itself linear, is *a fortiori* C-integrable!). But this finding also opened a new perspective, thanks to an observation of *Eckhaus* [2]. The idea goes as follows: the vanishing of the nonlinear contribution in the NLS equation indicates that, in this case, due to a cancellation, nonlinear effects are not sufficiently strong to play a significant role on the time and space scales corresponding to the *ansatz* (1.16–23, 30) with $p = 1$ [see (1.46)]; hence it is appropriate to try the same *ansatz* with a *larger* value of the parameter p, namely, by using more coarse-grained and slower space and time variables. And indeed, by making the choice $p = 2$, Eckhaus obtained from (1.52), instead of the NLS equation, the equation

$$i\psi_t + \psi_{xx} + [2(|\psi|^2)_x + |\psi|^4]\psi = 0 . \tag{1.55}$$

We have written this equation (now appropriately called "the Eckhaus equation" [13]) in the standard form that is obtained after an appropriate "cosmetic" rescaling of the (independent and dependent) variables.

On the basis of the reasoning outlined above, the Eckhaus equation should be C-integrable. In fact the following change of dependent variable [2, 13] linearizes it:

$$\varphi(x, t) = \psi(x, t) \exp\left\{ \int^x dx' |\psi(x', t)|^2 \right\} , \tag{1.56}$$

$$i\varphi_t + \varphi_{xx} = 0 . \tag{1.57}$$

Note the similarity of the transformations (1.53, 56). Indeed it is easy to *derive* (1.56) from (1.53) via the asymptotic *ansatz* (1.16–23, 30) with $p = 2$, and an analogous *ansatz* for $v(x, t)$ (this is left as an instructive exercise for the diligent reader).

As noted in [2], the Eckhaus equation (1.55), although originally obtained from the specific C-integrable equation (1.52), has, in analogy with the NLS equation, a certain *universal character*, since it describes the leading behavior under appropriate circumstances of a whole class of nonlinear evolution PDEs (see the following section). Moreover, it turns out to offer a most interesting "theoretical laboratory" to investigate the properties of nonlinear evolution equations [13–15]. Let us note incidentally, in this connection, that the Eckhaus equation is *not* S-integrable, if one makes precise the definition of S-integrability by identifying it with the existence of an infinity of *local* conservation laws; indeed, while the C-integrability of the Eckhaus equation implies of course the existence of an infinity of conservation laws, the nonlocal character of the linearizing transformation (1.56) (as well as its inverse [2, 13]) causes these conservation laws to have a nonlocal character. On the other hand, the solutions of the Eckhaus equation display a very rich solitonic phenomenology, including features that are characteristic of S-integrable equations, such as the *elastic* nature of solitonic collisions (which turns out to be, *in the context of the Cauchy problem*, a *generic* characteristic of soliton-bearing solutions of the Eckhaus equation) [13].

These findings suggest a more systematic investigation of the cases in which the derivation of the NLS equation, as outlined above, breaks down because some of the inequalities (say, $\eta \neq 0$, or $b_0 \neq 0$, or $A_2 \neq 0$) which were instrumental to obtain it, are invalid. This generally requires looking at larger, or smaller, space and time scales than those characterizing the emergence of the NLS equation. Such an analysis may be performed within two different, if interlocking, frameworks:

(i) It can be done in the context of the general class of nonlinear evolution PDEs (1.1) (or of other general classes to which this methodology is applicable), to obtain model equations that are presumably worth investigating, since they are likely to be *applicable and integrable*. This kind of study is particularly appealing inasmuch as it identifies a relatively small number of model equations as worthy of investigation, out of the boundless universe of nonlinear evolution equations. Let us emphasize in this connection that the vastness of the class of nonlinear evolution equations has always been a fundamental difficulty, making the study of nonlinearity unappealing due to the apparent need to chose between two alternatives: either undertake a general treatment that could not hope to gain much understanding about the specific behavior of solutions, or investigate special equations whose solutions could be analyzed in detail, but that might be (and were considered for a long time to be) just flukes. Hence, the importance of these findings, that indicate a possibility to identify, through a systematic analysis (which can be done easily, also in $n + 1$ dimensions with $n > 1$), a limited number of nonlinear evolution PDEs having a *universal character* likely to make them important both in applicative contexts and from a theoretical point of view.

(ii) The analysis can on the other hand be done in order to obtain integrable equations: S-integrable, or C-integrable, as the case may be. Here the procedure is, to start from an equation known to be integrable, and to apply to it the asymptotic expansion technique outlined above, perhaps focusing attention on special cases identified by the vanishing of key parameters, thereby obtained new equations, which of course are then, generally, also integrable (and indeed whose integrability can generally be explicitly demonstrated). An advantage of this approach is that it generally produces integrable model equations that are relatively simple and possess a universal character that makes them likely to be of applicative relevance. In this connection it should be recalled that there exist several techniques to manufacture integrable equations (both S-integrable and C-integrable); but the outcome of such an exercise may be unwieldy and therefore rather uninteresting. Hence, a technique of asymptotic expansion such as that described above that generally distills from a complicated equation a simpler, and in some sense universal, model equation, preserving in the process the integrability properties, is often fruitful. Let us note in this connection that the reason why the model equations yielded by these kinds of procedures are generally simpler and possess a certain universal character originates from the use of coarse-grained and slow variables that cause many detailed features of the original equation to be smoothed away (indeed the technique tends to replace derivatives with multiplication by constants; see the remark after (1.31)). On the other hand, the nontrivial nature of the procedure should also be emphasized, in particular, the fact that it involves a correct asymptotic limit (rather than just an arbitrary truncation); this accounts for the inheritance of integrability properties (from the parent equation, that serves as starting point of the analysis, to the model equation, that is generated by it). Note, however, that the nature of the model equation, although generally simpler than that of the parent equation, may differ from it quite substantially; it might even involve higher nonlinearities or it might have the form of a system of coupled PDEs even though the starting point of the treatment was a single equation for a single dependent variable (see examples below).

In the following section we briefly survey the results that have been obtained from both points of view, referring to the literature for more details on their derivation [2, 3, 10]; and in Sect. 3 we report several C-integrable equations. A more detailed study of the solutions of some of these equations will be published elsewhere [16], as well as an analysis of the results that are obtained by applying to them techniques of asymptotic expansion such as those described above [17].

But before ending this section we would also like to emphasize that the technique described above "may allow one to conclude that a given nonlinear evolution equation is not integrable, thereby providing a general and easily applicable *necessary condition for integrability*" [4]. This is a consequence of the inheritance of integrability properties through the limit procedure, implying that if the model equation yielded by this procedure is not integrable (and, more specifically, not S-integrable or not C-integrable, as the case may be), then the parent equation from which it has been obtained is itself not integrable (and, more specifically, not S-integrable or not C-integrable, as the case may be).

Let us illustrate this conclusion just on the specific example treated above, referring to [4] for its applicability in more general contexts. Consider a nonlinear evolution equation of the class (1.1) with (1.2) and (1.9, 10), and define the quantity $\eta(k)$ via (1.2, 27, 39, 48). (Note that these equations provide an *explicit* definition of $\eta(k)$ in terms of the parameters that characterize the original equation.) It may then be concluded that *the nonlinear evolution equation in question is not C-integrable if $\eta(k)$ does not vanish, and that it is not S-integrable if $\eta(k)$ is not real* [note that the conclusion of nonintegrability is obtained provided $\eta(k)$ is not vanishing, or not real, for some real value of k; provided that for that value of k, $\omega^2(k)$ is positive and $B_1(k)$ does not vanish; see (1.3, 28)]. This conclusion is of course implied by the notion that the NLS equation (1.47) (with B_1 and ω real and nonvanishing) is not C-integrable if $\eta \neq 0$, and is not S-integrable if $\eta \neq \eta^*$.

The ease of applicability of this kind of necessary conditions for integrability should be emphasized; and note that they only involve the constants that characterize the *quadratic* and *cubic* terms in the rhs of (1.9).

2. Survey of Model Equations

In this section we report, with a minimum of detail and no proofs, the main results of [2, 3, 10]. The general approach is of course that described in the preceding section (possibly with some minor adaptation); in particular, the value of the parameter p sets the space and time scale appropriate for each model equation, as implied by (1.21, 22, 30). All the quantities that appear as coefficients in the model equations written below are defined by explicit formulae, given in the cited papers. They depend generally on the wave number k and on the coefficients that define the quantities that provide the starting point of the analysis.

Consider first the class of evolution equations [2]

$$u_t + \sum_{j=0}^{J}(-1)^j a_j \frac{\partial^{2j+1}u}{\partial x^{2j+1}} = \sum_{m=2} \varepsilon^{m-1} F^{(m)} , \tag{2.1}$$

where $u \equiv u(x, t)$ is of course the (real) dependent variable, the coefficients a_j are real parameters (that characterize the linear part of the equation; note that the operator characterizing this part of the equation is odd, namely, it contains only derivatives of odd order), and $F^{(m)}$ is a (real) homogeneous polynomial of degree m in u and its derivatives.

Then the first relevant scale to display a nonlinear effect (in the $\varepsilon \to 0$ limit; according to the approach described, in a slightly different context, in the preceding section) corresponds to $p = 2/3$; the corresponding model equation reads as follows:

$$-i\psi_t + B_1\psi_{\xi\xi} = \alpha_1\varphi\psi , \tag{2.2a}$$

$$\varphi_\xi = \frac{\alpha_0}{w_0}|\psi|^2 . \tag{2.2b}$$

Here (and below) $\psi \equiv \psi(\xi, \tau)$, $\varphi \equiv \varphi(\xi, \tau)$.

Of course this model equation is applicable only if the (real) constant w_0 does not vanish. If instead $w_0 = 0$, a circumstance that may occur only for some special value of k, if at all [2], then the appropriate value of p is $\frac{1}{2}$, and the corresponding model equation reads

$$-i\psi_\tau + B_1\psi_{\xi\xi} = \alpha_1\varphi\psi , \tag{2.3a}$$

$$\varphi_\tau = \alpha_0|\psi|^2 + \alpha_2\varphi^2 . \tag{2.3b}$$

If, instead, $\alpha_0 = 0$ (so that (2.2b) implies that φ is ξ-independent, and (2.2a) becomes a linear equation for ψ), a more appropriate scale is characterized by $p = 1$, and the corresponding model equation reads

$$-i\psi_\tau + B_1\psi_{\xi\xi} = \left(\beta_1 + \frac{\alpha_5\alpha_6}{A_2}\right)|\psi|^2\psi + \alpha_1\varphi\psi , \tag{2.4a}$$

$$\varphi_\xi = 2\mathrm{Re}\left\{\frac{\alpha_3\psi\psi_\xi^*}{w_0}\right\}$$

$$= \left[\mathrm{Re}\left\{\frac{\alpha_3}{w_0}\right\}\right](|\psi|^2)_\xi + 2\left[\mathrm{Im}\left\{\frac{\alpha_3}{w_0}\right\}\right]\mathrm{Im}\{\psi^*\psi_\xi\} . \tag{2.4b}$$

Here we are of course assuming again that w_0 does not vanish. Note that, if either one of the two conditions $\alpha_1 = 0$ or $\alpha_3 = \alpha_3^*$ holds, then (2.4) becomes the NLS equation

$$-i\psi_\tau + B_1\psi_{\xi\xi} = \eta|\psi|^2\psi , \tag{2.5}$$

$$\eta = \beta_1 + \frac{\alpha_5\alpha_6}{A_2} + \frac{\alpha_1\alpha_3}{w_0} . \tag{2.6}$$

If instead both α_0 and w_0 vanish, $\alpha_0 = w_0 = 0$, then the appropriate scale is characterized by $p = \frac{2}{3}$ and the model equation reads

$$-i\psi_\tau + B_1\psi_{\xi\xi} = \alpha_1\varphi\psi , \tag{2.7a}$$

$$\varphi_\tau = \alpha_2\varphi^2 + 2\mathrm{Re}\{\alpha_3\psi\psi_\xi^*\} . \tag{2.7b}$$

If neither α_0 nor w_0 vanish but $\alpha_1 = 0$, then the appropriate choice for p is $p = 1$ and the model equation reads

$$-i\psi_\tau + B_1\psi_{\xi\xi} = \eta_1|\psi|^2\psi + \eta_2\varphi^2\psi + \alpha_9\varphi\psi_\xi , \tag{2.8a}$$

$$\varphi_\xi = \frac{\alpha_0}{w_0}|\psi|^2 + \frac{\alpha_2}{w_0}\varphi^2 . \tag{2.8b}$$

Note that this again reproduces the NLS equation if both η_2 and α_9 vanish.

Let us return to the (standard) case with $\alpha_0 = 0$, $\alpha_1 \neq 0$, $\alpha_3 = \alpha_3^*$, $w_0 \neq 0$, $A_2 \neq 0$ (a situation that is obtained "structurally" for a large subclass of (2.1) [2]), yielding the NLS equation (2.5) with (2.6); but assume now that $\eta = 0$, due to a cancellation (which might be "accidental" and occur only for some special value of k, or it might be "structural", hold for all values of k and, possibly, reflect the C-integrable nature of the original equation). Then the appropriate scale corresponds to the assignment $p = 4/3$, and the corresponding model equation reads

$$-i\psi_\tau + \psi_{\xi\xi} = \alpha_1\varphi\psi , \tag{2.9a}$$

$$\varphi_\xi = \theta|\psi|^4 . \tag{2.9b}$$

If, moreover, $\theta = 0$ (a situation that again prevails, as it were "structurally", for a large subclass of (2.1) [2]), then the appropriate scale is characterized by $p = 2$, and the model equation reads

$$-i\psi_\tau B_1\psi_{\xi\xi} = \alpha_1\varphi\psi + \varrho_0|\psi|^4\psi + \varrho_1(|\psi|^2)_\xi\psi + \varrho_2|\psi|^2\psi_\xi , \tag{2.10a}$$

$$\varphi_\xi = [\sigma_0|\psi|^4 + \sigma_1(|\psi|^2)_\xi + 2\sigma_2\mathrm{Im}\{\psi\psi_\xi^*\}]_\xi$$
$$+ \sigma_3|\psi|^2\varphi + \sigma_4|\psi|^6 + 2\sigma_5|\psi|^2\mathrm{Im}\{\psi\psi_\xi^*\} + \sigma_6|\psi_\xi|^2 . \tag{2.10b}$$

Moreover, if $\alpha_1 = 0$, or if the quantities σ_j satisfy the three conditions $\sigma_4 + \sigma_0\sigma_3 + \sigma_1\sigma_3/2 = \sigma_5 + \sigma_2\sigma_3 = \sigma_6 = 0$ (again, these conditions hold "structurally" for large classes of nonlinear PDEs of type (2.1)), then this equation may be recast in the simpler form

$$-i\psi_\tau + B_1\psi_{\xi\xi} = H_0|\psi|^4\psi + H_1(|\psi|^2)_\xi\psi + H_2|\psi|^2\psi_\xi . \tag{2.11}$$

Whenever the coefficient H_0 is real and does not vanish, i.e., $H_0 = H_0^* \neq 0$ and the real coefficient B_1 does not vanish, this model equation can be cast, via a "cosmetic" rescaling of dependent and independent variables, in the canonical form

$$i\psi_\tau + \psi_{\xi\xi} + s|\psi|^4\psi + iL_1(|\psi|^2)_\xi\psi + iL_2|\psi|^2\psi_\xi = 0 , \tag{2.12}$$

with

$$s = -\mathrm{sign}(B_1 H_0) , \quad L_j = i\frac{H_j}{B_1}\left|\frac{B_1}{H_0}\right|^{1/2} , \quad j = 1, 2 . \tag{2.13}$$

Let us note that (2.12) is S-integrable if L_1 and L_2 (are real and) satisfy the relation [2]

$$L_1 = L_1^* , \quad L_2 = L_2^* , \quad L_1(L_1 - L_2) = 4s . \tag{2.14}$$

Equation (2.12) is C-integrable if $s = +$, L_2 vanishes and L_1 is a complex number whose modulus equals two [13]:

$$s = + , \quad |L_1| = 2 , \quad L_2 = 0 . \tag{2.15}$$

Indeed, it appears that these conditions are not only *sufficient* for integrability, but also *necessary* [18, 19]. Note that the Eckhaus equation (1.55) is the special case of (2.12) with $s = +$, $L_1 = -2i$ and $L_2 = 0$; hence it satisfies the conditions (2.15) (and in fact it is C-integrable), but it does not satisfy the conditions required for S-integrability [18]. Indeed, in this case L_1 is not real, and moreover, the third condition (2.14) is not satisfied (although only due to a sign).

Let us proceed next to report the results relevant for the class of evolution equations with "even" linear part, already introduced in the preceding section, see (1.1) with (1.2,9) [3]. Some results have already been reported there, in particular, the standard result that is obtained with the scale characterized by $p = 1$ and yields the NLS equation (1.47). Let us recall that this result is obtained provided the quantity η in (1.48) exists (for which it is clearly required that neither b_0 nor A_2 vanish, $b_0 \neq 0$, $A_2 \neq 0$), and it is relevant provided η does not vanish.

If instead η does vanish, then the appropriate scale is characterized by $p = 2$, and the model equation that is obtained is essentially (2.11) [3].

If instead η does not exist because $A_2 = 0$ [see (1.27)] while $b_0 \neq 0$, then the relevant scale is characterized by $p = \frac{2}{3}$, and the corresponding model equation reads

$$-i\psi_\tau + B_1\psi_{\xi\xi} = \alpha\varphi\psi^* , \tag{2.16a}$$

$$i\varphi_\xi = \beta\psi^2 . \tag{2.16b}$$

This case may be characterized as being caused by a "resonance" affecting the first harmonic of the dominant mode [this corresponds to the condition $A_2 = 0$, see (1.26)]. If instead, with $b_0 \neq 0$ and $A_2 \neq 0$ [so that η does exist, see (1.48)], a resonance occurs in the *second* harmonic [namely, $A_3 = 0$; see (1.26, 27)], then the "standard" result is unaffected: namely, the appropriate scale is characterized by $p = 1$ and the relevant model equations is the NLS equation (1.47). If however, with $b_0 \neq 0$, $A_2 \neq 0$, and $A_3 \neq 0$, the quantity η vanishes, then the appropriate scale is characterized by $p = 4/3$ (rather than $p = 2$, which is instead appropriate if $\eta = 0$ but $A_3 \neq 0$), and the corresponding model equation reads

$$-i\psi_\tau + B\psi_{\xi\xi} = \alpha\varphi(\psi^*)^2 , \tag{2.17a}$$

$$i\varphi_\xi = \beta\psi^3 . \tag{2.17b}$$

All these cases are obtained provided that $b_0 \neq 0$ [see (1.2)]; note that this condition is necessary and sufficient to exclude that the frequencies $\omega(k)$ yielded by the linearized dispersion relation, see (1.3), vanish at $k = 0$, $\omega(0) \neq 0$. If instead $b_0 = 0$, implying $\omega(0) = 0$, then the first scale of interest is characterized by $p = \frac{1}{2}$, and the corresponding model equation reads

$$-i\psi_\tau + B\psi_{\xi\xi} = \alpha\varphi\psi , \tag{2.18a}$$

$$\varphi_{\xi\xi} = \beta_1|\psi|^2 + \beta_2\varphi^2 . \tag{2.18b}$$

If, in addition, the two quantities β_1 and β_2 vanish (a circumstance that occurs "structurally" for a large subclass of (1.1) [3]), then the appropriate scale is characterized by $p = \frac{2}{3}$, and the model equation coincides (up to trivial rescalings) with (2.2).

In the next stage, which becomes relevant if in addition one of the coefficients of the nonlinear terms on the right-hand sides of (2.2) vanishes, and which is then characterized by $p = 1$, the NLS equation would once more reappear. That this is the case has been shown in [3], rather than by a direct computation, by taking an altogether different approach: namely, by studying solutions of the original equation that are close to long traveling waves (rather than being close to a dispersive wave). The nonlinear evolution equations (1.1) (with (1.2) and $b_0 = 0$) may then be treated as a perturbed wave equation, by using a limit procedure appropriate to this situation. This leads, in many cases, just to the class of nonlinear evolution equations with *odd* linear part discussed above (in this section), and in [2]. Hence the NLS equation reappears merely via the application of the results reported above, and, moreover, an important link is established between the two classes of nonlinear evolution equations having *odd* linear part [see (2.1)] and having *even* linear part [see (1.1, 2)] *with* $b_0 = 0$, which both feature a linearized dispersion relation such that the frequency $\omega(k)$ vanishes at $k = 0$.

The results surveyed so far have originated from the first of the two points of view mentioned in the preceding section: namely, the identification of the "universal" model equations that are produced by the limit procedure under different circumstances, and that are therefore likely to become relevant in some applicative contexts corresponding to *weakly* nonlinear circumstances. Clearly a more systematic study than that reported here is still missing; it should, of course, also deal with nonlinear PDEs in more than $1 + 1$ dimensions (the extension necessary to treat such cases is straightforward; see below). Let us, however, note that if one is interested in applying the limiting procedure to a specific equation, it may well be more convenient to rederive the appropriate results by going through the asymptotic expansion than to use the general formulas valid for a whole class of nonlinear PDEs; for instance, it is easier to derive the Eckhaus equation (1.55) from (1.52) by working directly through the procedure of asymptotic expansion than by specializing to the special case of (1.52) the explicit, but cumbersome, formulas given in [2] for the entire class of evolution equations (2.1) [which clearly contains (1.52) as a special case].

For the reasons indicated in the preceding sections, the model equations identified in this manner (of which a partial list has just been reported) are also likely to be integrable. This provides, of course, an additional motivation to single them out as worthy of study. Note that each of these model equations should be seen as the limiting case of an entire class of nonlinear PDEs. This suggests an obvious methodology to demonstrate their integrability: to find just one integrable equation within the class of their "parent" equations. Moreover, this also implies that if they turn out *not* to be integrable (S-integrable or C-integrable, as the case may be), then the same negative conclusion applies to the entire class of their parent equations (see the discussion at the end of the preceding section).

Let us now proceed and report (from [2] and [10]) results that fit the second of the points of view mentioned in the preceding section; namely, to take as a starting point of the analysis an equation known to be integrable, and to derive from it, via the procedure of asymptotic expansion, another equation, generally "of nonlinear Schrödinger type", that is, therefore, also integrable; indeed, whose integrability can be generally demonstrated explicitly by applying the technique of asymptotic expansion not only to the evolution equation itself, but also to the mathematical structure that underlies its integrability (be it a Lax pair and a linear *Spectral* problem, in the case of S-integrability; or a linearizing *Change* of variable, in the case of C-integrability). Let us again recall in this connection the pioneering work of *Zakharov* and *Kuznetsov* [9], who indeed mentioned the possibility to discover, in this manner, novel integrable equations (although, in their paper, they focused mainly on the relations among nonlinear evolution PDEs already known to be integrable). Note that, in order to achieve such a goal, it is generally essential to push the approach beyond its more standard application; namely, to focus on the special cases characterized by the vanishing of some key parameter, as exemplified by the instances discussed in this section and by (some of) the examples reported below (in this section).

The nonlinear evolution equation

$$u_t - a_1 u_{xxx} + a_2 u_{xxxxx} = - 6a_1 \varepsilon u u_x$$
$$+ 10a_2(\varepsilon u u_{xxx} + 2\varepsilon u_x u_{xx} - 3\varepsilon^2 u^2 u_x) \qquad (2.19)$$

is S-integrable [5]; it clearly belongs to the class (2.1). From this equation, by the standard technique of asymptotic expansion described above (with $p = 1$ and $q = 2p = 2$; see (1.21, 22) and (1.30)) [2], one obtains the NLS equation

$$-i\psi_\tau + B_1 \psi_{\xi\xi} = \eta |\psi|^2 \psi , \qquad (2.20)$$

with

$$B_1 = -k(3a_1 + 10a_2 k^2) , \qquad (2.21a)$$

$$\eta = \frac{2B_1}{k^2} . \qquad (2.21b)$$

But suppose now that the values of the three (*a priori* arbitrary) constants a_1, a_2, and k^2 are so tuned as to produce a vanishing result for B_1 and hence η as well, see (2.21b); this of course requires that a_1 and a_2 have opposite signs, see (2.21a); or rather, let us assume that the values of the three constants a_1, a_2 and k imply [see (2.21a)]

$$B_1 = \varepsilon \nu + o(\varepsilon) , \qquad (2.22)$$

(so that B_1 vanishes in the $\varepsilon \to 0$ limit). It is then possible to apply the procedure with the same assignment for the space scale but a slower assignment for the time scale, namely, $p = 1$ and $q = 3$ [see (1.21, 22)], obtaining thereby the model equation [2]

$$-i\psi_\tau + \nu(\psi_{\xi\xi} - \lambda^2 |\psi|^2 \psi) - i\mu(\psi_{\xi\xi\xi} - 3\lambda^2 |\psi|^2 \psi_\xi) = 0 , \qquad (2.23)$$

where we have set for notational convenience

$$\mu = 2a_1 , \tag{2.24a}$$

$$\lambda^2 = \frac{20a_2}{3a_1} . \tag{2.24b}$$

Note that the three real parameters μ, ν, and λ my be arbitrarily modified by rescaling ψ, ξ, and τ.

Since the model equation (2.23) has been obtained from the S-integrable equation (2.19), one expects it to be S-integrable as well. Indeed, this equation coincides (up to trivial transformations) with the *Hirota* equation which is known to be S-integrable (see, e. g., (1.8–20) of [5]).

As noted in [2], this result is actually more general since it may be obtained by starting from the more general S-integrable equation [5]

$$u_t + \sum_{j=1}^{J} (-1)^j a_j L^j u_x = 0 , \tag{2.25}$$

where L is the integrodifferential operator

$$L = \frac{\partial^2}{\partial x^2} - 2\varepsilon \left[2u + u_x \left(\frac{\partial}{\partial x} \right)^{-1} \right] . \tag{2.26}$$

[Note that, for $J = 2$, (2.25) reduces to (2.19).] The standard procedure (with $p = 1$, $q = 2p = 2$) would then reproduce the NLS equation (2.20) with

$$B_1 = - \sum_{j=1}^{J} j(2j + 1)a_j k^{2j-1} , \tag{2.27a}$$

$$\eta = \frac{2B_1}{k^2} . \tag{3.27b}$$

Note that in this case η is again proportional to B_1; indeed, (2.27b) is identical to (2.21b)! If the J real constants a_j and the parameter k are so fine-tuned as to imply

$$B_1 = \varepsilon\nu + o(\varepsilon) \tag{2.28}$$

so that as ε vanishes B_1 and η vanish as well, then the procedure can be applied with the assignment $p = 1$, $q = 3$ and, after a rather tedious computation, one again obtains the Hirota equation (2.23) with

$$\mu = -\frac{1}{3} \sum_{j=1}^{J} (2j + 1)(2j - 1)a_j k_0^{2(j-1)} , \tag{2.29a}$$

$$\lambda^2 = \frac{2}{k_0^2} . \tag{2.29b}$$

Here k_0 is the value of the parameter k that implies the vanishing of B_1, see (2.27a), in the limit $\varepsilon = 0$. For instance, if $J = 2$, then $k_0^2 = -3a_1/10a_2$, so that (2.29b) reproduces (2.24b).

Note that the procedure might be iterated again. If the parameters ν and μ were themselves of order ε, one might apply the procedure with $p = 1$ and $q = 4$, thereby obtaining another Schrödinger-like equation that might be considered the next one of a hierarchy of S-integrable equations of which the NLS equation (2.20) and the Hirota equation (2.23) are, respectively, the first and second members; just as the class (2.25) constitutes a hierarchy of S-integrable equations of which the Korteweg–de Vries equation $u_t - a_1(u_{xxx} - 6\varepsilon u_x u) = 0$ is the first member ($J = 1$) and (2.19) the second ($J = 2$).

Let us proceed and consider another example [2]. The evolution equation

$$u_t - u_{xxx} = \varepsilon A u_x u + \varepsilon^2 B u_x u^2 \tag{2.30}$$

is S-integrable for arbitrary values of A and B. Note that for $B = 0$ it reduces to the Korteweg–de Vries equation, and for $A = 0$ to the modified Korteweg–de Vries equation. By applying to it the standard procedure [with $p = 1$ and $q = 2p = 2$; note that (2.30) belongs to the class (2.1)], one obtains the NLS equation (2.20) with

$$B_1 = -3k , \tag{2.31a}$$

$$\eta = Bk - \frac{A^2}{6k} . \tag{2.31b}$$

But suppose now that the constants A, B, and k are so adjusted as to yield a vanishing value for η, or rather a value of order ε^2, say

$$k = k_0 + O(\varepsilon^2) , \tag{2.32a}$$

$$k_0 = A(6B)^{-1/2} , \tag{2.32b}$$

so that

$$\eta = \varepsilon^2 \alpha . \tag{2.33}$$

It is then justified to apply the procedure with $p = 2$ and $q = 2p = 4$, thereby obtaining the model equation

$$i\psi_\tau + \beta \psi_{\xi\xi} + \alpha |\psi|^2 \psi + \frac{1}{2} \frac{\lambda^4}{\beta} |\psi|^4 \psi + i\lambda^2 \psi^2 \psi_\xi^* = 0 , \tag{2.34}$$

where we have set for notational convenience

$$\beta = 3k_0 , \tag{2.35a}$$

$$\lambda^2 = \frac{(A/k_0)^2}{3} . \tag{2.35b}$$

This equation can be transformed into the canonical form (2.12) with $s = +$, $L_1 = L_2 = \sqrt{2}$. (This is achieved by performing the following transformation from the unprimed to the primed variables,

$$\psi'(\xi',\tau') = 2^{-1/4}\frac{\lambda}{\beta}\psi(\xi,\tau)\exp\left(i\alpha\frac{(\lambda^2\xi + \alpha\beta\tau)}{\lambda^4}\right), \qquad (2.36a)$$

$$\xi' = \beta\left(\xi + \frac{2\alpha\beta\tau}{\lambda^2}\right), \qquad \tau' = \beta^3\tau, \qquad (2.36b)$$

and then dropping all primes for notational simplicity.) Note that these values of s, L_1, and L_2 satisfy the condition (2.14), thereby confirming the S-integrability of (2.34). This is as expected, since this equation has been obtained from the S-integrable equation (2.30).

Let us now proceed and report some examples from [10].

The Boussinesq equation

$$u_{tt} - u_{xx} - u_{xxxx} + 3(u^2)_{xx} = 0 \qquad (2.37)$$

is S-integrable [5]. This equation belongs to the class (1.1), but with $b_0 = 0$ [see (1.2)]. The standard treatment with $p = 1$ and $q = 2p = 2$ and $r = 1$ [see (1.16–22)] yields the standard nonlinear Schrödinger equation with the coefficient of the nonlinear term proportional to $(2k^2 - 3)/(4k^2 - 3)$. Hence, for the special values $k = \pm(3/4)^{1/2}$, a different assignment is required, which turns out to be $p = 2/3$, $q = 4/3$, $r = 1/3$. After an appropriate "cosmetic" rescaling of dependent and independent variables this leads to the model equation

$$i\psi_\tau + \psi_{\xi\xi} = \varphi\psi, \qquad (2.38a)$$

$$\varphi_\tau = \pm(|\psi|^2)_\xi. \qquad (2.38b)$$

As expected, this equation is S-integrable and of applicative interest [20].

Next, let us consider the "standard", "cylindrical" resp. "spherical" KdV equations, namely,

$$u_t - u_{xxx} + cu/t = 6u_x u, \qquad (2.39)$$

with $c = 0$, $c = 1/2$ resp. $c = 1$. Let us recall that the cylindrical KdV ($c = 1/2$) is S-integrable (as well, of course, as the ordinary KdV with $c = 0$).

This equation is not autonomous; nevertheless the same procedure as described above can be easily applied, yielding (after an appropriate "cosmetic" rescaling) the "standard", "cylindrical" resp. "spherical" NLS equations, namely

$$i\psi_\tau + \psi_{\xi\xi} + ic\psi/\tau = |\psi|^2\psi, \qquad (2.40)$$

with $c = 0$, $c = 1/2$ resp. $c = 1$. Hence one should expect (2.40) to be S-integrable for $c = 1/2$ ("cylindrical" case), as well as, of course, for $c = 0$; and this is indeed the case [21].

Next we apply the standard procedure (with obvious modifications) to the *matrix* KdV equation,

$$U_t + U_{xxx} = 3(UU_x + U_xU) = 3(U^2)_x ,$$ (2.41)

and thereby obtain (after an appropriate cosmetic rescaling) the *matrix* NLS equation

$$i\Psi_\tau + \Psi_{\xi\xi} = \Psi\Psi^*\Psi .$$ (2.42)

Here $U(x,t)$ and $\Psi(\xi,\tau)$ are square matrices of (arbitrary) order N; Ψ^* is the matrix complex conjugate to Ψ, so that $(\Psi^*)_{jk} = (\Psi_{jk})^*$. Of course both (2.41, 42) are S-integrable. Note that the matrices Ψ and Ψ^* need not commute. Several special cases of (2.41, 42) may be obtained by matrix reduction, including [from (2.42)] the standard, scalar, NLS equation with an optional sign in front of the nonlinear term.

Let us report now some results in $2 + 1$ dimensions [10].

The Kadomtsev–Petviashvili (KP) equation [22],

$$(u_t + u_{xxx})_x + su_{yy} = 3(u^2)_{xx} , \quad s = \pm ,$$ (2.43)

is known to be integrable [5, 6]. By applying to it the technique of asymptotic expansion, taking as starting point a dispersive wave characterized by a wave momentum $k = (k_1, k_2)$ and a corresponding frequency $\omega = -k_1^3 + sk_2^2/k_1$ [see the linear part of (2.43)], one obtains [10] (after an appropriate "cosmetic" rescaling) the model equation

$$i\psi_\tau + L_1\psi + \chi\psi = 0 ,$$ (2.44a)

$$L_2\chi = 2L_1|\psi|^2 ,$$ (2.44b)

where

$$L_1 = \frac{1}{4}(1 - s\lambda^2)\frac{\partial^2}{\partial\xi^2} + s\lambda\frac{\partial^2}{\partial\xi\,\partial\eta} - s\frac{\partial^2}{\partial\eta^2} ,$$ (2.44c)

$$L_2 = \frac{1}{4}(1 + s\lambda^2)\frac{\partial^2}{\partial\xi^2} + s\lambda\frac{\partial^2}{\partial\xi\,\partial\eta} - s\frac{\partial^2}{\partial\eta^2} ,$$ (2.44d)

and of course $\psi \equiv \psi(\xi,\eta,\tau)$, $\chi \equiv \chi(\xi,\eta,\tau)$ and $\lambda^2 = k_2^2/(3k_1^4)$ [10]. For $k_2 = \lambda = 0$ the treatment and result coincide with those of [9], and (2.44) yields the *Davey–Stewartson* (DS) equation [23], which is known to be S-integrable [5, 6, 24]. The nonlinear evolution equation (2.44) with arbitrary real λ is also S-integrable. The corresponding Lax pair has been obtained from that of the KP equation (2.43) [10]. But the additional generality of the treatment and the results of [10] relative to those of [9] is more apparent than substantial, since the rotation $\xi' = \xi + \lambda\eta/2$, $\eta' = \eta$ eliminates the λ from (2.44), that is, it reduces (2.44) to the standard DS equation.

Consider next the S-integrable equation in $2 + 1$ dimensions [25]:

$$u_t + u_{xxx} = 3(uv)_x , \tag{2.45a}$$

$$v_y = u_x . \tag{2.45b}$$

From this there easily obtains a model equation in $2+1$ dimensions, which is written below in the most natural form, obtained after a "cosmetic" rescaling [10]:

$$i\psi_\tau + \psi_{\xi\xi} = \varphi\psi , \tag{2.46a}$$

$$\varphi_\eta = 2s(|\psi|^2)_\xi , \tag{2.46b}$$

where s is a sign, $s^2 = 1$. The S-integrability and applicability of this equation were already known [20]. An explicit proof of S-integrability is actually provided at the end of this section.

Consider next the following S-integrable equation in $(2+1)$ dimensions [25]:

$$u_t + u_{xxx} + 3u_{xyy} = 6uu_x + 3[u_x(v^{(-)} - v^{(+)}) \\ + u_y(v^{(-)} + v^{(+)}) + 2u(v^{(-)} + v^{(+)})] , \tag{2.47a}$$

$$v_x^{(\pm)} \pm v_y^{(\pm)} = u_y . \tag{2.47b}$$

From this there obtains [10]:

$$i\psi_\tau + L_1\psi + \chi\psi = 0 , \tag{2.48a}$$

$$L_2\chi = -2sL_1|\psi|^2 , \tag{2.48b}$$

with

$$L_1 = \frac{\partial^2}{\partial\xi^2} + \frac{\partial^2}{\partial\eta^2} + 2\lambda\frac{\partial^2}{\partial\xi\,\partial\eta} , \tag{2.48c}$$

$$L_2 = -\frac{\partial^2}{\partial\xi^2} + \frac{\partial^2}{\partial\eta^2} , \tag{2.48d}$$

where λ is a parameter that can be chosen freely. In [10], the Lax pair for this equation is explicitly obtained from that of (2.47), thereby demonstrating the S-integrability of (2.48). This finding, however, is hardly new, since by a suitable rotation of the spatial variables it can be shown that this equation belongs to the following *Zakharov class*, whose integrability has been demonstrated in [20] (see also [26]):

$$i\psi_\tau + L_1\psi + \chi\psi = 0 , \tag{2.49a}$$

$$L_2\chi = \pm 2L_1|\psi|^2 , \tag{2.49b}$$

$$L_1 = a^2\frac{\partial^2}{\partial\eta^2} + 2(b-a)a\frac{\partial^2}{\partial\xi\,\partial\eta} + (b^2 - 2ba - a^2)\frac{\partial^2}{\partial\xi^2} , \tag{2.49c}$$

$$L_2 = a^2\frac{\partial^2}{\partial\eta^2} + 2(2b+1)a\frac{\partial^2}{\partial\xi\,\partial\eta} + b(b+1)\frac{\partial^2}{\partial\xi^2} , \tag{2.49d}$$

where b and a are arbitrary *real* constants and α is an arbitrary *complex* constant. The standard DS equation corresponds to $a = b = 1/2$ and $\alpha = 1$ or $\alpha = i$.

Next consider the S-integrable "sinh-Gordon" equation [27]

$$\{\exp(-q)[\exp(q)(q_{xt} + \sinh q)]_x\}_x$$
$$= \pm\{[\exp(-q)u_y]_{yt} - (1/2)[\exp(-2q)u_y^2]_{xt}\} , \qquad (2.50a)$$

$$\exp(q) = 1 + u_x . \qquad (2.50b)$$

From this there obtains [10] the model equation

$$i\psi_\tau + L_1\psi + \chi\psi = 0 , \qquad (2.51a)$$

$$L_2\chi = \pm2L_1|\psi|^2 , \qquad (2.51b)$$

with

$$L_1 = a\frac{\partial^2}{\partial\xi^2} + b\frac{\partial^2}{\partial\xi\,\partial\eta} + c\frac{\partial^2}{\partial\eta^2} , \qquad (2.51c)$$

$$L_2 = \frac{\partial^2}{\partial\xi^2} + B\frac{\partial^2}{\partial\xi\,\partial\eta} + C\frac{\partial^2}{\partial\eta^2} , \qquad (2.51d)$$

where a, b, c, B, and C are given by simple formulas in terms of a single free real parameter [10]. The S-integrability of (2.51) is explicitly demonstrated in [10] by exhibiting the corresponding Lax pair, obtained from that of (2.50) [27]; but this finding is not a novelty, since (2.51) can be reduced, by an appropriate rotation of the space variables, once again to the Zakharov class (2.49).

Next consider the following equation in $2 + 1$ dimensions of integrated KdV type:

$$u_t - u_{xxx} - u_{yyy} = 3(wu)_x , \qquad (2.52a)$$

$$w_{xy} = u_{xx} + u_{yy} . \qquad (2.52b)$$

From this one gets (after an appropriate "cosmetic" rescaling)

$$i\psi_\tau + L_1\psi + \varphi\psi = 0 , \qquad (2.53a)$$

$$L_2\varphi = 2L_1|\psi|^2 , \qquad (2.53b)$$

$$L_1 = s_1\frac{\partial^2}{\partial\xi^2} + s_2\frac{\partial^2}{\partial\eta^2} , \qquad (2.53c)$$

$$L_2 = 2s_1s_2\frac{\partial^2}{\partial\xi\,\partial\eta} , \qquad (2.53d)$$

where s_1 and s_2 are arbitrary signs, i.e., $s_1^2 = s_2^2 = 1$. Both (2.52, 53) are known to be S-integrable [28].

Finally, consider the *matrix* KP equation

$$U_t + U_{xxx} - 3W_y = 3(U^2)_x + 3i[U, W] , \qquad (2.54a)$$

$$U_y = W_x \ . \tag{2.54b}$$

Here $U \equiv U(x, y, t)$ and $W \equiv W(x, y, t)$ are square matrices of order N. Here and below we use the notation $[A, B] \equiv AB - BA$, $\{A, B\} \equiv AB + BA$. The following *matrix* DS equation is then obtained:

$$i\Psi_\tau + L_1\Psi = [\Psi, \Omega] + \lambda[\Phi, \Psi] + \{\Phi, \Psi\} - \{\Psi^2, \Psi^*\} \ , \tag{2.55a}$$

$$\Phi_\eta = \Omega_\xi \ , \tag{2.55b}$$

$$L_2\Phi - \Omega_\eta = \{\Psi, \Psi^*\}_x + [\Psi, \Psi_\eta] + [\Psi^*, \Psi_\eta] - \lambda\{[\Psi^*, \Psi_\xi] + [\Psi, \Psi_\xi^*]\} \ , \tag{2.55c}$$

with

$$L_1 = (1 + \lambda^2)\frac{\partial^2}{\partial\xi^2} - 2\lambda\frac{\partial^2}{\partial\xi\,\partial\eta} + \frac{\partial^2}{\partial\eta^2} \ , \tag{2.55d}$$

$$L_2 = (1 - \lambda^2)\frac{\partial}{\partial\xi} + 2\lambda\frac{\partial}{\partial\eta} \ , \tag{2.55e}$$

λ being a *real* constant that can be chosen freely.

Let us end this section by mentioning a possible generalization of the basic approach [10]. If the original equation that serves as starting point for the analysis has a linear part that is not too simple, it may well happen that two or more Fourier modes have the *same* group velocity. In such a case it may be of interest to focus on a solution that in the linear limit ($\varepsilon = 0$) is a superposition of these modes. Then the weak nonlinear effects induce a modulation of the amplitudes of these modes, that accounts also for the possibility that they interact with each other.

We limit our discussion of this possibility to two examples [10], one of which leads to no new results, while the other yields a nontrivial (S-integrable) generalization of (2.46).

The first example, in $1+1$ dimensions, takes as starting point the S-integrable equation (2.19); but it leads (contrary to what is implied in [10]) merely to two uncoupled NLS equations.

The second example takes as a starting point the S-integrable equation in $2 + 1$ dimensions (2.45), and yields the following generalization of (2.46):

$$i\psi_\tau^{(1)} + \psi_{\xi\xi}^{(1)} = \varphi\psi^{(1)} \ , \tag{2.56a}$$

$$i\psi_\tau^{(2)} + \psi_{\xi\xi}^{(2)} = \varphi\psi^{(2)} \ , \tag{2.56b}$$

$$\varphi_\eta = 2(s_1|\psi^{(1)}|^2 + s_2|\psi^{(2)}|^2)_\xi \ , \tag{2.56c}$$

where the parameters s_j are signs (namely, $s_j^2 = 1$; of course additional constants may be reintroduced by trivial rescalings). The S-integrability of this model equation is demonstrated by exhibiting the corresponding "Lax–Manakov triad", namely, the following three matrix differential operators L, A, and B such that the compatibility condition of the two linear equations for the (column) vector $\Phi(\xi, \eta, \tau)$,

$$L\Phi = 0 , \tag{2.57a}$$

$$\Phi_\tau + A\Phi = 0 , \tag{2.57b}$$

namely

$$L_\tau = LA - BL , \tag{2.58}$$

correspond to (2.56):

$$L = \begin{pmatrix} i\frac{\partial}{\partial\eta} & -s_1\psi^{(1)} & -s_2\psi^{(2)} \\ -s_1\psi^{(1)^*} & -is_1\frac{\partial}{\partial\xi} & 0 \\ -s_2\psi^{(2)^*} & 0 & -is_2\frac{\partial}{\partial\xi} \end{pmatrix} , \tag{2.59a}$$

$$A = \begin{pmatrix} -i\frac{\partial^2}{\partial\xi^2} + i\varphi & 0 & 0 \\ \psi^{(1)^*}\frac{\partial}{\partial\xi} - \psi_\xi^{(1)^*} & 0 & 0 \\ \psi^{(2)^*}\frac{\partial}{\partial\xi} - \psi_\xi^{(2)^*} & 0 & 0 \end{pmatrix} , \tag{2.59b}$$

$$B = \begin{pmatrix} -i\frac{\partial^2}{\partial\xi^2} + i\varphi & \psi^{(1)}\frac{\partial}{\partial\xi} + 2\psi_\xi^{(1)} & \psi^{(2)}\frac{\partial}{\partial\xi} + 2\psi_\xi^{(2)} \\ 0 & 0 & 0 \\ 0 & 0 & 0 \end{pmatrix} . \tag{2.59c}$$

Except for a minor generalization, that is, the introduction of the sign s_1 and some minor notational changes (including the correction of a misprint; the first term in the right-hand side of eq. (2.6) of Ref. [10] should read $-iD_1\,\partial^2/\partial\xi^2$ rather than $iD_1\,\partial^2/\partial\xi^2$), these results coincide with those of [10].

An obvious generalization of these findings is the model equation

$$i\psi_\tau^{(n)} + \psi_{\xi\xi}^{(n)} = \varphi\psi^{(n)} , \quad n = 1, 2, \ldots, N , \tag{2.60a}$$

$$\varphi_\eta = 2\left(\sum_{n=1}^{N} s_n|\psi^{(n)}|^2\right)_\xi , \tag{2.60b}$$

whose S-integrability is explicitly demonstrated by noticing that it may be put in the "Lax–Manakov" matrix form (2.58c) with the square matrices L, A, and B, of order $N + 1$, defined as follows:

$$L_{00} = i\frac{\partial}{\partial\eta} , \quad L_{0n} = -s_n\psi^{(n)} , \quad L_{n0} = L_{0n}^* , \quad L_{nn} = -is_n\frac{\partial}{\partial\xi} , \tag{2.61a}$$

$$A_{00} = -i\frac{\partial^2}{\partial\xi^2} + i\varphi , \quad A_{n0} = \psi^{(n)^*}\frac{\partial}{\partial\xi} - \psi_\xi^{(n)^*} , \tag{2.61b}$$

$$B_{00} = A_{00} , \quad B_{0n} = \psi^{(n)}\frac{\partial}{\partial\xi} + 2\psi_\xi^{(n)} . \tag{2.61c}$$

Here of course the matrix indices run from 0 to N, the index n runs from 1 to N, and all the matrix elements that are not explicitly defined vanish. The cases (2.46) resp. (2.56) correspond to $N = 1$ resp. $N = 2$.

3. C-Integrable Equations

In this section we briefly discuss a number of C-integrable equations. In contrast to the findings reported in the preceding sections, these results have not yet been published, nor submitted for publication elsewhere, as this chapter is being completed, in December 1987; although undoubtedly some of the equations reported below have already appeared in the literature. Most of these C-integrable equations are suited to serve as starting points for the application of the asymptotic limiting procedure discussed above. This line of research is now being pursued in collaboration with *Maccari* and *Levi* [17]. Some of these equations are also worthy of direct study; the Eckhaus equation has indeed shown that even C-integrable equations may exhibit quite an interesting phenomenology [13–15]. This is being pursued in collaboration with *De Lillo* [16].

A class of C-integrable equations is obtained via the change of dependent variable

$$v(x,t) = u(x,t) \exp \left\{ \int_{a(t)}^{x} dx' F[u(x',t)] \right\} \tag{3.1}$$

which provides a convenient technique to associate a *nonlinear* evolution PDE for $u(x,t)$ to a *linear* evolution PDE satisfied by $v(x,t)$. Here $F(u)$ is a given function that may be chosen in an arbitrary manner. Of course, more general transformations than (3.1) can be easily invented, but the equations yielded by (3.1) (see below) are sufficiently interesting, in our opinion, to deserve some study.

Before displaying the nonlinear equations for $u(x,t)$ induced by various linear equations for $v(x,t)$, let us discuss the transformation (3.1). It implies the relations

$$u[a(t),t] = v[a(t),t] \tag{3.2}$$

and

$$\frac{u_x(x,t)}{u(x,t)} + F[u(x,t)] = \frac{v_x(x,t)}{v(x,t)} . \tag{3.3}$$

Note that, once $a(t)$ and $F(u)$ are given, the formula (3.1) is the appropriate one to compute $v(x,t)$ from $u(x,t)$ by a quadrature, while (3.3), with (3.2), is the appropriate equation to calculate $u(x,t)$ from $v(x,t)$, which requires solving the nonlinear nonautonomous first-order ODE (3.3) for $u(x,t)$ with the boundary condition (3.2). Note that here the variable t plays merely the role of a parameter.

There are two choices of the function $F(u)$ for which (3.3) can be solved, thereby also reducing the calculation of $u(x,t)$ from a given $v(x,t)$ merely to a quadrature. The first of these choices sets

$$F(u) = Au^\lambda , \tag{3.4}$$

yielding

$$u(x,t) = v(x,t) \left\{ 1 + \lambda A \int\limits_{a(t)}^{x} dx' [v(x',t)]^{\lambda} \right\}^{-1/\lambda} . \tag{3.5}$$

The second choice sets

$$F(u) = A \ln(u) , \tag{3.6}$$

yielding

$$u(x,t) = v(x,t) \exp \left\{ -A \exp(-Ax) \int\limits_{a(t)}^{x} dx' \exp(Ax') \ln[v(x',t)] \right\} . \tag{3.7}$$

On the other hand, in the special case in which the x-dependence of $v(x,t)$ is purely exponential, $v(x,t) = v_0(t) \exp[p(t)x]$, so that

$$\frac{v_x}{v} = p , \tag{3.8}$$

the evaluation of $u(x,t)$ can be performed by a quadrature, followed by an inversion, for *any* choice of $F(u)$, since (3.3) with (3.8) clearly imply

$$\int\limits^{u(x,t)} du' \{u'[p - F(u')]\}^{-1} = x - \xi(t) . \tag{3.9}$$

Note that this case, with p a constant, generally yields a function $u(x,t)$ whose x-profile remains unchanged (in shape) over time, although it may move.

Let us now discuss a few instances of nonlinear evolution equations that obtain, via the transformation (3.1), from linear PDEs satisfied by $v(x,t)$; hence, all these equations are, by construction, C-integrable.

Let $v(x,t)$ satisfy the ("heat") equation

$$v_t - v_{xx} = 0 . \tag{3.10}$$

Then $u(x,t)$ satisfies the ("generalized Burgers") equation

$$u_t - u_{xx} + fu = 2u_x F(u) , \tag{3.11a}$$

$$f_x + uF'(u)f = -u_x^2 F''(u) . \tag{3.11b}$$

Here of course $u = u(x,t)$, $f = f(x,t)$ and the primes denote differentiation with respect to the argument, i.e., $F'(u) = dF(u)/du$, etc.

The consistency of (3.1, 10) with (3.11) requires, moreover, that $a(t)$ satisfy the ODE

$$\{F\dot{a}(t) + F^2 + F'u_x + f\}|_{x=a(t)} = 0 \tag{3.12a}$$

or, equivalently [see (3.2, 3)]

$$\{F\dot{a}(t) + F^2 + F'(v_x - Fv) + f\}|_{x=a(t)} = 0 . \tag{3.12b}$$

Here, of course, the dot indicates time differentiation; and note that the variable x is evaluated at $x = a(t)$ so that, for instance, F stands for $F\{u[a(t), t]\} = F\{v[a(t), t]\}$ [see (3.2)].

If the special choice (3.4) for F is made, then (3.11a, b) read

$$u_t - u_{xx} + fu = 2Au_x u^\lambda , \tag{3.13a}$$

$$f_x + \lambda Au^\lambda f = \lambda(1 - \lambda)Au_x^2 u^{\lambda-2} . \tag{3.13b}$$

This equation admits the "soliton"(kink-like) solution

$$u(x, t) = \left(\frac{p}{A}\right)^{1/\lambda} , \tag{3.14a}$$

$$f(x, t) = p^2 \left[g(t) + \left(\frac{1 - \lambda}{1 + E}\right)\right] \frac{E}{1 + E} , \tag{3.14b}$$

with

$$E \equiv \exp\{-\lambda p[x - \xi(t)]\} , \tag{3.14c}$$

$$\xi(t) = -p + tg(t) . \tag{3.14d}$$

Here $g(t)$ is an *arbitrary* function of t. Note that if λ and p are positive, for this solution

$$u(-\infty, t) = 0 , \quad u(+\infty, t) = \left(\frac{p}{A}\right)^{1/\lambda} , \tag{3.14e}$$

$$f(-\infty, t) = p^2 g(t) , \quad f(+\infty, t) = 0 . \tag{3.14f}$$

Of course, this solution is real, for all positive values of λ, only if A is positive. If instead A is negative, then the solution is still real provided p is also negative, in which case the behaviors of $u(x, t)$ as x tends to negative or positive infinity are exchanged, and likewise for f. Note that, if the nonvanishing asymptotic value of f is set to a constant value, say (for $p > 0$) $f(-\infty, t) = c$ [see (3.14f)], then the x-profiles of u and f move with the constant speed $c/p - p$ [see (3.14c, d)]. If instead one were to set a constant value for f at a given point, say

$$f(0, t) = 0 \tag{3.15a}$$

implying [see (3.14b)]

$$g(t) = \frac{(\lambda - 1)}{\{1 + \exp[\lambda p\xi(t)]\}} , \tag{3.15b}$$

then, by integrating (3.14d), one would obtain a more complicated behavior for $\xi(t)$ (except for $\lambda = 1$, of course) whose investigation is left as an amusing exercise for the diligent reader.

Another avatar of (3.13) obtains by setting

$$w(x,t) = A[u(x,t)]^\lambda = F[u(x,t)] . \tag{3.16}$$

It reads

$$w_t - w_{xx} - f_x = 2w_x w , \tag{3.17a}$$

$$f_x + \lambda w f = \frac{(1-\lambda)}{\lambda} \frac{w_x^2}{w} . \tag{3.17b}$$

The special case with $\lambda = 1$ deserves special attention, due to its simplicity and close resemblance to the usual Burgers equation. It reads

$$u_t - u_{xx} + fu = 2Au_x u , \tag{3.18a}$$

$$f_x + uf = 0 . \tag{3.18b}$$

For the special class of solutions with $f(x,t) = 0$, this reduces to the standard Burgers equation. In the general case it may be rewritten as a single equation via the position

$$z_x(x,t) = Au(x,t) . \tag{3.19}$$

Then the equation for $z(x,t)$ reads

$$z_t - z_{xx} = z_x^2 + h_1(t)\exp(-z) + h_2(t) , \tag{3.20}$$

with $h_1(t)$ and $h_2(t)$ two arbitrary functions. Note that if $h_1(t) = 0$, this is essentially again the standard Burgers equation [for $w(x,t) = z_x(x,t)$]. Also note that (3.20) admits the "soliton" solution

$$z(x,t) = \zeta(t) + \ln(1 + \exp\{p[x - \xi(t)]\}) , \tag{3.21a}$$

$$\zeta(t) = \ln\left\{\exp(\zeta_0) + \int_0^t dt' h_1(t') \exp\left[\int_{t'}^t dt'' h_2(t'')\right]\right\} , \tag{3.21b}$$

$$\xi(t) = \xi_0 - pt + p^{-1}\left[\zeta(t) - \zeta_0 - \int_0^t dt' h_2(t')\right] . \tag{3.21c}$$

This solution contains only the three arbitrary constants p, ξ_0, and ζ_0. If p is positive, it has the following asymptotic properties:

$$z(-\infty, t) = \zeta(t) , \tag{3.21d}$$

$$\lim_{x \to +\infty} \{z(x,t) - p[x - \xi(t)]\} = \zeta(t) . \tag{3.21e}$$

If p is negative, the two limiting behaviors are exchanged.

Let us now return to the general equation (3.11), to insert into it the choice (3.6). Then the equation takes the form

$$u_t + u_{xx} + fu = 2Au_x \ln u , \tag{3.22a}$$

$$f_x + Af = A \left(\frac{u_x}{u}\right)^2 . \tag{3.22b}$$

This equation possesses the following property of "Galilei" invariance: if the pair $u(x, t)$, $f(x, t)$ is a solution of (3.22), then the pair

$$\tilde{u}(x, t) = \exp\left(\frac{-c}{2A}\right) u(x - ct, t) , \tag{3.23a}$$

$$\tilde{f}(x, t) = f(x - ct, t) , \tag{3.23b}$$

is also a solution.

The "soliton" solution of (3.22) reads

$$u(x, t) = \exp(-\exp\{-A[x - \xi(t)]\}) , \tag{3.24a}$$

$$f(x, t) = A\xi(t)\exp\{-A[x - \xi(t)]\} - A^2 \exp\{-2A[x - \xi(t)]\} , \tag{3.24b}$$

with $\xi(t)$ an arbitrary function. If A is positive, this solution has the following asymptotic properties:

$$u(-\infty, t) = 0 , \quad f(-\infty, t) = -\infty , \tag{3.25a}$$

$$u(+\infty, t) = 1 , \quad f(+\infty, t) = 0 . \tag{3.25b}$$

If A is negative, the two limits are exchanged. Note that this solution, as all others, can be "boosted" by using the property of Galilei invariance (3.23).

But this case is not very interesting since it can be reduced to the standard Burgers equation

$$w_t - w_{xx} = 2w_x w \tag{3.26}$$

by introducing the new dependent variable

$$w(x, t) = A \ln[u(x, t)] + \frac{u_x(x, t)}{u(x, t)} = \left(A + \frac{\partial}{\partial x}\right) \ln[u(x, t)] . \tag{3.27a}$$

Let us now proceed and report the class of nonlinear evolution equations for $u(x, t)$ that correspond, via (3.1), to the following linear PDE for $v(x, t)$:

$$v_t - \alpha v_{xx} - \beta v_{xxx} = 0 . \tag{3.27b}$$

This might be considered as the next equation of a hierarchy of "generalized higher Burgers equations":

$$u_t - \alpha u_{xx} - \beta u_{xxx} + fu = (2\alpha F + 3\beta F^2)u_x + 3\beta F u_{xx}$$
$$+ 3\beta(F + uF''/2)u_x^2 , \tag{3.28a}$$

$$f_x + uF'f = -\left[\alpha + 3\beta\left(F - \frac{uF'}{2}\right)\right]F''u_x^2 + \frac{\beta}{2}F'''u_x^3 . \tag{3.28b}$$

The consistency of this equation with (3.27) via (3.1), requires, moreover, that the lower limit of integration $a(t)$ in the right-hand side of (3.1) satisfy the ODE

$$[\dot{a}F + f + (\alpha u_x + \beta u_{xx})F' + \alpha F^2 + \beta F^3 - \tfrac{1}{2}\beta F'' u_x^2]|_{x=a(t)} = 0 . \tag{3.29}$$

Let us also display the form that this equation takes when the choice (3.4) is made for F:

$$u_t - \alpha u_{xx} - \beta u_{xxx} + fu = A(2\alpha + 3\beta Au^\lambda)u^\lambda u_x + 3\beta Au^\lambda u_{xx}$$
$$+ (3\beta/2)\lambda(\lambda + 1)u^{\lambda-1}u_x^2 , \tag{3.30a}$$

$$f_x + \lambda Au^\lambda f = \lambda(1 - \lambda)A\left[\alpha + 3\beta\left(1 - \frac{\lambda}{2}\right)Au^\lambda\right]u^{\lambda-2}u_x^2$$
$$+ \lambda(1 - \lambda)\left(1 - \frac{\lambda}{2}\right)Au^{\lambda-3}u_x^3 , \tag{3.30b}$$

For $\lambda = 1$, this simplifies as follows:

$$u_t - \alpha u_{xx} - \beta u_{xxx} + fu = A(\alpha u^2 + \beta Au^3)_x + \tfrac{3}{2}\beta A(u^2)_{xx} , \tag{3.31a}$$

$$f_x + Auf = 0 . \tag{3.31b}$$

A subclass of solutions of this equation is characterized by an identically vanishing f ($f = 0$) and by a $u(x,t)$ that satisfies the usual Burgers equation if β vanishes and the next equation of the usual Burgers hierarchy if $\beta \neq 0$ [i.e., (3.31a) with $f = 0$].

Another interesting subclass of (3.30) obtains for $\lambda = 2$ and $\alpha = 0$ (for notational simplicity, we also set $A = \beta = 1$):

$$u_t - u_{xxx} + fu = 3u^4 u_x + 3u^2 u_{xx} + 9uu_x^2 , \tag{3.32a}$$

$$f_x + 2u^2 f = 0 . \tag{3.32b}$$

A subclass of solutions of this equation is characterized by an identically vanishing f ($f = 0$) and by a $u(x,t)$ that satisfies the nonlinear PDE studied in some detail in [11].

Let us now proceed to the class of nonlinear evolution equations satisfied by $u(x,t)$ when $v(x,t)$, related to it by (3.1), evolves according to the linear equation

$$v_{tt} - v_{xx} + \alpha v = 0 . \tag{3.33}$$

The corresponding equation for $u(x,t)$ reads

$$u_{tt} - u_{xx} + \alpha u = -2u_t g - ug_t - ug^2 + 2u_x F + uu_x F' + uF^2 , \tag{3.34a}$$

$$g_x = u_t F' = F_t . \tag{3.34b}$$

Here, of course, $u \equiv u(x,t)$, $g \equiv g(x,t)$, $F \equiv F[u(x,t)]$, and $F' = dF/du$. The consistency of (3.33) with (3.34) requires that the lower limit of integration in the right-hand side of (3.1) be time independent,

$$a(t) = a , \quad \dot{a}(t) = 0 , \tag{3.35}$$

and moreover that the function $g(x,t)$ satisfy the boundary condition

$$g(a,t) = 0 , \tag{3.36a}$$

implying [see (3.34b)]

$$g(x,t) = \int_a^x dx' F'[u(x',t)]u_t(x',t) = \frac{d}{dt} \int_a^x dx' F[u(x',t)] . \tag{3.36b}$$

If the special choice (3.4) for F is made, then (3.34a, b) read

$$u_{tt} - u_{xx} + \alpha u = -2u_t g - ug_t - ug^2 + (\lambda + 2)Au_x u^\lambda + A^2 u^{2\lambda+1} , \tag{3.37a}$$

$$g_x = \lambda A u_t u^{\lambda-1} . \tag{3.37b}$$

Finally, let us display the equation implied for $u(x,t)$, via (3.1), by the following linear equation for $v(x,t)$:

$$v_{xt} - \gamma v = 0 . \tag{3.38}$$

It reads

$$u_{xt} - \gamma u = -(uF + u_x)g_t - (F + F'u)u_t , \tag{3.39a}$$

$$g_x = F , \tag{3.39b}$$

with the boundary condition

$$g[a(t), t] = 0 , \tag{3.40a}$$

so that [see (3.39b)]

$$g(x,t) = \int_{a(t)}^x dx' F[u(x',t)] . \tag{3.40b}$$

Note that in this case there is no restriction on $a(t)$. Of course, in (3.39) $u \equiv u(x,t)$, $g \equiv g(x,t)$, $F \equiv F[u(x,t)]$, and $F' \equiv dF/du$.

If the special choice (3.4) for F is made, then (3.39a, b) read

$$u_{xt} - \gamma u = -(u_x + Au^{\lambda+1})g_t - (\lambda + 1)Au^\lambda u_t , \tag{3.41a}$$

$$g_x = Au^\lambda . \tag{3.41b}$$

This may be rewritten as a single equation for $g(x,t)$:

$$g_{xxt} - \lambda\gamma g_x = \left(\frac{\lambda-1}{\lambda}\right)\frac{g_{xx}g_{xt}}{g_x} - g_{xx}g_t - (\lambda+1)g_x g_{xt} - \lambda g_x^2 g_t . \tag{3.42}$$

Here we have set $A = 1$ for notational simplicity.

Note that, for $\gamma = 0$, this equation has the *general solution*

$$g(x,t) = \lambda^{-1} \ln \left\{ h_1(t) + \int^x dx'[h_2(t) + h_3(x')]^\lambda \right\} , \tag{3.43}$$

containing the three *arbitrary* functions h_1, h_2, and h_3. In particular, for $\lambda = 1$ (in which case the right-hand side of (3.42) is polynomial), this formula yields

$$g(x,t) = \ln[h_1(t) + x h_2(t) + H(x)] , \tag{3.44}$$

while for $\lambda = 2$ it gives

$$g(x,t) = \frac{1}{2} \ln \left\{ h_1(t) + x[h_2(t)]^2 + 2H(x)h_2(t) + \int^x dx'[H'(x')]^2 \right\} . \tag{3.45}$$

Here h_1, h_2, and H are three arbitrary functions.

Let us note that the equation for $u(x,t)$ can also be recast in a form different from (3.39), which should presumably be more convenient to treat the Cauchy problem. It reads as follows:

$$u_{xt} - \gamma u = -f(u_x + Fu) - [F + F'u - (u_x + Fu)h]u_t , \tag{3.46a}$$

$$f_x + fh(u_x + Fu) = \gamma uh , \tag{3.46b}$$

$$h_x = -h^2 u_x + (1 - hu)(hF - F') , \tag{3.46c}$$

with the additional conditions

$$f[a(t), t] = 0 , \tag{3.46d}$$

$$[\dot{a}F - hu_t]|_{x=a(t)} = 0 . \tag{3.46e}$$

These equations may be simplified by considering the special solution $h = 1/u$, that clearly satisfies (3.46c). Then the equations take the simpler form

$$u_{xt} - f_x u = \left(\frac{u_x}{u} - F'u \right) u_t , \tag{3.47a}$$

$$f_x = \gamma - \left(\frac{u_x}{u} + F \right) f , \tag{3.47b}$$

$$f[a(t), t] = 0 \tag{3.47c}$$

$$\left(\dot{a}F - \frac{u_t}{u} \right)\Big|_{x=a(t)} = 0 . \tag{3.47d}$$

An even simpler form obtains via the position

$$f(x,t) = q(x,t) + \frac{u_t(x,t)}{u(x,t)} ,$$

namely,

$$u_{xt} - \gamma u = -q(u_x + Fu) - (F + F'u)u_t , \tag{3.48a}$$

$$q_x = F'u_t , \tag{3.48b}$$

$$q[a(t), t] = -\frac{u_t[a(t), t]}{u[a(t), t]} , \tag{3.48c}$$

$$\dot{a}(t)F[u(a(t), t)] + q[a(t), t] = 0 . \tag{3.48d}$$

This becomes particularly simple when $F(u) = A/u$:

$$u_{xt} - \gamma u = -q(u_x + A) , \tag{3.49a}$$

$$q_x = -A\frac{u_t}{u^2} = \left(\frac{A}{u}\right)_t , \tag{3.49b}$$

$$q[a(t), t] = -\frac{u_t[a(t), t]}{u[a(t), t]} , \tag{3.49c}$$

$$A\dot{a}(t) = u_t[a(t), t] . \tag{3.49d}$$

Let us now turn to a class of nonlinear evolution equations for a complex function $\psi(x, t)$, that are natural generalizations of the Eckhaus equation (1.55). They are closely related to those discussed above, since they are generated by the transformation

$$\varphi(x, t) = \psi(x, t)\exp\left[\frac{1}{2}\int\limits_{a(t)}^{x} dx' F(|\psi(x', t)|^2)\right] , \tag{3.50}$$

where we assume F to be a real, but otherwise a priori arbitrary, function. Note that this relation implies that φ and ψ have the same phase, while their squared moduli, say $|\varphi(x, t)|^2 = v(x, t)$ and $|\psi(x, t)|^2 = u(x, t)$, are related by (3.1). Hence the discussion about the explicit invertibility of this transformation, as given at the beginning of this section, remains applicable in the present case.

Assume now that $\varphi(x, t)$ satisfies the linear Schrödinger equation

$$i\varphi_t + \varphi_{xx} = 0 . \tag{3.51}$$

The corresponding evolution equation for $\psi(x, t)$ then reads

$$i\psi_t + \psi_{xx} + if\psi = -\left[(|\psi|^2)_x F' + \frac{F^2}{4}\right]\psi + (|\psi|^2 F' - F)\psi_x , \tag{3.52a}$$

$$f_x + |\psi|^2 F'f = \text{Im}\{\psi_x\psi^*\}[(|\psi|^2)_x F'' + 2(|\psi|^2 F' - F)] . \tag{3.52b}$$

Here, of course, $\psi \equiv \psi(x, t)$ is complex, while $f \equiv f(x, t)$ and $F \equiv F(|\psi|^2)$ are real. The compatibility of (3.51, 52) via (3.50) requires, moreover, that the lower limit of the integral in the right-hand side of (3.50) satisfy the ODE

$$[F\dot{a} + 2f - 2\,\text{Im}\{\psi_x\psi^*\}F']|_{x=a(t)} = 0 . \tag{3.53}$$

The evolution equation (3.52) possesses the following property (*Galilei invariance*): if the pair $\psi(x,t)$, $f(x,t)$ is a solution of (3.52), then the pair

$$\tilde{\psi}(x,t) = \exp\left[\left(\frac{i}{2}\right)c\left(x - \frac{ct}{2}\right)\right]\psi(x - ct, t), \qquad (3.54a)$$

$$\tilde{f}(x,t) = f(x - ct, t) + c[|\psi(x - ct, t)|^2 F'(|\psi(x - ct, t)|^2) \\ - F(|\psi(x - ct, t)|^2)] \qquad (3.54b)$$

is also a solution.

If the special choice of (3.4) for F is made, then (3.52a, b) read

$$i\psi_t + \psi_{xx} + if\psi = -\left[\lambda A(|\psi|^2)_x|\psi|^{2(\lambda-1)} + \left(\frac{A}{2}\right)^2|\psi|^{4\lambda}\right]\psi$$

$$+ (\lambda - 1)A|\psi|^{2\lambda}\psi_x, \qquad (3.55a)$$

$$f_x + \lambda A|\psi|^{2\lambda}f = (\lambda - 1)A\,\mathrm{Im}\{\psi_x\psi^*\}[\lambda(|\psi|^2)_x|\psi|^{2(\lambda-2)} + 2|\psi|^{2\lambda}]. \quad (3.55b)$$

In the special case that $\lambda = 1$ this simplifies considerably to

$$i\psi_t + \psi_{xx} + if\psi + [|\psi|^4 + 2(|\psi|^2)_x]\psi = 0, \qquad (3.56a)$$

$$f_x + 2|\psi|^2 f = 0, \qquad (3.56b)$$

where we have set $A = 2$ for convenience. A subclass of the solutions of this equation is characterized by an $f(x, t)$ that vanishes identically and a $u(x, t)$ that obeys the Eckhaus equation (1.55).

Another avatar of (3.52) that is worth recording obtains from the position

$$\psi(x,t) = [\sigma(x, t)]^{1/2}\exp[i\theta(x, t)], \quad |\psi(x, t)|^2 = \sigma(x, t), \qquad (3.57)$$

with σ and θ real, and reads

$$\sigma_t + 2\sigma_x\theta_x + 2\sigma\theta_{xx} = -2f\sigma + 2\sigma\theta_x(\sigma f' - F), \qquad (3.58a)$$

$$4\sigma\theta_t - 2\sigma_{xx} + \frac{\sigma_x^2}{\sigma} + 4\sigma\theta_x^2 = \sigma F^2 + 2\sigma_x(\sigma F' + F), \qquad (3.58b)$$

$$f_x + 2\sigma f = \sigma\theta_x[\sigma_x F'' + 2(\sigma F' - F)], \qquad (3.58c)$$

where, of course, $F \equiv F(\sigma)$. In view of the possible application of the technique of asymptotic expansion, the following position may prove useful: $\sigma = 1 + \varepsilon\mu$, $\theta = \varepsilon\nu$, $f = \varepsilon^2\varrho$, $F(\sigma) = \varepsilon H(1 + \varepsilon\mu)$, yielding

$$\mu_t + 2\nu_{xx} = -2\varepsilon\{(\mu\nu_x)_x$$
$$+ (1 + \varepsilon\mu)\varrho + (1 + \varepsilon\mu)[H - (1 + \varepsilon\mu)H']\nu_x\}, \qquad (3.59a)$$

$$4\nu_t - 2\mu_{xx} = \varepsilon\left\{-4\mu\nu_t - \frac{\mu_x^2}{(1 + \varepsilon\mu)} - 4(1 + \varepsilon\mu)\nu_x^2\right.$$
$$\left. + (1 + \varepsilon\mu)H^2 + 2\mu_x[H + (1 + \varepsilon\mu)H']\right\}, \qquad (3.59b)$$

$$\varrho_x + 2\varrho + 2\nu_x(H - H') = \varepsilon\{2\mu\varrho + (\mu_x H'' + 4\mu H' - 2\mu H)\nu_x\}. \qquad (3.59c)$$

Let us moreover note that a nontrivial extension of the transformation (3.50) reads as follows

$$\varphi(x,t) = \psi(x,t)\exp\left\{\frac{1}{2}\int_{a(t)}^{x} dx' F[|\psi(x',t)|^2]\right.$$

$$\left. +i\int_{b(t)}^{x} dx' G[|\psi(x',t)|^2]\right\}, \tag{3.60}$$

with F and G two *real* functions that may be freely chosen. Note that this leads to the *same* relation as (3.50) among the moduli of φ and ψ, so that the previous analysis of the invertibility of the transformation remains applicable. The nonlinear evolution equation for ψ that corresponds via (3.60) to the linear Schrödinger equation (3.51) for φ reads as follows:

$$i\psi_t + \psi_{xx} + if\psi + g\psi = -[U + iV + (F' + 2iG')\sigma_x]\psi$$
$$- (M + iN)\psi_x, \tag{3.61a}$$

$$f_x + \sigma F' f = \sigma V F' - \text{Im}\{\psi_x\psi\}(MF - F''\sigma_x), \tag{3.61b}$$

$$g_x - 2\sigma G' f = -2\sigma V G' + 2\,\text{Im}\{\psi_x\psi^*\}(MG' - G''\sigma_x), \tag{3.61c}$$

with $f \equiv f(x,t)$, $g \equiv g(x,t)$ both real and

$$\sigma(x,t) \equiv |\psi(x,t)|^2, \tag{3.61d}$$

$$U \equiv U[\sigma(x,t)] = [F(\sigma)/2]^2 - \sigma^2[G'(\sigma)]^2 + 4\int_{\sigma_1}^{\sigma} d\sigma'\,\sigma'[G'(\sigma')]^2, \tag{3.61e}$$

$$V \equiv V[\sigma(x,t)] = F(\sigma)G(\sigma)$$
$$- \int_{\sigma_2}^{\sigma} d\sigma'\,F'(\sigma')[G(\sigma') + \sigma'G'(\sigma') - \sigma'^2 G''(\sigma')], \tag{3.61f}$$

$$M \equiv M[\sigma(x,t)] = F(\sigma) - \sigma F'(\sigma), \tag{3.61g}$$

$$N \equiv 2[G(\sigma) - \sigma G'(\sigma)]. \tag{3.61h}$$

Here σ_1 and σ_2 are two (arbitrary) real constants. The compatibility of (3.61) with (3.51) via (3.60) requires, moreover, that $a(t)$ and $b(t)$ satisfy the following ODEs:

$$[F\dot{a} + 2f + Q - 2\,\text{Im}\{\psi_x\psi^*\}F]|_{x=a(t)} = 0, \tag{3.62a}$$

$$[G\dot{b} + g + R + 2\,\text{Im}\{\psi_x\psi^*\}G']|_{x=a(t)} = 0, \tag{3.62b}$$

$$Q \equiv Q[\sigma(x,t)] = -2 \int_{\sigma_2}^{\sigma} d\sigma' F'(\sigma')[G(\sigma') + \sigma' G'(\sigma') - \sigma'^2 G''(\sigma')] , \quad (3.62c)$$

$$R \equiv R[\sigma(x,t)] = [G(\sigma)]^2 - \sigma^2 [G'(\sigma)]^2 + 4 \int_{\sigma_1}^{\sigma} d\sigma' \sigma' [G'(\sigma')]^2 . \quad (3.62d)$$

These formulas become rather simple if

$$F(\sigma) = 2A\sigma^\lambda , \quad (3.63a)$$

$$G(\sigma) = B\sigma^\mu , \quad (3.63b)$$

in which case (we have also set for simplicity $\sigma_1 = \sigma_2 = 0$),

$$U(\sigma) = A^2 \sigma^{2\lambda} + \mu(2-\mu)B^2 \sigma^{2\mu} , \quad (3.64a)$$

$$V(\sigma) = 2AB\mu(1 - 2\lambda + \lambda\mu)(\lambda + \mu)^{-1} \sigma^{\lambda+\mu} , \quad (3.64b)$$

$$M(\sigma) = 2A(1 - \lambda)\sigma^\lambda , \quad (3.64c)$$

$$N(\sigma) = 2B(1 - \mu)\sigma^\mu . \quad (3.64d)$$

In particular, if $\lambda = \mu = 1$ (3.61) becomes

$$i\psi_t + \psi_{xx} + if\psi + g\psi + [|\psi|^4 + 2\exp(i\alpha)(|\psi|^2)_x]\psi = 0 , \quad (3.65a)$$

$$f_x + 2\cos(\alpha)|\psi|^2 f = 0 , \quad (3.65b)$$

$$g_x - 2\sin(\alpha)|\psi|^2 f = 0 , \quad (3.65c)$$

where we have set for notational simplicity $A = \cos\alpha$, $B = \sin\alpha$. The integrability of this equation, in the special case of $f(x,t) = g(x,t) = 0$, has already been noted [13]. The special solution with $g(x,t) = -\tan(\alpha)f(x,t)$ is also notable. In this case the system takes the reduced form

$$i\psi_t + \psi_{xx} + i\exp(i\alpha)h\psi + [|\psi|^4 + 2\exp(i\alpha)(|\psi|^2)_x]\psi = 0 , \quad (3.65d)$$

$$h_x + 2\cos(\alpha)|\psi|^2 h = 0 , \quad (3.65e)$$

where we have set $f(x,t) = \cos(\alpha)h(x,t)$.

The examples of C-integrable equations considered so far are all related to the changes of dependent variable (3.1) or (3.50) [or (3.60)], and they refer to equations in $1 + 1$ dimensions. Let us merely mention here that via the *same* changes of dependent variable it is also possible to generate C-integrable nonlinear evolution equations in $n + 1$ dimensions with $n > 1$ (whose interest remains to be investigated). Let us limit our presentation here to two examples.

Let

$$v(x,y,t) = u(x,y,t)\exp[g(x,y,t)] \quad (3.66)$$

with

$$g(x, y, t) = \int\limits_{a(y,t)}^{x} dx' \, F[u(x', y, t)] \tag{3.67}$$

and with $F(u)$ a given function (whose choice remains our privilege). It is then easily seen that to the linear equation for v,

$$a_{33}v_{tt} + a_{22}v_{yy} + a_{11}v_{xx} + 2a_{32}v_{ty} + 2a_{31}v_{tx} + 2a_{21}v_{yx}$$
$$+ a_{30}v_t + a_{20}v_y + a_{10}v_x + a_{00}v = 0 , \tag{3.68}$$

there formally corresponds the equation for u

$$a_{33}u_{tt} + a_{22}u_{yy} + a_{11}u_{xx} + 2a_{32}u_{ty} + 2a_{31}u_{tx} + 2a_{21}u_{yx}$$
$$+ a_{30}u_t + a_{20}u_y + a_{10}u_x + a_{00}u = -[a_{33}(g_{tt} + g_t^2) + a_{22}(g_{yy} + g_y^2)$$
$$+ a_{11}F^2 + 2a_{32}(g_{ty} + g_t g_y) + 2a_{31}(g_{tx} + Fg_t) + 2a_{21}(g_{yx} + Fg_y)$$
$$+ a_{30}g_t + a_{20}g_y + a_{10}F]u - 2(a_{33}g_t + a_{23}g_y + a_{13}F)u_t$$
$$- 2(a_{22}g_y + a_{23}g_t + a_{12}F)u_y$$
$$- [a_{11}(2F + F'u) + 2a_{13}g_t + 2a_{23}g_y]u_x , \tag{3.69a}$$

$$g_x = F . \tag{3.69b}$$

Of course the last equation must be supplemented with the "boundary condition"

$$g[a(y, t), y, t] = 0 \tag{3.69c}$$

[see (3.67)]. Note, however, that the choice of $a(y, t)$ remains unconstrained. An analogous treatment may be based on the change of variable

$$\varphi(x, y, t) = \psi(x, y, t) \exp[g(x, y, t)] , \tag{3.70}$$

with

$$g(x, y, t) = \frac{1}{2} \int\limits_{a(y,t)}^{x} dx' \, F[|\psi(x', y, t)|^2] , \tag{3.71}$$

where again F is a function that may be chosen arbitrarily. Then, for instance, the linear Schrödinger equation in $2 + 1$ dimensions for φ,

$$i\varphi_t + \varphi_{xx} + \varphi_{yy} = 0 \tag{3.72}$$

yields for ψ the following nonlinear evolution equation:

$$i\psi_t + \psi_{xx} + \psi_{yy} = -\left[ig_t + g_{yy} + g_y^2 + (|\psi|^2)_x \frac{F'}{2} + \frac{F^2}{4} \right] \psi$$
$$- F\psi_x - 2g_y\psi_y , \tag{3.73a}$$

$$g_x = \frac{F}{2} , \tag{3.73b}$$

supplemented of course by the "boundary condition" [see (3.71)]

$$g[a(y,t), y, t] = 0 ,\tag{3.74}$$

with the choice of $a(y,t)$ unconstrained.

So far we have only discussed C-integrable equations that can be linearized by a change of dependent variable (and even within this class, we have looked only at certain special changes of dependent variable). A much richer phenomenology obtains if changes of variables are considered that involve both *dependent and independent* variables. Here we limit our treatment to the exhibition of a few examples, merely to illustrate the potential usefulness of this approach.

Suppose that $u(x,t)$ satisfies the Liouville equation

$$u_{xt} = a \exp(u) ,\tag{3.75}$$

whose *general solution* is given by the explicit formula

$$u(x,t) = f(x) + g(t) - 2\ln\left\{\frac{p}{2}\int_{x_0}^{x} dx' \exp[f(x')] + \frac{a}{p}\int_{t_0}^{t} dt' \exp[g(t')]\right\}\tag{3.76}$$

in terms of the two arbitrary functions $f(x)$ and $g(t)$ and the three constants p, x_0 and t_0. Now set

$$u(x, \gamma\tau) = v(\xi, \tau) + \frac{\xi}{\gamma} ,\tag{3.77a}$$

$$u_x(x, \gamma\tau) = \exp\left(\frac{\xi}{\gamma}\right) .\tag{3.77b}$$

It is then easily seen that $v(\xi, \tau)$ satisfies the *generalized Liouville equation*

$$v_{\xi\tau} = a \exp(v) + b[\exp(v)]_{\xi\xi}\tag{3.78a}$$

with

$$b = -a\gamma^2 .\tag{3.78b}$$

Note incidentally that the two constants a and b in the right-hand side of (3.78a) can be multiplied by arbitrary factors via a trivial rescaling of the independent variables (and the addition of a constant to the dependent variable); but this does not permit us to set $b = 0$, indeed it is seen from (3.77, 78b) that the limit $b \to 0$ is nontrivial.

Let us emphasize that given $u(x,t)$, the transformation (3.77) yields $v(\xi, \tau)$ merely by inverting the function u_x [to obtain x as a function of ξ and τ, use (3.77b); then insert this expression of x into (3.77a)]. If on the other hand $v(\xi, \tau)$ is given, then to obtain $u(x,t)$ one quadrature is required, in addition to algebraic operations (indeed, first one must solve (3.77a) to get ξ as a function of u, and then one must solve the autonomous first order ODE for u that results from (3.77b) when ξ is expressed in terms of u). Note that only the space variables x and ξ, as well as the dependent variables u and v, are involved in the transformation (3.77a); the time variable plays instead a merely parametric role (except for a "cosmetic" rescaling).

Let us now proceed with a few other C-integrable equations obtainable via a change of variables analogous to (3.77), that we choose to write in the following canonical form:

$$u(x,t) = v(\xi,t) , \tag{3.79a}$$

$$u_x(x,t) = \xi(x,t) . \tag{3.79b}$$

It is then easily seen that the nonlinear evolution equation for $u(x,t)$

$$u_t = \frac{a u_{xxx}}{u_{xx}^3} + \frac{f(u_x)}{u_{xx}} + u g(u_x) + h(u_x) \tag{3.80}$$

gets transformed into the following *linear* equation for $v(\xi,t)$:

$$\begin{aligned}
v_t = a v_{\xi\xi\xi} &- \left[\frac{3a}{\xi} + f(\xi)\right] v_{\xi\xi} \\
&+ \left[\frac{3a}{\xi^2} + 2\frac{f(\xi)}{\xi} - f'(\xi) - \xi g(\xi)\right] v_\xi \\
&+ [g(\xi) - \xi g'(\xi)]v + h(\xi) - \xi h'(\xi) .
\end{aligned} \tag{3.81}$$

Here a is an arbitrary constant, and $f(\xi)$, $g(\xi)$ and $h(\xi)$ are arbitrary functions (all these quantities might, moreover, depend on t). The C-integrability of (3.80) was already announced in [11]. The special case $f(\xi) = -3a/\xi$, $g(\xi) = b\xi$, $h(\xi) = c\xi$ is worth noting. Then (3.80) reads

$$u_t = a \left(\frac{u_{xxx}}{u_{xx}^3} - \frac{3}{u_x u_{xx}}\right) + b u_x u + c u_x \tag{3.82}$$

and (3.81) becomes simply

$$v_t = a v_{\xi\xi\xi} - b\xi^2 v_\xi . \tag{3.83}$$

Another special case worth noting obtains for $f(\xi) = -b$, $g(\xi) = h(\xi) = 0$. Then, via the additional change of variable $r(x,t) = [u_{xx}(x,t)]^{-1}$, (3.80) becomes

$$r_t = a(r^3 r_{xxx} + 3r^2 r_x r_{xx}) + b r^2 r_{xx} , \tag{3.84a}$$

$$r_t = a(r^3 r_{xx})_x + b r^2 r_{xx} , \tag{3.84b}$$

$$r_t = r^2 \left[\frac{a}{2}(r^2)_{xxx} + b r_{xx}\right] , \tag{3.84c}$$

$$s_t = -\left[\frac{a}{2}(s^{-2})_x + b s^{-1}\right]_{xx} , \tag{3.84d}$$

$$p_t = a p^{3/2} p_{xxx} + 2b p^{3/2} (p^{1/2})_{xx} . \tag{3.84e}$$

Of course, in the last two equations, $s(x,t) = [r(x,t)]^{-1} = u_{xx}(x,t)$, $p(x,t) = [r(x,t)]^2 = [s(x,t)]^{-2} = [u_{xx}(x,t)]^{-2}$. The C-integrability of (3.84) with $b = 0$ was already announced in [11].

Yet another special case of (3.80) worth noting obtains for $f(\xi) = b\xi$, $h(\xi) = 0$, $g(\xi) = c\xi$:

$$u_t = a\frac{u_{xxx}}{u_{xx}^3} + buu_x , \tag{3.85a}$$

$$u_t = -\frac{a}{2}[(u_{xx})^{-2}]_x + \frac{b}{2}(u^2)_x , \tag{3.85b}$$

or equivalently, via the position $u(x,t) = w_x(x,t)$,

$$w_t = -\frac{a}{2}(w_{xxx})^{-2} + \frac{b}{2}(w_x)^2 + c . \tag{3.85c}$$

It is also easily seen that the change of variable (3.79) transforms the nonlinear equation for $u(x,t)$

$$u_{xt} = \frac{f(u_x)u_{xxx}}{u_{xx}^2} + [ug(u_x) + h(u_x)]u_{xx} + m(u_x) \tag{3.86}$$

into the *linear* equation for $v(\xi,t)$

$$v_{\xi t} = f(\xi)v_{\xi\xi\xi} + \left[f'(\xi) - 2\frac{f(\xi)}{\xi} - m(\xi)\right]v_{\xi\xi}$$
$$+ \left[2\frac{f(\xi)}{\xi^2} - \frac{f'(\xi)}{\xi} + \frac{m(\xi)}{\xi} - m'(\xi) + \xi g(\xi)\right]v_\xi - \xi g'(\xi)v - \xi h'(\xi) .$$
$$\tag{3.87}$$

The C-integrability of (3.86) was already announced in [11].

There are also several special cases of (3.86) that are worth noting. The first obtains for $f(\xi) = 0$ and $g(\xi) = a$ and reads

$$u_{xt} = u_{xx}[au + h(u_x)] + m(u_x) . \tag{3.88}$$

Note that this equation has the following property (*Galilei invariance*): if $u(x,t)$ is a solution of (3.88), then $\tilde{u}(x,t) = u(x - ct, t) - c/a$ is also a solution. Note, moreover, that the corresponding equation for $v(\xi,t)$ becomes the following *first order linear* PDE for $w(\xi,t) = v_\xi(\xi,t)$:

$$w_t + m(\xi)w_\xi = \left[\frac{m(\xi)}{\xi} - m'(\xi) + a\xi\right]w - \xi h'(\xi) . \tag{3.89}$$

Another special case of (3.86) worth noting obtains for $f(\xi) = -a$, $g(\xi) = G'(\xi)$, $h(\xi) = H'(\xi)$, and $m(\xi) = \xi G(\xi)$:

$$u_t = a(u_{xx})^{-1} + uG(u_x) + H(u_x) . \tag{3.90a}$$

If, moreover, $G(\xi) = b$ and $H(\xi) = c\xi$, then via the positions $u_{xx}(x,t) = s(x,t)$ and $[u_{xx}(x,t)]^{-1} = r(x,t)$, (3.90a) becomes simply

$$s_t = a(s^{-1})_{xx} + bs + cs_x , \tag{3.90b}$$

$$r_t = -ar^2 r_{xx} - br + cr_x \ . \tag{3.90c}$$

The transformation (3.79) also proves expedient to transform a linear equation for $u(x,t)$ into a nonlinear equation for $v(\xi,t)$. For instance, to the equation

$$u_t = Au + Bu_x + Cu_{xx} + Du_{xxx} \ , \tag{3.91}$$

there corresponds the equation

$$v_t = A(v - \xi v_\xi) + C\frac{\xi^2 v_{\xi\xi}}{v_\xi^2} + D\left(\frac{\xi}{v_\xi^2}\right)^2 (\xi v_\xi v_{\xi\xi\xi} - 3\xi v_{\xi\xi}^2 + 3v_\xi v_{\xi\xi}) \ . \tag{3.92}$$

This equation becomes autonomous (and somewhat neater) via the transformation

$$y = \ln \xi \tag{3.93a}$$

$$v_\xi(\xi,t) = [w(y,t)]^{-1} \tag{3.93b}$$

which yields

$$w_t = -Aw_y + Cw^2(w_y + w_{yy})$$
$$+ D[w^3(2w_y + w_{yyy}) + 3w^2(w_y^2 + ww_{yy} + w_y w_{yy})] \tag{3.94a}$$

or equivalently (but in a form, obtained by setting $w(y,t) = 1 + \varepsilon z(y,t)$ and $C = -3D + \varepsilon\gamma D$, that is more suitable to apply the expansion technique)

$$z_t + (A+D)z_y - Dz_{yyy} = \varepsilon D\{(z_y + z_{yy})[\gamma - (3 - \varepsilon\gamma)z]$$
$$+ (2z_y + 3z_{yy} + z_{yyy})z(3 + 3\varepsilon z + \varepsilon^2 z^2)$$
$$+ 3(z_y^2 + z_y z_{yy})(1 + \varepsilon z)^2\} \ . \tag{3.94b}$$

Another change of variables that is also suitable to identify interesting C-integrable equations is the following one:

$$u(x,t) = \xi(x,t) \ , \tag{3.95a}$$

$$u_x(x,t) = v(\xi,t) \ . \tag{3.95b}$$

Note that in this case, as in the previous one, if $u(x,t)$ is given, $v(\xi,t)$ may be obtained by purely algebraic manipulations, while if $v(\xi,t)$ is given, the evaluation of $u(x,t)$ requires one quadrature [to solve the first order ODE

$$u_x = v(u,t) \tag{3.95c}$$

implied by (3.95a, b)]. Note that also in this case, as in the previous one, the change of variables involves only the "spatial" independent variables x and ξ (in addition to the dependent variables u and v), while the time variable t plays a purely parametric role (but of course, the different significance of the time differentiation in u_t and v_t should be emphasized; in the case of u, the subscripted t indicates partial t-differentiation at x fixed, in the case of v it indicates partial t-differentiation at ξ fixed).

The transformation (3.95) may be used to relate a *nonlinear* PDE satisfied by u to a *linear* PDE satisfied by v (or by some other function related to v), or vice versa.

A few examples of the former approach read as follows:

The nonlinear evolution PDE satisfied by $u(x, t)$

$$u_t = \frac{f(u)u_{xx}}{u_x^2} + g(u)u_x + h(u) \tag{3.96}$$

becomes, via (3.95),

$$v_t = f(\xi)\left(v_{\xi\xi} - 2\frac{v_\xi^2}{v}\right) + [f'(\xi) - h(\xi)]v_\xi + g'(\xi)v^2 + h'(\xi)v, \tag{3.97}$$

which is then linearized by the substitution

$$v(\xi, t) = [w(\xi, t)]^{-1}, \tag{3.98}$$

yielding

$$w_t = f(\xi)w_{\xi\xi} + [f'(\xi) - h(\xi)]w_\xi - h'(\xi)w - g'(\xi). \tag{3.99}$$

The C-integrability of (3.96) was already announced in [11].

The nonlinear evolution PDE satisfied by $u(x, t)$

$$u_t = f(u)\left[\frac{u_{xxx}}{u_x^3} - 3\left(\frac{u_{xx}}{u_x^2}\right)^2\right] + g(u)\frac{u_{xx}}{u_x^2} + h(u), \tag{3.100}$$

becomes, via (3.95) followed by (3.98), the linear equation

$$w_t = [f(\xi)w_{\xi\xi} + g(\xi)w_\xi - h(\xi)w]_\xi. \tag{3.101}$$

The C-integrability of (3.100) was also already announced in [11]. The special case of (3.100) corresponding to $f(u) = -3a$, $g(u) = b$, $h(u) = 0$ is worth noting; it reproduces (3.84a, b, c) via the position $r(x, t) = [u_x(x, t)]^{-1}$.

The nonlinear evolution PDE satisfied by $u(x, t)$

$$u_t = f(u)\left[\frac{u_{xxx}}{u_x^3} - \frac{9}{4}\left(\frac{u_{xx}}{u_x^2}\right)^2\right] + \frac{1}{2}f'(u)\frac{u_{xx}}{u_x^2} + h(u) \tag{3.102}$$

becomes, via (3.95) followed by the substitution

$$v(\xi, t) = [z(\xi, t)]^{-2}, $$

the linear equation

$$z_t = f(\xi)z_{\xi\xi\xi} + \frac{3}{2}f'(\xi)z_{\xi\xi} + [\tfrac{1}{2}f''(\xi) - h(\xi)]z_\xi - \tfrac{1}{2}h'(\xi)z. \tag{3.103}$$

The C-integrability of (3.102) was already announced in [11].

As an example of the second approach [namely, to relate via (3.95) a *linear* equation for $u(x, t)$ to a *nonlinear* equation for $v(\xi, t)$], we report the following example:

$$u_t = Au + Bu_x + Cu_{xx} + Du_{xxx} , \tag{3.104}$$

$$v_t = A(v - \xi v_\xi) + Cv^2 v_{\xi\xi} + Dv^2[3v_\xi v_{\xi\xi} + vv_{\xi\xi\xi}] , \tag{3.105a}$$

$$w_t = -A(\xi w)_\xi - C(w^{-1})_{\xi\xi} - \frac{D}{2}(w^{-2})_{\xi\xi\xi} . \tag{3.105b}$$

[Equation (3.105b) is related to (3.105a) by the simple change of variable $w(\xi, t) = [v(\xi, t)]^{-1}$; note that for $A = 0$, and up to trivial notational changes, (3.105a) coincides with (3.84a), and (3.105b) with (3.84d).]

Of course the transformation (3.95) [or (3.79)] may be used to relate two *nonlinear* equations to each other. For instance via (3.95) the nonlinear equation

$$u_t = f(u)u_{xxx} + g(u)u_{xx} + h(u)u_x + M(u) \tag{3.106}$$

yields

$$v_t = v^2 \left[\frac{1}{2}f(\xi)(v^2)_{\xi\xi} + g(\xi)v_\xi + h(\xi) + \frac{M(\xi)}{v}\right]_\xi , \tag{3.107}$$

while

$$u_t = f(u_x)u_{xxx} + g(u_x)u_{xx} + H(u_x) + m(u_x)u \tag{3.108}$$

yields

$$v_t = v^2 \left[\frac{1}{2}f(v)(v^2)_{\xi\xi} + g(v)v_\xi + \frac{H(v)}{v} + \frac{\xi m(v)}{v}\right]_\xi . \tag{3.109}$$

Hence, for instance, the equation [see (2.30)]

$$u_t = u_{xxx} + Auu_x + Bu^2 u_x \tag{3.110}$$

yields

$$v_t = v^3 v_{\xi\xi\xi} + 3v^2 v_\xi v_{\xi\xi} + (A + 2B\xi)v^2 . \tag{3.111}$$

Therefore this last equation, for any arbitrary value of the two constants A and B (note that, if $B \neq 0$, it is not autonomous), should be considered S-integrable [being related to the S-integrable equation (3.110) by the change of variables (3.95)]; and it is of course C-integrable if $A = B = 0$ [indeed, it is then a special case of the C-integrable equation (3.84)].

Let us also report an example where the transformation (3.95) produces a neat (C-integrable) evolution equation, after being applied to a *nonlinear* (C-integrable) PDE itself obtained, via a change of variable involving only the dependent variable, from a linear PDE. Indeed let $w(x, t)$ satisfy the *linear* PDE

$$w_t = aw + bw_{xxx} ,$$
(3.112)

and via the change of dependent variable

$$u_x(x,t) = [w(x,t)]^2 ,$$
(3.113)

obtain for $u(x,t)$ the nonlinear C-integrable PDE

$$u_t = 2au + b\left(u_{xxx} - \frac{3}{4}\frac{u_{xx}^2}{u_x}\right) .$$
(3.114)

Now apply the transformation (3.95), thereby obtaining for $v(\xi,t)$ the neat equation

$$v_t = 2(v - \xi v_\xi) + b(v^3 v_{\xi\xi\xi} + \tfrac{3}{2}v^2 v_\xi v_{\xi\xi}) .$$
(3.115)

Note the factor 3/2 in the right-hand side of (3.115), in contrast to the factor 3 in the right-hand side of (3.105a).

These examples should have convinced the reader of the potential usefulness of these approaches to generate potentially interesting nonlinear evolution equations. Of course, many more results than have been reported here obtain by repeated application of changes of variables of the kind considered above, or by more general and systematic analogous approaches. Here we have merely surveyed a few simple transformations that are expedient to identify C-integrable equations. Our presentation has mainly focused on the exhibition of examples. In this same vein, we now show in compact (user-friendly!) form, a list of very simple C- and S-integrable equations, for a single real dependent variable, followed by a synthetic key to their integrability. An asterisk affixed to the equation number indicates that the equation is S-integrable (although it may of course also be C-integrable for some *special* value of the arbitrary constants it features). All the equations listed below without an asterisk are C-integrable (for arbitrary values of the constants and arbitrary choices of the functions that appear in them, which might also arbitrarily depend on the time variable, without spoiling the property of integrability).

$$u_t = u_{xx} + uu_x ,$$
(3.116)

$$u_t = u_{xx} + u_x^2 + a\exp(u) + b ,$$
(3.117)

$$u_t = u^2 u_{xx} + au ,$$
(3.118)

$$u_t = u^2 u_{xx} + au^2 u_x = u^2(u_{xx} + au_x) ,$$
(3.119)

$$u_t = u^2 u_{xx} + a(u - xu_x) = u^2\left[u_{xx} + a\left(\frac{x}{u}\right)_x\right] ,$$
(3.120)

$$u_t = \left(\frac{1}{u}\right)_{xx} + au ,$$
(3.121)

$$u_t = \left(\frac{1}{u}\right)_{xx} + a\left(\frac{1}{u}\right)_x ,$$
(3.122)

$$u_t = \left(\frac{1}{u}\right)_{xx} + a(xu)_x \,, \tag{3.123}$$

$$u_t = \left(\frac{1}{u_x}\right)_x + \frac{a}{u_x} \,, \tag{3.124}$$

$$u_t = \left(\frac{1}{u_x}\right)_x + axu_x \,, \tag{3.125}$$

$$u_t = \frac{f(u_x)}{u_{xx}} + h(u_x) + ug(u_x) \,, \tag{3.126}$$

$$u_{xt} = a\exp(u) + b[\exp(u)]_{xx} \,, \tag{3.127}$$

$$u_{xt} = u_x u_t + (au_x^2 + buu_{xx})\exp(-u) \,, \tag{3.128}$$

$$u_t = u_{xxx} + 2auu_x + 3bu^2 u_x = (u_{xx} + au^2 + bu^3)_x \,, \tag{3.129}*$$

$$u_t = u_{xxx} + 3u^4 u_x + 9uu_x^2 + 3u^2 u_{xx} \,, \tag{3.130}$$

$$u_t = u_{xxx} + au_x^2 + bu_x^3 \,, \tag{3.131}*$$

$$u_t = a(u_{xx} + 2uu_x) + b[u_{xxx} + 3(u^2 u_x + u_x^2 + uu_{xx})]$$
$$= a(u_x + u^2)_x + b(u_{xx} + u^3 + 3uu_x)_x \,, \tag{3.132}$$

$$u_t = a(u_{xx} + u_x^2) + b(u_{xxx} + u_x^3 + 3u_x u_{xx}) \,, \tag{3.133}$$

$$u_t = u^3 u_{xxx} \,, \tag{3.134}*$$

$$u_t = u^3 u_{xxx} + \frac{3}{2}u^2 u_x u_{xx} + a(u - xu_x)$$
$$= u^2 \left[uu_{xx} + \left(\frac{u_x}{2}\right)^2 + \frac{ax}{u}\right]_x \,, \tag{3.135}$$

$$u_t = u^3 u_{xxx} + 3u^2 u_x u_{xx} + (a + bx)u^2 = u^2[\tfrac{1}{2}(u^2)_{xxx} + a + bx] \,, \tag{3.136}*$$

$$u_t = u^3 u_{xxx} + 3u^2 u_x u_{xx} + au^2 u_{xx} + b(u - xu_x)$$
$$= u^2 \left[\left(\frac{1}{2}\right)(u^2)_{xxx} + au_{xx} + b\left(\frac{x}{u}\right)_x\right] \,, \tag{3.137}$$

$$u_t = u^3 u_{xxx}$$
$$\quad + 3u^2(u + u_x)u_{xx}$$
$$\quad + u^2(2u + 3u_x)u_x$$
$$\quad + au^2(u_x + u_{xx}) \tag{3.138}$$

$$u_t = u^{3/2}(u_{xxx} + a + bx) \,, \tag{3.139}*$$

$$u_t = u^{3/2}[u_{xxx} + a(u^{1/2})_{xx} + b(xu^{-1/2})_x] \,, \tag{3.140}$$

$$u_t = u^2 u_{xxx} + \tfrac{1}{9}u_x^3 + b(3u - 2xu_x) \,, \tag{3.141}$$

$$u_t = u^{12/5}u_{xxx} + \tfrac{3}{5}u^{7/5}u_x u_{xx} + b(5u - 4xu_x)$$
$$= u^{7/5}\left[uu_{xx} - \tfrac{1}{5}u_x^2 + 5bxu^{-4/5}\right]_x \,, \tag{3.142}$$

$$u_t = (u^{-1/2})_{xxx} , \tag{3.143)*}$$

$$u_t = (u^{-2})_{xxx} + a + bx , \tag{3.144)*}$$

$$u_t = (u^{-2})_{xxx} + a(u^{-1})_{xx} + b(xu)_x , \tag{3.145}$$

$$u_t = (u^{-2})_{xxx} + 3(u^{-2})_{xx} + 2(u^{-2})_x + a[(u^{-1})_{xx} + (u^{-1})_x] , \tag{3.146}$$

$$u_t = (u^{-6}u_{xx} - 3u^{-7}u_x^2 + cxu^2)_x , \tag{3.147}$$

$$u_t = [(u_x)^{-1/2}]_{xx} , \tag{3.148)*}$$

$$u_t = [(u_x)^{-2}]_{xx} + a + bx + cx^2 , \tag{3.149)*}$$

$$u_t = [(u_x)^{-2}]_{xx} + a[(u_x)^{-1}]_x + bxu_x , \tag{3.150}$$

$$u_t = [(u_x)^{-2}]_{xx} + 3[(u_x)^{-2}]_x + 2(u_x)^{-2} + a\{[(u_x)^{-1}]_x + (u_x)^{-1}\} , \tag{3.151}$$

$$u_t = u_x^{-6}u_{xxx} - 3u_x^{-7}u_{xx}^2 + cxu_x^2 , \tag{3.152}$$

$$u_t = [(u_{xx})^{-1/2}]_x , \tag{3.153)*}$$

$$u_t = [(u_{xx})^{-2}]_x + a_0 + a_1 x + a_2 x^2 + a_3 x^3 , \tag{3.154)*}$$

$$u_t = [(u_{xx})^{-2}]_x + \frac{f(u_x)}{u_{xx}} + h(u_x) + ug(u_x) , \tag{3.155}$$

$$u_t = [(u_{xx})^{-2} + au^2]_x , \tag{3.156}$$

$$u_t = (u_{xxx})^{-1/2} , \tag{3.157)*}$$

$$u_t = (u_{xxx})^{-2} + b_0 + b_1 x + b_2 x^2 + b_3 x^3 + b_4 x^4 , \tag{3.158)*}$$

$$u_t = (u_{xxx})^{-2} + au_x^2 . \tag{3.159}$$

Key to integrability:

(3.116)	"Burgers"; (3.18) with $f = 0$
(3.117)	see (3.20)
(3.118)	see (3.90c)
(3.119)	see (3.94) with $A = D = 0$, rescaled y
(3.120)	see (3.105a) with $D = 0$

[The C-integrability of (3.118), (3.119) and (3.120) was already announced in [11]; but unfortunately these 3 equations were misprinted there; the correct form to ensure C-integrability is that given here.]

(3.121–123)	see (3.118–120) with u replaced by $1/u$
(3.124, 125)	see (3.122, 123) with u replaced by u_x and integrating over x
(3.126)	see (3.80) with $a = 0$
(3.127)	see (3.78a)
(3.128)	see (3.88) with $h(\xi) = 0$, $m(\xi) = c\xi^2$, and u replaced by $\exp(-u)$
(3.129)	"KdV" + "MKdV", see (2.30)
(3.130)	see (3.32) with $f(u) = 0$, and [11]
(3.131)	see (3.129) with u replaced by u_x and integrating over x
(3.132)	"Burgers" + "higher Burgers"; see (3.31) with $f(u) = 0$
(3.133)	see (3.132) with u replaced by u_x and integrating over x (or,

more directly, u replaced by $\ln u$)

(3.134)	"Harry Dym", see [5]
(3.135)	see (3.115)
(3.136)	see (3.111)
(3.137)	see (3.105a)
(3.138)	see (3.94)
(3.139, 140)	see (3.136, 137) with u replaced by $u^{1/2}$
(3.141, 142)	see (3.135) with u replaced by $u^{2/3}$ and $u^{4/5}$
(3.143)	see (3.134) with u replaced by $u^{-1/2}$
(3.144–146)	see (3.136–138) with u replaced by $1/u$
(3.147)	see (3.142) with u replaced by $u^{-5/2}$
(3.148–152)	see (3.143–147) with u replaced by u_x and integrating over x
(3.153, 154)	see (3.148, 149) with u replaced by u_x and integrating over x
(3.155)	see (3.80)
(3.156)	see (3.85b)
(3.157–159)	see (3.153, 154, 156) with u replaced by u_x and integrating over x

Let us end this section by reporting just one more example of application of the transformation (3.95), but now in a $(2+1)$-dimensional context:

$$u(x, y, t) = \xi(x, y, t) , \tag{3.160a}$$

$$u_x(x, y, t) = v(\xi, y, t) . \tag{3.160b}$$

Note that this transformation involves again only the independent variables x and ξ and, of course, the dependent variables u and v. Hence, as in the previous case, not only can the evaluation of v from a given u be performed by purely algebraic operations, but also the evaluation of u from a given v requires merely one quadrature. This would not be, generally, the case for transformations involving simultaneously more independent variables (in which case, incidentally, a more appropriate transformation to use would seem to be (3.79), which clearly has a natural generalization to deal simultaneously with several spatial variables; but, let us reiterate, at the cost of making the computation of u from a given v unwieldy). Of course, transformations that involve, more than parametrically, several dependent variables, but still retain the desirable feature of being invertible by quadratures, may be realized by sequential applications of several transformations, each dealing with one old and one new independent variable.

The change of variables (3.160) transforms the following linear evolution equation for $u(x, y, t)$,

$$u_t = a_{00}u + a_{10}u_x + a_{01}u_y + a_{20}u_{xx} + a_{11}u_{xy} + a_{02}u_{yy}$$
$$+ a_{30}u_{xxx} + a_{21}u_{xxy} + a_{12}u_{xyy} + a_{03}u_{yyy} , \tag{3.161}$$

into the following nonlinear evolution equation for $v(\xi, y, t)$,

$$v_t = a_{00}(v - \xi v_\xi) + a_{01}v_y + a_{20}v^2 v_{\xi\xi} + a_{11}v^2 w_{\xi\xi} + a_{02}(vw_{\xi y} + vww_{\xi\xi})$$
$$+ a_{30}(v^3 v_{\xi\xi\xi} + 3v^2 v_\xi v_{\xi\xi})$$
$$+ a_{21}(2vv_y v_{\xi\xi} + v^2 v_{\xi\xi y} + vv_\xi v_{y\xi} + 3vwv_\xi v_{\xi\xi} + v^2 wv_{\xi\xi\xi})$$
$$+ a_{12}[(w_{\xi\xi}v^2 + w_\xi v_\xi v)_y + (w_{\xi\xi}v^2 + w_\xi v_\xi v)_\xi w$$
$$- (w_{\xi y}v + w_\xi v_y + w_{\xi\xi}vw + v_\xi w_\xi w)v_\xi]$$
$$+ a_{03}[(w_{\xi y}v + w_\xi v_y + w_{\xi\xi}vw + w_\xi v_\xi w)_y$$
$$+ (w_{\xi y}v + w_\xi v_y + w_{\xi\xi}vw + w_\xi v_\xi w)_\xi w$$
$$- (w_{yy} + 2w_{\xi y}w + w_\xi w_y + w_{\xi\xi}w^2 + w_\xi^2 w)v_\xi], \tag{3.162a}$$

$$w_\xi = \frac{wv_\xi + v_y}{v}. \tag{3.162b}$$

Note that this equation is autonomous iff $a_{00} = 0$ (and all the other quantities a are constant; although, to derive (3.162) from (3.161) via (3.160), it is only required that the a variables be x-independent, hence also ξ-independent). This equation may be recast in a form suitable to application of the technique of asymptotic expansion, via the formal substitution $v \to 1 + \varepsilon v$, $w \to \varepsilon w$, $a_{00} \to \varepsilon a_{00}$, $a_{20} \to \varepsilon a_{20}$, $a_{11} \to \varepsilon a_{11}$, $a_{02} \to \varepsilon a_{02}$.

4. Envoi

The results reported in this paper outline a research strategy to identify nonlinear evolution equations that are presumably worthy of study and to actually investigate them. The possibility to discover in this manner *integrable* equations that are also likely to be *applicable* is particularly appealing. This methodology is also useful to study equations generated in some applicative context, by providing straightforward and reliable techniques to test their integrability and to reduce them to simpler forms that might retain applicative relevance (in a regime of weak nonlinearity), while being susceptible to deeper and more revealing study from an analytical point of view (for instance, they might themselves be, in some sense, "integrable", even though the original equations did not posses this property). Finally (last but not least), this methodology, by providing various possibilities to uncover nontrivial relations among integrable equations, contributes to our understanding of this field of research; it supports the hunch that all integrable equations are somehow related to each other, all being manifestations of a single underlying mathematical structure.

Our presentation here has focused on the exhibition of explicit examples rather than a systematic treatment (the best test of a pie comes from the eating). Clearly, much remains to be done, especially in the context of multidimensional problems; hence the character of this presentation is essentially a report on work in progress; as it is indeed appropriate for a chapter with an interrogative title of a book with an interrogative title.

Addendum

In the time that elapsed since my contribution for this book was written, some interesting developments, closely related to the results reported above, have occurred. This *addendum* outlines concisely these findings and provides some relevant references.

The main point of departure, to obtain the results of Sects. 2 and 3, is to take a nonlinear evolution PDE whose *linear* part is *dispersive*, to focus attention on the solution of the *linear* part of the equation that represents *a single dispersive wave*, and then to investigate how such a solution gets modified due to the *nonlinear* part of the equation. The main finding is that, in a regime of *weak nonlinearity*, the main effect is accounted (exactly – in an asymptotic sense) by a *slow modulation* of the amplitude of the dispersive wave; a modulation that, in appropriate ("slow" and "coarse-grained") variables, is generally governed by *universal* nonlinear PDEs (of which the nonlinear Schrödinger equation is a prototype). These universal nonlinear PDEs are important, because their very origin implies that they are likely to be *both widely applicable and integrable*.

An analogous analysis can be performed, taking as point of departure *a superposition of several dispersive waves* (as solution of the *linear* part of the original nonlinear PDE), rather than a single dispersive wave. In this manner one arrives at a coupled system of nonlinear (generally first-order) PDEs, that describe the *slow modulation of several, weakly nonlinearly interacting, dispersive waves*. This phenomenon is again governed by *universal* equations, which are quite interesting, because they are again likely to be *both applicable and integrable*. The derivation of such equations, by a technique closely analogous to that of Sect. 2 above, is given in [29]. And a particular coupled system of nonlinear evolution PDEs of this type, that turn out to be (nontrivially) C-integrable, and that gives rise to a novel solitonic phenomenology, is treated in [30].

A definition of C-integrability somewhat more precise than that given in Sect. 1 reads as follows: "A nonlinear PDE is C-integrable if its solution can be obtained by solving: (i) a finite system of *nondifferential* (possibly nonlinear, i.e., algebraic or transcendental) equations; and (ii) a finite system of *linear PDEs* (including ODEs and quadratures)". The corresponding definition of S-integrability would be closely analogous, except for the replacement (extension) of (ii) as follows: "(ii) a finite system of *linear PDEs* (including ODEs and quadratures), as well as *linear integral equations* (Fredholm, Volterra)".

A nontrivial aspect of this definition has to do with the precise meaning of the term "solving". We mean, as a heuristic definition of this term, "the capability to manufacture a large class of solutions (perhaps all solutions)"; note however that this need not coincide with the capability to solve (if applicable) the initial value ("Cauchy") problem (see below for an additional discussion of this important point).

A technique to manufacture C-integrable nonlinear PDEs, that is closely related to some of the results of Sect. 4, works as follows. Let

$$Dw(x,t) = 0 \tag{A.1}$$

be a linear PDE that implies the "conservation law"

$$h_t(x,t) = g_x(x,t) , \tag{A.2}$$

where h and g are *homogeneous functions of degree p in w, w_x and w_t*. Then set

$$u(x,t) = w(x,t)[F(x,t)]^{-1/p} , \tag{A.3}$$

with

$$F_x(x,t) = h(x,t) , \tag{A.4a}$$

$$F_t(x,t) = g(x,t) . \tag{A.4b}$$

Note that these two equations are compatible thanks to (A.2).

Logarithmic differentiation of (A.3) yields for

$$\phi(x,t) = \frac{w_x(x,t)}{w(x,t)} , \tag{A.5a}$$

$$\chi(x,t) = \frac{w_t(x,t)}{w(x,t)} , \tag{A.5b}$$

the following expressions [via (A.4)]:

$$\phi = \frac{u_x}{u} + H(u,\phi,\chi) , \tag{A.6a}$$

$$\chi = \frac{u_t}{u} + G(u,\phi,\chi) , \tag{A.6b}$$

where

$$H(u,\phi,\chi) = \frac{h}{p}\left(\frac{u}{w}\right)^p , \tag{A.7a}$$

$$G(u,\phi,\chi) = \frac{g}{p}\left(\frac{u}{w}\right)^p . \tag{A.7b}$$

Note that the fact that H and G are functions of u, ϕ and χ is implied, via the definitions (A.3, 4), by the above hypothesis stipulating h and g to be *homogeneous functions of degree p in w, w_x and w_t*.

Solving (A.6) for ϕ and χ, one can express these quantities in terms of u, u_x and u_t.

On the other hand, by applying the linear differential operator D to (A.3), by using (A.1, 3, 4, 5), and by taking again advantage of the fact that h and g are homogeneous functions of degree p, one generally obtains for u a PDE of the form

$$Du = M , \tag{A.8}$$

where M is a function of u, ϕ, χ and the derivatives of these three functions. But since, as we just saw, ϕ and χ can themselves be considered functions of u, u_x, and u_t, the quantity M in the right-hand side of (A.6) can finally be considered a (nonlinear) function of u and its derivatives. Hence (A.6) is a *nonlinear PDE* satisfied by u. And this PDE is *C-integrable*, since it can be transformed into the linear PDE (A.1) via (A.3); or, more precisely, since a very large class of solutions of this equation can be manufactured inserting solutions of the linear PDE (A.1) into the formula

$$u(x,t) = w(x,t) \left[C(t) + \int\limits_{y(t)}^{x} dx' h(x',t) \right]^{-1/p}$$
(A.9a)

with

$$C_t(t) = y_t(t) h[y(t), t] + g[y(t), t] ,$$
(A.9b)

which is clearly implied by (A.3, 4). Note that (A.9b) can be solved by quadratures, since only one of the two functions $C(t)$, $y(t)$ has to be determined (in terms of the other, which can be assigned at one's convenience). Note, on the other hand, that the possibility to solve the Cauchy problem for the nonlinear PDE (A.8) hinges on the possibility to invert the transformation (A.3), namely to be also able to express w in terms of u; this might be possible by quadratures, or might require solving a nonlinear ODE, depending on the structure of h.

A class of C-integrable nonlinear PDEs can be generated in this manner taking for D the odd differential operator

$$D = \frac{\partial}{\partial t} + \sum_{j=0}^{J} a_j \frac{\partial^{2j+1}}{\partial x^{2j+1}} ,$$
(A.10)

and for h the quadratic expression

$$h = \alpha w^2 + \beta w_x^2 ,$$
(A.11)

with α and β arbitrary constants, since a corresponding quadratic expression for g can then be always found. Some examples generated by this technique will be reported elsewhere (note, incidentally, that in this case the function χ need not be introduced at all).

Here we report explicitly only two simple examples; the first of which, incidentally, is likely to pose a difficult challenge for whoever hopes that the "Painlevé approach" may provide a general method to identify integrable equations.

First example. Let (A.1) read

$$w_{xt} = 0 ,$$
(A.12)

and set

$$h(x, t) = \alpha(x)[w_x(x, t)]^p \,, \tag{A.13a}$$

$$g(x, t) = \beta(t)[w_t(x, t)]^p \,, \tag{A.13b}$$

with $\alpha(x)$ and $\beta(t)$ arbitrary functions and p an arbitrary (nonvanishing) number.

One then obtains for $v = u^{-p}$ the nonlinear PDE (written below in two *equivalent* forms; see (A.15a, b)):

$$v_{xt} = -p[\chi v_x + \phi v_t + (p-1)\chi\phi v] \,, \tag{A.14a}$$

$$v_{xt} = -p[\alpha\phi^p\chi + \beta\chi^p\phi - (p+1)\chi\phi v] \,, \tag{A.14b}$$

where ϕ and χ are related to v, v_x and v_t by the (nondifferential) equations

$$\alpha\phi^p - p\phi v - v_x = 0 \,, \tag{A.15a}$$

$$\beta\chi^p - p\chi v - v_t = 0 \,. \tag{A.15b}$$

Note that these equations can be *explicitly* solved for ϕ and χ if p takes one of the following values:1, 2, 3, 1/2, 1/3. As implied by its derivation, the nonlinear PDE (A.14) is C-integrable; indeed, its *general solution* is given by the explicit formula

$$v(x, t) = [f_1(x) + f_2(t)]\left\{\left[\frac{f_1'(x)}{\alpha(x)}\right]^{1/p} + \left[\frac{f_2'(t)}{\beta(t)}\right]^{1/p}\right\}^{-p} \,, \tag{A.16}$$

where $f_1(x)$ and $f_2(t)$ are arbitrary functions.

Second example. Let (A.1) read

$$w_{xt}(x, t) - a(x, t)\, w(x, t) = 0 \,, \tag{A.17}$$

where $a(x, t)$ satisfies the restriction

$$[\alpha(x)\, a(x, t)]_x = [\beta(t)\, a(x, t)]_t \tag{A.18}$$

with $\alpha(x)$ and $\beta(t)$ a priori arbitrary (except for this restriction; which, incidentally, is identically satisfied if α, β and a are three arbitrary constants); and set

$$h = \alpha w_x^2 + a\beta w^2 \,, \tag{A.19a}$$

$$g = \beta w_t^2 + a\alpha w^2 \,. \tag{A.19b}$$

There obtains then for $u(x, t)$ the C-integrable nonlinear PDE

$$\begin{aligned} u_{xt} - au = -\tfrac{1}{2}u^3[(a\alpha)_x \\ + (\alpha\phi + \beta\chi)(\tfrac{3}{2}a + \phi\chi) \\ - \tfrac{3}{2}u^2(\alpha\phi^2 + a\beta)(\beta\chi^2 + a\alpha)] \end{aligned} \tag{A.20}$$

with

$$\phi = \frac{1 - s_1[1 - \alpha u(\alpha\beta u^3 + 2u_x)]^{1/2}}{\alpha u^2} , \tag{A.21a}$$

$$\chi = \frac{1 - s_2[1 - \beta u(\alpha\alpha u^3 + 2u_t)]^{1/2}}{\beta u^2} , \tag{A.21b}$$

and $s_1^2 = s_2^2 = 1$.

An apparently simpler avatar of this equation is obtained setting $v = u^{-2}$, so that (A.20) is replaced by

$$v_{xt} = (a\alpha)_x - 2v(a - 3\phi\chi) \\ \tfrac{1}{2}(\alpha\phi + \beta\chi)(3a + 4\phi\chi) , \tag{A.22}$$

and (A.21) by

$$\phi = \frac{v - s_1 \left[v^2 - \alpha(a\beta - v_x)\right]^{1/2}}{\alpha} , \tag{A.23a}$$

$$\chi = \frac{v - s_2 \left[v^2 - \beta(a\alpha - v_t)\right]^{1/2}}{\beta} , \tag{A.23b}$$

again with $s_1^2 = s_2^2 = 1$. Of course (A.22, 23) coincide with (A.14, 15) if $p = 2$ and $a = 0$.

Extensions of the technique outlined above, which also yield interesting results, include the possibility to take as point of departure, rather than the linear PDE (A.1), a nonlinear PDE (itself C-integrable, or S-integrable), and/or to consider nonhomogeneous conservation laws. Another possible extension is to let h and g depend also on higher derivatives of w than the first. Examples of integrable nonlinear PDEs manufactured in this manner will be reported elsewhere.

References

1   Galileo Galilei: *Il Saggiatore*, 1623; *Opere di Galileo Galilei* Vol. VI (Edizione nazionale, Barbèra, Firenze 1890–1909) [see p. 232]
2   F. Calogero, W. Eckhaus: "Nonlinear Evolution Equations, Rescalings, Model PDEs and Their Integrability. I", Inverse Problems 3, 229–262 (1987)
3   F. Calogero, W. Eckhaus: "Nonlinear Evolution Equations, Rescalings, Model PDEs and Their Integrability. II", Inverse Problems 4, 11–33 (1988)
4   F. Calogero, W. Eckhaus: "Necessary Conditions for Integrability of Nonlinear PDEs", Inverse Problems 3, L27–L32 (1987)
5   F. Calogero, A. Degasperis: *Spectral Transform and Solitons: Tools to Solve and Investigate Nonlinear Evolution Equations. I* (North-Holland, Amsterdam 1982)
6   S. P. Novikov, S. V. Manakov, L. P. Pitaevskii, V. E. Zakharov: *Theory of Solitons* (Plenum, New York 1984) [Original Russian title: *Teoria solitonov*]
7   Suppl. Progr. Theor. Phys. 55, 1974 (issue devoted to the "Reductive Perturbation Method for Nonlinear Wave Propagation")

8   W. Eckhaus: "The Long-Time Behaviour for Perturbed Wave-Equations and Related Problems", preprint *404*, Department of Mathematics, University of Utrecht, December 1985, published in part in Lect. Notes Phys. (Springer, Berlin–Heidelberg–New York 1986)

9   V. E. Zakharov, E. A. Kuznetsov: "Multiscale Expansion in the Theory of Systems Integrable by the Inverse Scattering Transform", Physica 18 D, 455–463 (1986)

10   F. Calogero, A. Maccari: "Equations of Nonlinear Schrödinger Type in $1+1$ and $2+1$ Dimensions, Obtained from Integrable PDEs", in *Inverse Problems: an Interdisciplinary Study* (Proceedings of the Meeting on Inverse Problems held in Montpellier, November 1986), ed. by P. C. Sabatier, Advances in Electronics and Electron Physics 19 (Academic, London 1987) 463–480

11   F. Calogero: "The Evolution PDE $u_t = u_{xxx} + 3(u_{xx}u^2 + 3u_x^2 u) + 3u_x u^4$", J. Math. Phys. 28, 538–555 (1987)

12   V. V. Sokolov, A. B. Shabat: "Necessary Conditions on Nontrivial Lie-Bäcklund Algebras and Existence of Conservation Laws", preprint (in Russian), Department of Physics and Mathematics, Bashkirian Section of the Soviet Academy of Sciences, Ufa, 1982

13   F. Calogero, S. De Lillo:"The Eckhaus PDE $i\psi_t + \psi_{xx} + 2(|\psi|^2)_x \psi + |\psi|^4 \psi = 0$", Inverse Problems 3, 633–681 (1987)

14   F. Calogero, S. De Lillo: "On the Eckhaus Equation", in: *Nonlinear Evolutions*, Proceedings of the 4th NEEDS Workshop held in Baraluc-les-Bains, June 1987, ed. by J. Leon (World Scientific, Singapore 1988) 691–696

15   F. Calogero, S. De Lillo: "Cauchy Problems on the Semiline and on a Finite Interval for the Eckhaus Equation", Inverse Problems 4, L33–L37 (1988)

16   F. Calogero, S. De Lillo: to be published

17   F. Calogero, D. Levi, A. Maccari: to be published

18   A. V. Mikhailov, A. B. Shabat, R. I. Yamilov: "Extension of the Module of Invertible Transformations. Classification of Integrable Systems", Commun. Math. Phys. 115, 1 (1988)

19   A. V. Mikhailov: private communication

20   V. E. Zakharov: "The Inverse Scattering Method", in *Solitons*, ed. by R. K. Bullough, J. P. Caudrey, Top. Curr. Phys. Vol. 17 (Springer, Berlin–Heidelberg–New York 1980) 243–285

21   G. Leebert, C. Karney, A. Bers, D. Kaup: "Two-dimensional Self-modulation of Lower Hybrid Waves in Inhomogeneous Plasmas", Phys. Fluids 22, 1545–1553 (1979)

22   B. B. Kadomtsev, B. I. Petviashvili: "On the Stability of Solitary Waves in Weakly Dispersive Media", Sov. Phys. Dokl. 15, 539–541 (1970) [Russian original: Dokl. Akad. Nauk SSSR 192, 753–756 (1970)]

23   A. Davey, K. Stewartson: "On Three-Dimensional Packets of Surface Waves", Proc. Roy. Soc. London A 338, 101–110 (1974)

24   D. Anker, N. C. Freeman: "On the Soliton-Solutions of the Davey–Stewartson Equation for Long Waves", Proc. Roy. Soc. London A 360, 529–540 (1978)

25   M. Boiti, J. J.–P. Leon, M. Manna, F. Pempinelli: "On the Spectral Transform of a Korteweg–de Vries Equation in Two Spatial Dimensions", Inverse Problems 2, 271–279 (1986)

26   E. I. Schulman: "On the Integrability of Equations of Davey–Stewartson Type", Theor. Math. Phys. 56, 720–724 (1983) [Russian original: Teor. Mat. Fiz. 56, 131–136 (1983)]

27   M. Boiti, J. J.–P. Leon, F. Pempinelli: "Integrable Two-Dimensional Generalization of the Sine and Sinh-Gordon Equations", Inverse Problems 3, 37–49 (1986)

28   L. P. Nizhnik: "Integration of Multidimensional Nonlinear Equations by the Method of the Inverse Problem", Sov. Phys. Dokl. 25, 706–708 (1980) [Russian original: Dokl. Akad. Nauk SSSR 254, 332–335 (1980)]

29   F. Calogero: "Universality and integrability of the nonlinear PDEs describing N-wave interactions", J. Math. Phys. 30, 28–40 (1989)

30   F. Calogero: "Solutions of certain integrable nonlinear PDEs describing nonresonant N-wave interactions", J. Math. Phys. 30, 639–654 (1989)

Painlevé Property and Integrability

N. Ercolani and E. D. Siggia

1. Background

A brief account of this work has appeared in [1] and an expanded version in [2]. This discussion is merely intended to illustrate, in the simplest possible terms, our approach to the Painlevé problem. Precise statements of theorems, in the few instances where we have them, as well as nontrivial examples are all reserved for [2].

We will consider only polynomial systems of equations ($\cdot \equiv d/dt$)

$$\dot{x}_i = f_i(\{x_j\}) \quad i, j = 1, \ldots n , \tag{1.1}$$

so that the right-hand side of (1.1) does not introduce any singularities into the time flows. In fact, the differential equations define the continuation of $x_i(t)$ from real to complex times. All variables will henceforth be understood as being complex. The Painlevé conjecture then asserts that if the singularities of $x_i(t)$ are no worse than poles for all t, (i.e., no branch points, logarithms, essential singularities, etc.) then (1.1) is integrable. We define "integrable" below. In other terms, a system whose solutions are globally meromorphic is said to possess the Painlevé property.

1.1 Motivation

There are several reasons for pursuing this subject that go beyond a specific interest in integrable systems. However integrability is defined, it is clearly a global property of the flows in phase space. Integrals are smooth functions that are defined *globally*. Any analytic system of equations, however, is *locally* integrable but the local patches of level sets do not in general fit together.

Singularity analysis is also purely local. Polynomial equations only have singularities when some variable blows up. The leading singularity can be guessed by assuming a form, e.g., bt^{-a}, and calculating a and b. When the leading singularity is a pole, it is fairly simple to continue the series. Logarithms may then appear, which means the system is not Painlevé. If it is, on the other hand, then formal Laurent series are obtained. With additional labor their convergence may be established. This is all local analysis, and can be made very explicit.

The problem, then, is to understand why local analysis should have global, geometric consequences.

A second general context into which the Painlevé conjecture falls is that of singularities in nonlinear systems. Poles are a particularly mild singularity, and for Painlevé systems their occurrence is merely an indication that one is working on too small a phase space. Most of this chapter will be devoted to constructing an augmented phase space on which the flows are analytic at all times. Other types of singularities may not admit such an interpretation, but it is interesting to understand the Painlevé case first, which is not altogether trivial.

A third intriguing aspect of the Painlevé conjecture is that it might be made into a tool for computing integrals. Currently, to prove a system integrable means displaying the integrals or solving the initial value problem. This requires ingenuity and insight. Any constructive algorithm would be useful. Partial results along these lines comprise the last section.

Lastly, a thorough understanding of the relationship between singularities and integrability could lead to an alternative to the KAM notion of being "near" to integrable.

1.2 History

Kowalevskaya was the first to use singularity analysis to screen Hamiltonian systems for integrability. Her reasoning was that all other integrable systems known at that time had meromorphic solutions. It was relatively simple to find a previously unknown set of parameter values for which the equations for a top in a gravitational field had only poles. She then explicitly solved the initial value problem in terms of hyperelliptic functions, thereby demonstrating integrability [3].

Some forty years later Painlevé considered all equations of the form

$$\ddot{x} = f(t, x, \dot{x})$$

which are analytic in t and rational in x and $\dot{x}$, and for which the only moveable singularities (those whose location depends on initial data) are poles. In addition to the known transcendental functions, six new ones were found, the so-called Painlevé transcendents [4]. Success in integrating several of these by inverse scattering methods lead *Ablowitz* et al to suspect a connection between the Painlevé property and integrability [5]. Many examples have been worked out since then [6].

2. Integrability

There is no single definition of integrability that seems appropriate in all circumstances. We mention two common ones (for autonomous systems).

Liouville integrability follows for an n-degree of freedom Hamilton system if there are n-independent integrals $K_i(q, p)$ in involution [7]. If the intersection of their level sets is compact, then it is topologically a complex n-torus.

For the system (1.1), we define ODE integrability to be existence of $n - 1$ independent analytic integrals. We note several pecularities of these definitions.

An integrable Hamiltonian system is generally not ODE integrable. If we insist on functional independence of the (real) integrals at infinity formed from the real and imaginary parts of $\mathcal{H}$, then $\mathcal{H} = p^2 - q^5 - 1$ is not Liouville integrable (i.e., the level surface is of genus 2, not a torus, and any vector field on it must vanish somewhere). Thus, although a solution exists locally for real and finite (q, p), this example is not Liouville integrable over the complexes. The converse pathology occurs for any repulsive and short ranged potential between pairs of particles in an n-particle system on the real line. These examples are always integrable since the momenta at $t = +\infty$ are constant and probably analytic in the initial data.

3. Riccati Example

The Riccati equation [4], though ancient, and not Hamiltonian, is an exceedingly instructive illustration of how the singularities serve to augment the original phase space in such a way that on the augmented manifold the flow is analytic. The equation reads

$$\dot{x} = a_0(t) + a_1(t)x + a_2(t)x^2 , \tag{3.1}$$

where $a_i(t)$ are entire in t. Clearly $x(t)$ is analytic whenever it is defined (i.e., finite). The only singularities are poles, near which $\dot{x} \sim a_2 x^2$ or $x \sim cst/(t - t_0)$. Consider the substitution, $\tilde{x} = 1/x$, under which (3.1) becomes

$$-\dot{\tilde{x}} = a_0 \tilde{x}^2 + a_1 \tilde{x} + a_2 . \tag{3.2}$$

Clearly (3.2) is also an analytic equation, and $\tilde{x}(t)$ is analytic whenever it exists, particularly around $\tilde{x} = 0$, where $\tilde{x}$ vanishes as $\sim (t - t_0)$. Therefore, we have proven that the Laurent series for x, which could be found by formally expanding (3.1) actually converges.

Define an *augmented phase space* as the set

$$M = \{x \in \mathbb{C}\} \cup \{\tilde{x} = 0\} . \tag{3.3}$$

Cover $\tilde{x} = 0$ with the open set (patch) $\{\tilde{x} \in \mathbb{C}\}$ and identify this path with the other one, $\{x \in \mathbb{C}\}$, by the *transition function* $\tilde{x} = x^{-1}$ for x and $\tilde{x} \neq 0$. This makes M into a manifold.

In this example M is compact, and is just the Riemann sphere. Consider the finite time map φ_t. It takes M 1:1 and onto itself and is biholomorphic since (3.1, 2) are analytic. Hence, it must be fractional linear, i.e.,

$$y(t) = \frac{\alpha y(t_0) + \beta}{\gamma y(t_0) + \delta}, \tag{3.4}$$

where α, β, γ and δ are entire functions of t and t_0. [Proof: Assume φ_t maps $y(t_0) = \Delta$ to infinity, write a Laurent series $y(t) = \sum a_n(y(t_0) - \Delta)^n$, then $a_{n>0} = 0$ since otherwise infinity would also map to infinity. Furthermore, $a_{n \leq -2} = 0$, since the map is uniquely invertible.]

One could imagine a multivariable generalization of (3.4) for which there would be no invariant integrals even though one has an explicit formula for the dependence on initial data. Such a system should surely be called solvable though it is not integrable in the technical sense defined above.

The main body of this chapter will be devoted to illustrating how the steps which were required to construct M apply to an arbitrary Hamiltonian and Painlevé system.

4. Balances

The information that one can obtain through a singularity analysis of the equations of motion for a polynomial Hamiltonian will now be considered. Solutions of entire analytic differential equations will only fail to be locally analytic in time and initial data when they blow up. A *balance* is defined to be the leading term in a formal asymptotic expansion about such a singularity. We say "formal" since the convergence of the series is not obvious. All balances for a Painlevé system must in fact be the first term in a Laurent expansion. Testing for the Painlevé property usually means showing that all formal solutions around any singular point are Laurent.

A *principal* balance will be a formal Laurent (or equivalently pole) series with the maximum number of free constants allowed by the dimension of the phase space. *Lower* balances can be ordered by the number of free constants. For instance, for

$$\mathcal{H} = \tfrac{1}{2}(p_1^2 + p_2^2) + q_1^2 q_2 + 2q_2^2 \tag{4.1}$$

one finds a principal balance

$$\begin{aligned}
q_1 &= c_1 t^{-1} + \tfrac{5}{12}c_1 t + c_2 t^2 - \tfrac{1}{72}c_1^5 t^3 - \tfrac{5}{6}c_2 t^4 - \tfrac{1}{5}c_1 c_3 t^5 + o(t^5) \\
q_2 &= -t^{-2} + \tfrac{1}{12}c_1^2 + \tfrac{1}{48}c_1^4 t^2 + \tfrac{1}{3}c_1 c_2 t^3 + c_3 t^4 + o(t^5).
\end{aligned} \tag{4.2}$$

In addition to c_1, c_2, and c_3, a fourth constant t_0 has been hidden in $t = $ time $-t_0$.

For any balance we define a *resonance* ϱ to be the lowest power of t at which a new constant enters the series measured from the leading one, i.e., $q_i = t^{-f_i}$ $(cst + \ldots + t^{\varrho})$. Thus in (4.2) the resonances occur for $\varrho = 0, 3, 6$, and -1 for t_0. Equation (4.1) also has two lower balances

$$q_1 = \pm 6i t^{-2}(1 + \ldots), \quad q_2 = -3t^{-2}(1 + \ldots)$$

with resonances $\varrho = -1, 6, 8$.

The principal balance series can be used as a formal variable change from $\{q, p\}$ to t_0 and the constants. Thus we find that the value of $\mathcal{H}$ may be expressed as

$$E = 14c_3 + \tfrac{35}{432}c_1^6 .$$

The 2-form $\omega^{(2)}$ can be rewritten and, of course, must be t independent;

$$\omega^{(2)} = \sum_i dp_i \wedge dq_i = dt_0 \wedge dE + 3 \, dc_2 \wedge dc_1 . \tag{4.3}$$

Note that (4.3) establishes a conjugate pairing between the constants and leads to relations among the associated ϱ [2].

5. Elliptic Example

We now consider how the construction of a manifold on which the flows exist, analytically, for all times may be extended to a Hamiltonian system. The general argument plus nontrivial examples are given in [2]. Consider as an example

$$2\mathcal{H}(q, p) = p^2 - 4q^3 - 2q .$$

The principal, and only, balance reads

$$q = (t - t_0)^{-2}(1 + \ldots), \quad p = -2(t - t_0)^{-2}(1 + \ldots) .$$

We will solve the Hamilton-Jacobi equation perturbatively around infinity for a canonical variable change analogous to $\tilde{x} = x^{-1}$ in the Riccati example. Thus

$$\mathcal{H}(q, \partial S/\partial q) = E \tag{5.1}$$

has the approximate solution

$$\tilde{S}(q, v) = 4/5q^{5/2} + q^{1/2} - vq^{-1/2} , \tag{5.2}$$

which satisfies (5.1) to $O(q^{-1})$. We have replaced E by v in (5.2) since we want to use v as a variable in the coordinate patch at infinity. Since $\tilde{S}$ is only approximate, v will not be constant and it would be misleading to call it E.

Define a variable change by

$$u = \partial \tilde{S}/\partial v \,, \quad p = \partial \tilde{S}/\partial q \,,$$

which implies

$$q = u^{-2}$$
$$p = -2u^{-3} - \tfrac{1}{2}u - \tfrac{1}{2}vu^3 \,. \qquad (5.3)$$

Since the variable change is canonical, the equation of motion in (u, v) variables is derived from a Hamiltonian,

$$\mathcal{H}(u, v) = v + \tfrac{1}{8}u^2 + \tfrac{1}{4}vu^4 + \tfrac{1}{8}v^2u^6 \,.$$

In particular, for small u (i.e., p, q near infinity)

$$\dot{u} = 1 + O(u^4) \,, \quad \dot{v} = O(u) \,.$$

Hence v is approximately constant and $u \sim (t - t_0)$. This could be seen equally well by comparing the transition functions (5.3) with the time series. Since $\mathcal{H}(u, v)$ is polynomial, the flows are analytic around infinity.

Define an augmented manifold by

$$M = \{q, p \in \mathbf{C}^2\} \cup \{u = 0, v \in \mathbf{C}\} \,.$$

Note that we have resolved the singularity, and "infinity" is nothing but $u = 0$, and v is an arbitrary complex number

If we were to take this curve and integrate "backwards" by $-t$ we would obtain another analytic curve consisting of all initial data that hits infinity in a time t. Finally, to make M into a manifold we have to cover infinity by a patch consisting of a tube around $u = 0$ which is narrow enough so that the transition functions are uniquely invertible for $u, v(q, p)$.

Note that M is not compact. It ignores points such as $p = $ infinity, $q = $ finite which are never reached by the flow.

6. Augmented Manifold

In [2] a general algorithm is presented for constructing an augmented manifold M for any polynomial Hamiltonian system with the properties:

1. $\{p, q\} \in \mathbf{C}^{2n}$ is a dense subset of M,
2. $M - \mathbf{C}^{2n}$ is a finite union of analytic hypersurfaces,
3. the time flows extend to M, exist for all times, and are analytic, and
4. the transition functions between the patches of M are canonical.

The principal balance(s) in local coordinates always correspond to $u = 0$, $\{v \in \mathbf{C}, i = 1, 2, \ldots, 2n - 1\}$; that is they are codimension one. If one initial point $\{p, q\}$ blows up in a time t_0 then an open set of neighboring points blow up in a time near to t_0 (i.e. integrate the u equation backward from $u = 0$). The lower balances add points to the boundary of the principal balance (i.e., certain v_i infinite). In appropriate local coordinates, the added sets are just given by $u_i = 0$ with the number of variables u_i equalling the codimension. The codimension 2 lower balances may be realized as a singularity in the principal balance equations in which certain v_i approach infinity as $u \to 0$. (Clearly any singularity in these equations for $u \neq 0$ is due to bad coordinates and disappears when transferred back to q, p variables.)

The Hamiltonian in local variables is always analytic and there is a coordinate patch for each balance, implying conditions 2, 3. Lastly, all computations are done with the Hamilton-Jacobi equation, guaranteeing that M is symplectic.

7. Argument for Integrability

Given a manifold M with the properties just described, there is a heuristic argument as to why the flow is simple. The assertion is not that it is integrable in the technical sense but rather that either the finite-time map is birational as in the Riccati example or a level surface exists in the form of a time dependent entire function $F(t, q, p)$ whose total time derivative is zero.

The argument, which is no better than intuitive exploits the characterization of "entire" functions on a manifold by rate of growth. The simplest example of this reasoning is Liouville's theorem which says that if the maximum modulus of an entire function grows algebraically then it is a polynomial. There is an extension to several variables which leads with some reasoning to (3.4) which applies when M is either compact, or the rate of growth of the finite time map ϕ_t is algebraic as its arguments tend towards the omitted regions (i.e., $p \to \infty$, q finite, for the elliptic example). In the latter case, M admits a formal compactification. Bianalytic maps such as ϕ_t between compact spaces can all be given explicit functional forms, as in the Riccati example. We consider all such examples to be solved.

Hence the only problematic case is when ϕ_t behaves essentially as its arguments tend toward the points required to compactify. Consider a trivial example,

$$\mathcal{H} = \tfrac{1}{2}(pq)^2$$
$$p_t = p_0 e^{-p_0 q_0 t}, \qquad q_t = q_0 e^{p_0 q_0 t}.$$

The map ϕ_t is clearly essential but the exponent depends only on an invariant $p_t q_t = p_0 q_0$. We believe something like this must happen if composition is not to result in an explosion of essentialness as in $\exp(\exp(\ldots))$. Thus, we would like to claim that an invariant surface results from the group property of ϕ_t plus essential growth.

The basic argument may be repeated if the first level set does not result in a compactifiable submanifold.

We also mention in this section a refinement of the Painlevé test which permits one to detect ODE integrability.

Conjecture. If there are no lower balances, then a Painlevé system is ODE integrable.

The converse is clearly trivial.

To argue in the direction stated, consider the set of all complex t poles for given initial conditions. If there is one pole, there must be an infinite number by Picard's Theorem. The assumption of no lower balances implies that these poles cannot coalesce or disappear as initial conditions are changed. Consider any two and use them to define a map from the data at infinity to itself by integrating from one to the other. This generates a bi-entire map from C^{2n-1} to C^{2n-1}. We would like to assert that integral invariants exist.

8. Separability

We first illustrate how to solve an integrable system by separating variables in the Hamilton-Jacobi equation. There are separable (and hence integrable) equations which are not Painlevé but this occurs in the known examples because one is working in too small a phase space. If a system is separable in a technical sense yet to be defined, then the local analysis embodied in the Painlevé test yields a good deal of information about how to perform the separation and the form of the other integrals in involution. First, as an example, we separate (4.1).

Let

$$q_1 = i\xi_1\xi_2 , \quad q_2 = \tfrac{1}{2}(\xi_1^2 + \xi_2^2) . \tag{8.1}$$

Re-express $\partial S/\partial q$ interms of $\partial S/\partial \xi_i = \eta_i$ in the equation

$$\mathcal{H}\left(q, \frac{\partial S}{\partial q}\right) = h_1 ,$$

and one finds a hyperelliptic curve γ:

$$\eta_i^2 = -\tfrac{1}{2}\xi_i^3 + 2h_1\xi_i^2 + h_2 \tag{8.2}$$

The action is

$$S = \sum_1^2 \int^{\xi_i} \eta_i d\xi_i .$$

The numbers h_i are the values of the two integrals in involution. Equation (8.2) written for $i = 1, 2$ can be simultaneously inverted to find $h_i(\xi, \eta)$ or $H_i(q, p)$. The "times" conjugate to h_i are given by $t_i = \partial S / \partial h_i$. If we differentiate this pair of equations we find an expression for the flows $\xi_i(t_j)$ on the level surfaces:

$$\delta_{ij} = \left(\frac{\partial^2 S}{\partial h_i \partial \xi_k} \right) \left(\frac{\partial \xi_k}{\partial t_j} \right) .$$

Definition. For a *hyperelliptically separable* system there is
a. a good variable change $q_i = q_i(\xi_j)$ where q_i is a symmetric function of $\{\xi_j\}$,
b. under which the Hamilton-Jacobi equation separates into n copies of a hyperelliptic curve γ with the equation,

$$\begin{aligned} \eta^2 &= \xi^d + \dots \\ d &\geq 2n + 1 , \end{aligned} \tag{8.3}$$

c. the integrals h_i occur as the coefficients of ξ^{α_i} with $\alpha_i + 1 - d/2 < 0$,
d. the set $\{t_i = \partial S / \partial h_i, \ i = 1, 2, \dots n\}$ modulo the periods of γ is a torus.

The fourth condition guarantees that the Hamiltonian in question is Painlevé. When $d = 2n + 1$, condition (d) is automatic and for larger d some symmetry is required of γ. Otherwise the period lattice would be of rank $> 2n$. The above example is hyperelliptically separable.
The following facts then follow:

a. The principal balance corresponds to $\xi_1 \sim t^{-1}$ or t^{-2} (depending on whether d is even or odd) and $\xi_{i>1} \sim cst$. The leading exponents f_i, g_i for $q \sim t^{-f_i}$, $p \sim t^{-g_i}$ are the same for all the n-flows in involution:
b. $H_i(q, p)$ is polynomial in q, p;
c. there is a lowest balance with just n free constants, the h_i, plus t_0;
d. from an expansion of the Hamilton-Jacobi equation at a principal balance, there follows the degree d of γ and bounds on the weighted degrees of $H_i(q, p)$. Here, each q, p is given the weight f_i, g_i defined in (a).

By comparing the series (4.2) with the separating variable change (8.1) it will be observed that property (a) reduces the calculation of (8.1) to checking only a few possibilities. In (d), the degree of $\eta^2(\xi)$ follows from expanding $S \sim \int^{\xi_1(t)} \eta_1 d\xi_1$ and comparing with the first (largest) term in $S(q(t), v)$. The degrees of H_i basically reflect the order of their occurrence, α_i, in η^2, but their calculation from the Hamilton-Jacobi expansion is laborious [2].
We believe similar results can be formulated when γ is replaced by a rational function in ξ or something higher order in η.

References

1   N. Ercolani, E.D. Siggia: Phys. Lett. A **119**, 112 (1986)
2   N. Ercolani, E.D. Siggia: Physica D **34**, 303 (1989)
3   V.V. Golubov: *Lectures on Integration of the Equations of Motion of a Rigid Body About a Fixed Point* (State Publishing House, Moscow 1953)
4   E.L. Ince: *Ordinary Differential Equations* (Dover, New York 1947)
5   M.J. Ablowitz, A. Ramani, H. Segur: J. Math. Phys. **21**, 715 (1980)
6   Y.F. Chang, M. Tabor, J. Weiss: J. Math. Phys. **23**, 531 (1982);
     T. Bountis, H. Segur, F. Vivaldi: Phys. Rev. A **25**, 1257 (1982);
     T. Bountis: in *Dynamical Systems and Chaos,* ed. by L. Garrido
7   V.I. Arnold: *Mathematical Methods of Classical Mechanics*, Graduate Texts in Mathematics, Vol. 60 (Springer, Berlin-Heidelberg 1980)

Integrability

H. Flaschka, A.C. Newell and M. Tabor

1. Integrability

A comprehensive definition of the term "integrable" is proving to be elusive. Rather, use of this term invokes a variety of intuitive notions (and not infrequently, some lively debate) corresponding to a belief that integrable systems are in some sense "exactly soluble" and exhibit globally (i.e., for all initial conditions) "regular" solutions. In contrast, the term "nonintegrable" is, generally, taken to imply that a system cannot be "solved exactly" and that its solutions can behave in an "irregular" fashion. Here the notion of irregular behavior corresponds to dynamics that are very sensitive to initial conditions, with neighboring trajectories in the phase space locally diverging on the average at an exponential rate. This characteristic is measured by Lyapunov exponents. A system with at least one positive exponent will display irregular motion. In contrast, regular motion is associated with no positive exponents. Unfortunately, the definition of the Lyapunov exponents involves long time averages, their existence is only guaranteed for a limited set of situations and their values are difficult to compute both analytically and numerically. It is unlikely, therefore, that an algorithm which tests a given system for Lyapunov exponents will be a successful test for integrability.

To justify our own ideas on "integrability", we discuss a variety of concepts for both finite and infinite dimensional systems that illustrate how difficult it is to make a "universal" definition. To begin, consider the range of possibilities for finite systems of ordinary differential equations. A traditional point of view is/was to talk about "integration by quadratures". Loosely speaking, this means that for an nth order system the identification of a sufficient number of analytic first integrals of motion reduces the system (by one order for each integral) to a form that can be explicitly integrated. For second order systems, such as elliptic differential equations, the identification of one integral reduces the solution to a single quadrature. For example, for

$$\ddot{x} = ax + bx^2 + cx^3 , \tag{1.1}$$

the integral is just

$$I = \tfrac{1}{2}\dot{x}^2 - \tfrac{1}{2}ax^2 - \tfrac{1}{3}bx^3 - \tfrac{1}{4}cx^4$$

and the resulting quadrature is

$$t - t_0 = \int \frac{dx}{\sqrt{2(I + ax^2/2 + bx^3/3 + cx^4/4)}} , \qquad (1.2)$$

and the inversion of this formula is achieved in terms of Jacobi elliptic functions.

As is well known, a quadrature type definition of integrability is far too restrictive. For example, the simple first order equation

$$\dot{x} = x^2 + t \qquad (1.3)$$

cannot be separated and therefore integrated by quadratures. However, such an equation is still "exactly soluble" in the sense that the direct linearization obtained by the substitution $x = -\dot{y}/y$ enables one to express the solution in terms of Airy functions. In a similar vein, the second order equation

$$\ddot{x} = tx + x^3 + y , \qquad (1.4)$$

namely the second Painlevé Equation PII, cannot be integrated by quadratures. Neither is there any direct linearization. Nonetheless, this equation, and the other five equations for the Painlevé transcendents I, III, IV, V, VI, are certainly not nonintegrable. A more sophisticated indirect linearization known as the Inverse Monodromy Transform [1, 2] (a relative of the Inverse Scattering Transform used for certain classes of PDEs) can be used to find the general solution.

For Hamiltonian systems the notion of integrability seems to be clear. Here, due to the special symplectic structure of Hamiltonian phase space, the identification of just N integrals of motion (for a $2N$-dimensional phase space) enables one to reduce the solution to a set of trivial integrations. Global existence of solutions of all commuting flows guarantees that the intersection E of the N level surfaces representing the motion constants is topologically equivalent to a product of circles and lines (Arnol'd–Liouville Theorem). If in addition E is compact, from which global existence automatically follows, it is an N-torus, namely a product of circles. In that case, the typical E can be coordinatized by action-angle variables and in those coordinates the flow is simply a straight line flow with constant velocity. Certainly in these cases, the Lyapunov exponents exist. They are clearly zero and the flow is regular. The Painlevé equations I through VI can also be cast in the form of Hamiltonian systems with two degrees of freedom and a second transcendental constant of motion can be found which is in involution with the Hamiltonian and which is an analytic function of the original coordinates as long as they remain finite. However, it is not so easy to prove that the dynamics in each of these cases are regular. In general, solutions blow up in finite time (the phase space is divided into regions of global existence and finite-time blow-up) and therefore one cannot expect Arnol'd–Liouville integrability to hold. Because the structure of the E is not well understood, and because we do not know how to coordinatize this surface, we cannot definitely say whether the motion is regular or irregular. Although it is likely that the flow is regular and in some sense integrable, there might be some complicated behavior of the kind illustrated by the following examples.

Let $H(q, p)$ be a real polynomial of q and p and set $q = x_1 + iy_2$, $p = y_1 + ix_2$. Then $F = \text{Re}\{H\}$ and $G = \text{Im}\{H\}$ are in involution with respect to the Poisson bracket $\{F, G\} = \sum_{i=1}^{2} (\partial F / \partial y_i)(\partial G / \partial x_i) - (\partial F / \partial x_i)(\partial G / \partial y_i)$ on R^4. The level surface E, the intersection of $F = c_1$, $G = c_2$, can be thought of as a complex one-dimensional manifold $H(q = x_1 + iy_2, p = y_1 + ix_2) = c_1 + ic_2$. Thus E is a Riemann surface with the points at infinity missing. If, for example, $H = p^2 - P_8(q)$ where P_8 is a polynomial of degree eight in q, then E can be pictured as a real two-dimensional surface with three holes and two tubes going off to infinity. Solutions of the Hamiltonian equations generated on E by F or G blow up in finite time. Orbits can wind around the holes in some, possibly complicated, way before shooting off to infinity along one of the tubes. Sensitivity is introduced by the number of times an orbit can wind around the holes before escaping off to infinity. From this picture, one might guess that nearby orbits separate at algebraic rather than exponential rates and that therefore these flows belong in the regular category. These kinds of subtleties, arising in situations which would appear to be widespread, are rarely addressed and suggest that a general definition of integrability may be difficult.

Notions of integrability are not restricted to volume preserving flows and can also apply to dissipative systems in which the phase volume contracts. For these systems one can sometimes find what are oxymoronically termed "time dependent integrals". For example, for the system

$$\dot{x} = \tfrac{1}{2}(y - x) \,,$$
$$\dot{y} = -xz - y \,,$$
$$\dot{z} = xy - z \,,$$

(1.5)

a special case of the Lorenz equations, one can identify the two quantities

$$I_1 = (x^2 - z)e^t \,,$$
$$I_2 = (y^2 + z^2)e^{2t} \,,$$

(1.6)

which enable one to reduce the third order system to a first order system which can be explicitly integrated in terms of elliptic functions.

From the discussion so far we see that there are various ways in which we can consider a system to be exactly soluble, e.g., integration by quadratures, direct linearization, Arnol'd–Liouville integrability, but we have also encountered many subtleties. In an analogous fashion, the same range of possibilities exists for systems with infinite degrees of freedom, e.g., partial differential equations, and any algorithm which we might suggest for testing integrability should also be able to handle these situations.

One of the earliest indications that certain nonlinear partial differential equations of evolution type might have special properties arose with the discovery of an infinite set of conservation laws for the Korteweg–de Vries (KdV) equation. These conservation laws implied that the flow was constrained by an infinite set of symmetries to be on a restricted manifold in the phase space. The existence of the infinite set of conservation laws persuaded *Gardner, Greene, Kruskal,* and *Miura* [3] to look for a simple way of generating the conserved quantities. This in turn led to the Miura–Gardner one parameter family of Bäcklund transformations between the solutions of KdV and of the modified KdV equation. From this the Lax pair was found, the inverse scattering transform for directly linearizing the equation was developed and the rest, as they say, is history. It even turned out that the KdV equation could be written in Hamiltonian form [4,5] and that the inverse scattering transform was a canonical transformation to action-angle variables [4]. In a very real sense, the integration of the KdV equation by IST was the infinite dimensional analogue of Arnol'd–Liouville integrability. While the existence of the infinite set of conservation laws was an important link in the chain of discoveries (the fact that there might be hidden symmetries in this equation was suggested by the discovery of *Kruskal* and *Zabusky* [6] of the remarkable soliton interaction properties), it was the discovery of the Bäcklund transformation which was the key step because it suggested the possibility of an algorithmic approach to check a given equation for symmetries.

Indeed, the nonexistence of an infinite number of conservation laws does not preclude integrability. There are equations which only have a finite number of conservation laws, which are not Hamiltonian, which are in fact dissipative, and which also can be directly integrated. The most notable of these is the Burgers equation which has only one conservation law, although an infinite number of symmetries. It can be integrated by directly transforming it to the linear heat equation. Again the transformation can be considered to be a one parameter family of Bäcklund transformation between solutions of the heat and Burgers equations.

Our philosophy then is that one parameter Bäcklund transformations are the main step to uncovering a given equation's integrability properties. These transformations occur between solutions of different equations or between solutions of the same equation. In the latter cases one appends the adjective auto. They can also be used to build multiparameter general solutions from simple ones. Each application of the transformation enriches the solution by adding new structure and one new parameter. This ladder process is the superposition principle of nonlinear equations and it suggests a definition of integrability which is entirely *local*. An infinite dimensional system (a PDE) is integrable if one can build an n-phase solution with n undetermined constants. A new constant is added on each application of the transformation. We emphasize for PDEs that a definition of integrability which is purely local in the independent variables is most important. It is simply not possible to decide in general whether evolution equations, posed as initial-boundary value problems (e.g., initial value in t, boundary value in x) are integrable, because the boundary conditions place such constraints, that a general solution constructed using a superposition principle may not be possible. For example, on the infinite line with the dependent variable $q(x,t)$ belonging to the class of function $\int (1 + x^2)|q(x)|\, dx < \infty$, the KdV equation is an exactly integrable Hamiltonian system. But how about the half line problem or the problems on finite intervals? Even though well posed, the dynamics may prove to be irregular in a "deterministic chaos" sense.

What ingredients should a successful algorithm for testing have? It should be compatible with the intuitive notions of regular and irregular flow discussed above. It should also be able to place in the integrable class all those systems whose general solutions exhibit regularity. In addition, we would like the method to make contact with exactly solvable models which arise in other areas of physics (e. g., the two-dimensional Ising model) but which are not evolution equations at all. The purpose of this paper is to suggest that the Painlevé algorithm is one of the most powerful for identifying integrable systems, although we emphasize that there is no proof that all these goals are achieved. Nevertheless, despite a few failures, mostly of pathological type, it has had some spectacular successes and has identified and provided the means to solve several systems about which nothing was previously known. (One of its successful applications, reproduced in Sect. 4, displayed, in a period of ten minutes during a lecture by Mikhailov, the Lax pair for a system that the Mikhailov–Shabat method had deemed should be integrable.)

The Painlevé method can be applied to systems of ordinary and partial differential equations alike. We will describe its application in detail in the following sections. Here we will simply point out the salient features, ask how this method makes contact with the other schemes (Zakharov–Schulman, Wahlquist–Estabrook, Mikhailov–Shabat: see other chapters in this volume for descriptions) for uncovering the integrable character of equations, conjecture about the properties of solutions of exactly solvable systems which are not evolution equations, and finally, point out how the Painlevé method can be used to understand something about the behavior of nonintegrable systems of a certain type.

The basic idea is to expand each dependent variable in the system of equations as a Laurent series about a pole manifold. If the equations are a set of ODEs then this simply means that one looks for solutions as Laurent series in the complex time variables $t - t_0$. In order that the equations for the coefficients of the various powers of $(t - t_0)$ in the Laurent expansion have self-consistent solutions, certain conditions on the structure of the given system of equations are required. If the system satisfies these conditions and the number of undetermined constants appearing in the Laurent series together with t_0 is equal to the order of the system of equations, then we say that the solution has the Painlevé property. A slight extension of these ideas and definitions, which will be described shortly, is needed when one is dealing with sets of partial differential equations with at least two independent variables.

The point is that in order for the general solution of a given equation to have a Laurent expansion about some pole in the space of complex independent variables, the equation has to have special properties. Indeed, most of the systems which are known to be integrable in the Arnol'd–Liouville sense have the Painlevé property and, in fact, *Kovalevskaya* [7] used this idea to construct the third exactly integrable model of rigid-body motion, for the top with the particular ratios of moments of inertia which now bears her name. A more appropriate name for the method would be, therefore, the Kovalevskaya method. Painlevé's name is attached because, a little later, he asked a related question: What is the set of second order equations $y'' = R(y, y', t)$ with R rational in y, y'; and analytic in t which has the property that the location of any algebraic, logarithmic, or essential singularities is independent of the initial conditions? This property means that only the location of the poles (t_0) can depend on the arbitrary constants of integration. The set of equations which possess this property was reducible either to a set of known equations which were linear, or could be integrated by quadratures, or to a sextet of new equations, the so-called Painlevé equations I, II, III, IV, V, VI, the solutions of which are known as Painlevé transcendents.

Let us now list some of the advantages of the method. The first is that when a system has the Painlevé property, one is led naturally to a Bäcklund transformation, the Lax Pair and the Hirota bilinear formulation. A second advantage is that the ideas make contact with exactly solvable models in other branches of physics. A discovery of the past ten years has been that there is a connection between exactly solvable models in statistical mechanics and exactly integrable PDEs. The earliest example of this connection is that the n-point correlation functions of the two-dimensional Ising model in the scaling limit satisfy partial differential equations which are exactly integrable in the local sense described above. Other exactly solvable models in one-dimensional quantum field theory also have this property. Because of this link, one is tempted to ask whether there is a distinction between solvable and unsolvable models in statistical mechanics, other than nobody smart enough has yet managed to write down closed form solutions. In other words, is there an analogue in statistical mechanics and quantum field theory to the intrinsically nature of solutions of perturbed integrable systems? If there is, what constitutes an irregular flow in the statistical mechanics context? One possible answer and point of contact with differential equations is that the calculated solutions, the free energy and the correlation functions, may have extremely complicated singularity structures in the complexified space of the independent variables, namely the separation vectors describing the relative geometry of the points involved in the correlation and the Boltzmann weights. The Boltzmann weights are the parameters which weight the various site-site interactions in the original Hamiltonian. If these singularity structures are close to the real axes, then one might find that the calculated functions are very complicated functions of the arguments. In order to test some of these ideas, it would be a valuable exercise to see whether the Yang–Baxter or Star–Triangle relations, which guarantee that there exists a one parameter family of commuting transfer matrices for a given model, can be related to the Painlevé conditions that the differential equations describing the correlation functions must satisfy.

A third advantage of the method is that it can reveal some of the geometric and algebraic structure that underlies many integrable equations. For example, factorization in Lie groups is now understood to provide a uniform approach to many soliton equations, and in some test cases the Painlevé analysis has identified both the Lie group and aspects of the geometric setting for factorization problems.

A fourth advantage of the method is that it can uncover information about nonintegrable systems. The movable singularities of these systems are, typically, complicated branch points. These appear to cluster recursively, leading to an overall singularity distribution of pathological complexity. Despite this complexity, a "rescaling" technique can sometimes be used to map the problem onto an "underlying" integrable part. This enables one to obtain an accurate picture of the singularity pattern and even, in certain cases, identify integrals of motion that may still exist.

It should be clear to the reader at this stage that a comprehensive definition of an algorithm for testing integrability is still an open challenge. In this article we focus attention on the Painlevé method whose advantages we have mentioned. However, we also emphasize that other methods are available. Each has had its successes and its failures. What is remarkable is that, on the surface, each of these methods looks so different. The Wahlquist–Estabrook method is based on the notion that in order to be integrable the given system should be the compatibility equation for a one (nontrivial) parameter system of linear equations (the Lax pair). This idea reminds one of the ideas of Baxter that in order to find solvable systems in statistical mechanics, one should construct a one parameter family of commuting transfer matrices. The Zakharov–Schulman method appears to be very different indeed. They rewrite the PDE of interest in Fourier coordinates as if they consider the system to have weak nonlinear couplings. The existence of an infinite set of motion invariants is then related to properties of the coefficients multiplying products of the Fourier amplitudes in the integrand of the convolution integrals which are the transforms of the nonlinear terms in the equations. These coefficients should have the property that the effects of small denominators, which arise in a Birkhoff expansion due to wave–wave resonances, are canceled.

It is clear that a better understanding of integrability will come not simply by understanding any one of these methods individually, but by understanding their interrelation as well. The Painlevé method appears to be able to reconstruct global properties of the given equation by revealing the behavior of its solutions near infinity. The Zakharov–Schulman method, on the other hand, looks in the neighborhood of the origin. The fact that two different regions in the phase space can supply global information is, presumably, due to the fact that local information about analytic constants of motion continues analytically to give global structure.

The organization of the paper is as follows. In Sect. 2, we introduce the method. In Sects. 3 and 4, we introduce two new results, the Lax pairs for an integrable Hénon–Heiles Hamiltonian and for an example of Mikhailov and Shabat. In Sect. 5, we summarize some of the results on the KdV hierarchy, the details of which can be found in [8]. In Sect. 6, we discuss how the Painlevé analysis can help uncover some of the algebraic structure of the equation system.

Finally in Sect. 7, we show how Painlevé analysis can be used to "integrate the nonintegrable".

2. Introduction to the Method

Weiss, Tabor, and *Carnevale* (henceforth referred to as WTC) [9] have constructed a powerful generalization of the Painlevé test for ordinary differential equations that is capable of uncovering both the integrable properties of many nonlinear systems and special classes of solutions of nonintegrable systems. Although originally designed to treat partial differential equations, their method has turned out to be capable of handling both nonlinear ordinary and partial differential equations in a highly unified way. The method involves expanding the dependent variable in a Laurent series about a *singular manifold* – the pole manifold – and gives rise to a suggestive and provocative formalism from which one may deduce Lax pairs and Bäcklund transformations. A variety of applications and extensions of the WTC method have been made in the impressive sequence of papers by *Weiss* [10–13], by *Tabor* and coworkers [14–16] and others. These papers have clearly demonstrated the effectiveness of the method even if, on occasion, various transformations and tricks are involved that are motivated by an a priori knowledge of the answer. Lacking, however, has been a more serious attempt to gain a deeper insight into how and why the method actually works. In this paper we report on our efforts in this direction and the relation of the WTC method to other work on algebraically integrable systems. The ultimate goal is to show that the generalized Laurent expansions can not only show that a system is integrable but that the expansions can also be used to provide an algorithm which successfully captures all its properties; namely, the Lax pairs, the Bäcklund transformations, the Hirota equations, the motion invariants, symmetries and commuting flows, the geometrical structure of the phase space and, finally, the algebraic properties (symmetries) which make the system's exact integrability transparent and inevitable.

The novel feature of the WTC approach is the flexibility contained in the singular manifold function. It is this flexibility that allows the Laurent expansions to be collapsed into Bäcklund transformations which give rise to the Lax pairs – the heart of a system's integrability. The Hirota formulation of the equation under study is also a straightforward consequence. In contrast to this approach, but in parallel with its goals, there have recently been several valuable investigations, notably by *Adler* and van *Moerbeke* [17, 18], *Haine* [19], *Ercolani* and *Siggia* [20] which address in depth how the traditional Laurent expansions (i.e., the standard ODE type) relate to and give information on algebraically completely integrable systems of ordinary differential equations. In particular, Ercolani and Siggia have shown for hyperelliptically separable, finite dimensional, Hamiltonian systems how the "lowest balance" leads to an algorithm that will, in principle, produce all the constants of motion.

2.1 The WTC Method for Partial Differential Equations

Let $q(x,t)$ satisfy the nonlinear evolution equation

$$q_t = k[q, q_x \ldots] \tag{2.1}$$

and let

$$\phi(x,t) = 0 \tag{2.2}$$

be the manifold (the "singular" or "pole" manifold) on which $q(x,t)$ is singular. At this stage, ϕ is simply a new coordinate and one wishes to construct a Laurent expansion for a solution of (2.1) in the neighborhood of its level surfaces, and in particular $\phi = 0$. The basic ideas of the WTC method is to ask that the expansion

$$q(x,t) = \frac{1}{\phi^\alpha} \sum_{j=0}^\infty u_j(x,t)\phi^j \tag{2.3}$$

be single valued. This requires that (i) α is an integer, (ii) ϕ is analytic in x and t and (iii) the equations for the coefficients u_j have self-consistent solutions. If all these conditions are satisfied, (2.3) can be considered to be a Laurent expansion of the solution in the neighborhood of the singular manifold. We shall say that an equation whose solutions have this property has the *Laurent property*. In order that the solution can be constructed in the neighborhood of (2.2) it is necessary to stipulate that *neither ϕ_x nor ϕ_t vanish on $\phi(x,t) = 0$*. This important requirement corresponds to the demand that the zeros are simple when we use the Implicit Function Theorem on (2.2) to express x as a function of t [i.e., $\phi(x,t) = x - \alpha(t)$ on $\phi = 0$ such that $\phi(\alpha(t), t) = 0$]. This condition is satisfied by most, but not all, points on the pole manifold (2.2). We will discuss later what happens at a confluence of poles; at this point we simply remark that the WTC method is also capable of capturing the most singular pole behavior of the system under investigation.

A specific illustration of the method is provided by the KdV equation

$$q_t + 6qq_x + q_{xxx} = 0 \; ; \tag{2.4}$$

direct substitution of the ansatz (2.3) into (2.4) quickly shows that the *leading order* is $\alpha = 2$ and that $u_0 = -2\phi_x^2$. At this point we remark that for equations with stronger nonlinearities, such as higher members of the KdV hierarchy, one may find a number of different leading orders – which we term *branches* (or *balances*) – and each will lead to a separate Laurent expansion which must be tested for single-valuedness. Continuing the analysis for (2.4) we find a set of recursion relations for the u_j of the form

$$(j+1)(j-4)(j-6)u_j = F_j(\phi_t, \phi_x, \ldots, u_k; \; k < j) \,. \tag{2.5}$$

We note that the left-hand side of (2.5) vanishes at $j = -1$, 4 and 6. These values of j are called *resonances*. At each such resonance, consistency demands that the right-hand side of (2.5) vanish, thereby ensuring the indeterminacy of the corresponding u_j. (The resonance at $j = -1$ corresponds to the arbitrariness of ϕ itself.) What does it mean for u_j to be indeterminate? If we were dealing with a system of ODE and ϕ were simply $x - x_0$, then the undetermined functions would be constants. If, in a PDE, ϕ were $x - \alpha(t)$, then the u_j could also be t-dependent but would not depend on $x - \alpha(t)$. Similarly, for a general ϕ, the undetermined functions are independent of ϕ.

If the equations for the "resonant" u_j are not consistent, we have to introduce terms of the type $\phi^j \log \phi$ into the series (2.3) thereby rendering it a multivalued psi-series. In general the number of positive resonances depends on both the shape of $k[q, q_x, \ldots]$ and the branch. The branch with most positive resonances is termed the "principal" branch and corresponds to the general solution of (2.1) if it has, in keeping with the Cauchy–Kovalevskaya theorem, as many undetermined functions as the order of the system. The other, "lower" branches are sometimes referred to as the singular branches. If all branches of the Laurent expansion are single valued for arbitrary ϕ and the principal branch has its full complement of undetermined functions, then we say that (2.1) has the *Painlevé property*. Sometimes, if we loosely refer to a particular branch as having the Painlevé property, we simply mean it is single valued and has the appropriate number of undetermined functions (e.g., the principal expansion will have the full complement).

Referring to the specifics of (2.4) we find that at

$$j = 0 : \quad u_0 = -2\phi_x^2 \,, \tag{2.5a}$$

$$j = 1 : \quad u_1 = 2\phi_{xx} \,, \tag{2.5b}$$

$$j = 2 : \quad \phi_x\phi_t + 4\phi_x\phi_{xxx} - 3\phi_{xx}^2 + 6\phi_x^2 u_2 = 0 \,, \tag{2.5c}$$

$$j = 3 : \quad \phi_{xt} + 6\phi_{xx}u_2 + \phi_{xxxx} - 2\phi_x^2 u_3 = 0 \,, \tag{2.5d}$$

$$j = 4 : \quad \frac{\partial}{\partial x}(\phi_{xt} + \phi_{xxxx} + 6\phi_{xx}u_2 - 2\phi_x^2 u_3) = 0 \,, \tag{2.5e}$$

which demonstrates that u_4 is indeed arbitrary. Further analysis also shows u_6 to be arbitrary. Thus the expansion

$$q(x,t) = 2\frac{\partial^2}{\partial x^2}\log\phi + \sum_{j=2}^{\infty} u_j\phi^{j-2} \tag{2.6}$$

possesses the Painlevé property as defined by WTC thereby identifying the KdV equation as integrable. Furthermore, WTC introduced the additional ideas of a *truncated expansion* by observing that the expansion (2.6) can be consistently truncated at $O(\phi^0)$ by

(i) setting the arbitrary functions u_4 and u_6 equal to zero,

(ii) requiring that $u_3 = 0$.

This results in the following system of equations:

$$q(x,t) = 2\frac{\partial^2}{\partial x^2} \log \phi + u_2 , \tag{2.7a}$$

$$\phi_x \phi_t - 3\phi_{xx}^2 + 4\phi_x \phi_{xxx} + 6u_2 \phi_x^2 = 0 , \tag{2.7b}$$

$$\phi_{xt} + \phi_{xxxx} + 6u_2 \phi_{xx} = 0 , \tag{2.7c}$$

$$u_{2t} + 6u_2 u_{2x} + u_{2xxx} = 0 . \tag{2.7d}$$

Because u_2 satisfies the KdV equation, (2.7a) is an auto-Bäcklund transformation. The remaining three equations (2.7b–d) are an overdetermined system for the two variables $\phi(x,t)$ and $u_2(x,t)$. However, this system is entirely self-consistent, with (2.7d) being the solvability condition for (2.7b) and (2.7c), which are found, after a certain transformation, to be precisely the Lax pair for the KdV equation (2.7d). Thus the truncated expansion (2.7a) is an auto-Bäcklund transformation for the solutions of (2.4). The key to the success of the WTC procedure, therefore, is that the flexibility in ϕ allows one to collapse an infinite Laurent series in $x - \alpha(t)$, or some other $\tilde\phi$ with the same zero manifold as ϕ, to a Bäcklund transformation.

To demonstrate this explicitly, consider

$$q_t + 6qq_x + q_{xxx} = \left\{ \left(\frac{2\phi_x}{\phi}\right)_t + \frac{(6\phi_{xx}^2 - 8\phi_x \phi_{xxx} - 12u\phi_{xx}^2)}{\phi^2} \right.$$
$$\left. + \frac{(2\phi_{xxxx} + 12u\phi_{xx})}{\phi} \right\}_x + u_t + 6uu_x + u_{xxx} , \tag{2.8}$$

where we have set $u_2 = u$. On making the substitution

$$\phi_x = \psi^2 , \tag{2.9}$$

(2.8) becomes

$$q_t + 6qq_x + q_{xxx} = \left\{ -\frac{4\psi^2}{\phi^2} \int^x \psi(\psi_t + 4\psi_{xxx} + 6u\psi_x + 3u\psi_x \psi)\, dx + \frac{4\psi}{\phi} \right.$$
$$\left. \times \left[(\psi_t + 4\psi_{xxx} + 6u\psi_x + 3u_x \psi) - 3\psi \left(\frac{\psi_{xx}}{\psi} + u\right)_x \right] \right\}_x$$
$$+ u_t + 6uu_x + u_{xxx} ,$$

from which one deduces the Lax pair

$$\psi_{xx} + (\lambda + u(x,t))\psi = 0 , \tag{2.10a}$$

$$\psi_t + 4\psi_{xxx} + 6u\psi_x + 3u_x \psi = 0 . \tag{2.10b}$$

2.2 The WTC Method for Ordinary Differential Equations

Now consider the stationary version of (2.4), namely the ordinary differential equation

$$q_{xx} + 3q^2 = 0 \tag{2.11}$$

which can be integrated in terms of Weierstrass elliptic functions. Equation (2.11) is easily shown to have the Painlevé property for ODEs and possesses the Laurent expansion (about some movable pole at $x = x_0$)

$$q(x) = \frac{1}{(x - x_0)^2} \sum_{j=0}^{\infty} q_j (x - x_0)^j , \tag{2.12}$$

with resonances at -1 and 6 (i.e., q_6 and x_0 are the arbitrary constants). We now generalize the above Laurent series to an expansion in terms of some singular manifold function $\phi(x)$, namely,

$$q(x) = \frac{1}{\phi^2} \sum_{j=0}^{\infty} u_j(x) \phi^j , \tag{2.13}$$

where the $u_j(x)$ are certain functions analytic in the neighborhood of the singular manifold $\phi(x) = 0$ which are nonzero on $\phi = 0$. It is assumed that $\phi_x \neq 0$ on the manifold. Clearly the choice $\phi(x) = x - x_0$ enables us to reclaim the original Laurent series (2.12).

Substitution of the ansatz (2.13) into (2.11) leads to an analysis that is, bar the t-dependent terms, identical to that just carried out for the KdV equation. Working with the truncated expansion

$$q = 2 \frac{\partial^2}{\partial x^2} \log \phi + u_2 \tag{2.14}$$

and the squared eigenfunction relation $\phi_x = \psi^2$ leads to

$$q_{xx} + 3q^2 = -\frac{4\psi^2}{\phi^2} \left\{ \int^x \psi [4\psi_{xxx} + 6u_2 \psi_2 + 3u_{2x} \psi] \, dx \right\}$$
$$+ \frac{4\psi}{\phi} \left\{ [4\psi_{xxx} + 6u_2 \psi_x + 3u_{2x} \psi] - 3\psi \left(\frac{\psi_{xx}}{\psi} + u_2 \right)_x \right\}$$
$$+ u_{2xx} + 3u_2^2 . \tag{2.15}$$

However, as it stands, the system of equations that one would obtain by setting each order of ϕ to zero, i.e.,

$$\psi_{xx} + u_2 \psi = \lambda \psi , \tag{2.16a}$$

$$4\psi_{xxx} + 6u_2 \psi + 3u_{2x} \psi = 0 , \tag{2.16b}$$

where

$$u_{2xx} + 3u_2^2 = 0 \tag{2.16c}$$

although self-consistent, is not sufficiently general to supply the ingredients through which (2.16c) can be integrated.

The crucial point is that the original WTC prescription of setting each order of ϕ equal to zero is too restrictive. Rather, one should think of each order of ϕ being zero *modulo some function of* ϕ which tends to zero as ϕ does. In this way the addition and subtraction of judiciously chosen quantities at various orders of ϕ can considerably strengthen the power of the method. For example, the correct Lax pair for (2.16c) requires the addition of the term $y\psi$ on the right-hand side of (2.16b). The presence of such a term is not surprising because if we think of $u_2(x)$ in (2.16c) as the t-dependent limit of the u_2 in (2.7c) and (2.10), then we can also argue that the corresponding $\psi(x, t; \lambda)$ can be separated as $\psi(x; \lambda) \exp(yt)$. In order to add this second free parameter y into the Lax pair of the stationary equation, we observe that we can

(i) add an amount $4\phi_x \phi y$ at ϕ^{-2} and
(ii) subtract the identical amount $4\phi_x y$ at order ϕ^{-1}.

Since $\phi = \int^x \psi^2 dx$, this modifies (2.15) to

$$q_{xx} + 3q^2 = -\frac{4\psi^2}{\phi^2} \left\{ \int^x \psi(4\psi_{xxx} + 6u_2\psi_x + 3u_{2x}\psi - y\psi)\, dx \right\}$$
$$+ \frac{4\psi}{\phi} \left\{ (4\psi_{xxx} + 6u_2\psi_x + 3u_{2x}\psi - y\psi) - 3\psi \left(\frac{\psi_{xx}}{\psi} + u_2 \right)_x \right\}$$
$$+ u_{2xx} + 3u_2^2 . \tag{2.17}$$

Equating the coefficients of powers of ϕ to zero yields the correct Lax pair

$$\psi_{xx} + u_2\psi = \lambda\psi , \tag{2.18a}$$

$$4\psi_{xxx} + 6u_2\psi_x + 3u_{2x}\psi = y\psi . \tag{2.18b}$$

In the next two subsections, we treat an integrable version of the Hénon–Heiles Hamiltonian and all the stationary equations for the KdV hierarchy using the same ideas. The Hénon–Heiles example is noteworthy because, before being treated by the WTC method, the Lax pair was not known.

In all these examples our modification of the WTC method has involved the incorporation of the extra free parameter which enables us to obtain Lax pairs and auto-Bäcklund transformations. In [21] we give a more detailed investigation of our modified procedure for ODEs and show, among other things, that it can also be used to obtain Bäcklund transformations between different equations and consequently as a technique for finding a variety of nontrivial special solutions. In [21], we treat the Painlevé second and fourth equations in some detail.

The results obtained by means of the WTC procedure, both for PDEs (previous subsection) and ODEs (this subsection) naturally give rise to a number of questions, including:

(i) What is the nature of the function ϕ associated with the *"truncated expansion"* and what is its relation to the eigenfunction of the Lax pair, its role in Bäcklund transformations and in helping to uncover the Hirota formulation for the given equation?

(ii) What is the relation between the truncated and nontruncated expansions?

(iii) What information do the Painlevé expansions for the other balances contain?

(iv) How general is the WTC procedure and how might it be used in noninte-grable situations?

A summary of comments on these questions and an outline of the rest of the paper is now given in Sects. 2.3–5.

2.3 The Nature of ϕ

We begin by considering the nature of the WTC function $\phi(x,t)$. As mentioned briefly, the $\phi(x,t)$ that appears in the infinite Laurent series (2.3, 6) is simply a new coordinate [one might think of it as $x - \alpha(t)$] in the space of independent variables and the expansions (2.3, 6) are Laurent series in the new variable. Provided they converge in some neighborhood of $\phi = 0$ and satisfy the Painlevé property defined in the paragraph before (2.5), any choice of this function will lead to a solution of the PDE with as many free parameters as there are resonances – just as (2.12) is a Laurent expansion for an exact solution of the ODE (2.11). On the other hand, the ϕ associated with the truncated expansion, namely, the ϕ appearing in (2.7a) which satisfies the overdetermined system of (2.7b–d) is not arbitrary at all but a very special function indeed. For the following discussion, let us designate this as $\bar{\phi}$. It will have the same zero manifold as the arbitrary ϕ which appears in (2.3, 6); $\bar{\phi}$ will be zero on $\phi = 0$ and will have the same properties on its derivatives, i.e., $\bar{\phi}_x$, $\bar{\phi}_t$ are nonzero on $\bar{\phi} = 0$. It is this $\bar{\phi}$ which we now discuss.

A valuable context for this discussion is the KdV hierarchy,

$$q_{t_{2n+1}} = \frac{\partial}{\partial x} L^n q \,, \tag{2.19}$$

where

$$L = -\frac{1}{4}\frac{\partial^2}{\partial x^2} - q + \frac{1}{2}\int^x dx\, q_x \,,$$

for which (2.4) with $t_3 = 4t$ is the first nontrivial member. In Sect. 3 of [8] we see that the truncated expansion (2.7a) is an auto-Bäcklund transformation for any member of this hierarchy and furthermore prove (and not simply assume) that the principal Painlevé expansion can indeed be self-consistently truncated in this manner for each member of the family (2.19). Since all the flows of this family commute, we can think of q, u_2, and ϕ in (2.7a) as functions of all the flow times $t_1 = x, t_3, t_5, \ldots, t_{2n+1}, \ldots$. Now, it is well-known [22–24] that all solutions of the KdV hierarchy can be characterized in terms of a single function, the Hirota τ-*function*, which is related to a solution $q(x, t_{2r+1})$ as follows:

$$q(x, t_3, \ldots, t_{2n+1}, \ldots) = 2\frac{\partial^2}{\partial x^2} \ln \tau(x, \ldots, t_{2n+1}, \ldots). \tag{2.20}$$

Indeed, it can be shown that

$$L^n q = 2\frac{\partial^2}{\partial x\, \partial t_{2n+1}} \ln \tau(x, \ldots, t_{2n+1}, \ldots). \tag{2.21}$$

The conservation laws for the t_{2n+1} flow are conveniently written as

$$\frac{\partial}{\partial t_{2n+1}} 2\frac{\partial^2 \ln \tau}{\partial t_1 \partial t_{2j+1}} = \frac{\partial}{\partial t_1} 2\frac{\partial^2 \ln \tau}{\partial t_{2n+1} \partial t_{2j+1}} \tag{2.22}$$

(with $t_1 = x$), thus identifying the conserved densities and fluxes of the $(2n+1)$st flow as

$$C_{2j+1,2n+1} = 2\frac{\partial^2}{\partial t_1 \partial t_{2j+1}} \ln \tau, \tag{2.23}$$

$$F_{2j+1,2n+1} = 2\frac{\partial^2}{\partial t_{2n+1} \partial t_{2j+1}} \ln \tau, \tag{2.24}$$

$j = 1, 2, \ldots$, respectively [24]. It is shown in [23] that $C_{2j+1,2n+1}$ and $F_{2j+1,2n+1}$ can be written as local expressions in q and its x derivatives. In short, every property of a common solution of every member of the KdV hierarchy can be deduced from a single τ function.

With these preliminary remarks, we now come to the main points. Denoting the τ function of the solutions q and u as τ_q and τ_u, respectively, we see that candidates for the pole function $\phi(x, t_3, \ldots, t_{2n+1}, \ldots)$ in a truncated Painlevé expansion are simply the ratios of two τ functions, i.e.,

$$\bar{\phi}(x, \ldots, t_{2n+1}, \ldots) = \frac{\tau_q(x, \ldots, t_{2n+1}, \ldots)}{\tau_u(x, \ldots, t_{2n+1}, \ldots)}, \tag{2.25}$$

with the additional property that the zero manifold

$$\bar{\phi}(x, \ldots, t_{2n+1}, \ldots) = 0 \tag{2.26}$$

is a manifold on which $q(x, \ldots, t_{2n+1}, \ldots)$ has pole singularities and $u(x, \ldots, t_{2n+1}, \ldots)$ does not. Equivalently, it is the manifold on which τ_q has zeros but on which τ_u does not vanish. Note that whereas $\bar{\phi}$ is a special function, it can have many degrees of freedom. All it must be is a ratio of two legitimate τ functions. We have seen already that it is related to the squared eigenfunctions with arbitrary λ of the scattering problem (2.10a) associated with the KdV family and will shortly explain that this connection is general.

The idea of thinking of the $\bar{\phi}$ function as a ratio of τ functions is very useful. We know that the τ function is the building block from which the most general solutions of the family of integrable equations is obtained. One begins, in general, with the simplest τ, namely, $\tau = 1$, and by applying a sequence of group operations from a certain infinite-dimensional symmetry group, one can build arbitrarily complicated solutions. Some group elements are exponentials of an operator X, called the vertex operator, which depends on $x, t_3, \ldots$ and arbitrary parameters $\lambda_j, j = 1, 2, \ldots$ [for the KdV hierarchy there is only one; for the Kadomtsev–Petviashvili (KP) hierarchy, there are two] such that

$$\tau_{\text{new}} = e^X \tau_{\text{old}} . \tag{2.27}$$

Equation (2.27) is an auto-Bäcklund transformation in the τ-function framework. The new τ function may be richer than the old one and contains new zeros which the old one does not. We might ask: can we factor τ_{new} algebraically, into a product $\bar{\phi}$ times τ_{old} so that $\bar{\phi}$ contains all the information about this new zero of τ_{new}? We have seen that this is indeed possible. Not only is $\bar{\phi}$ a ratio of legitimate τ functions but one now can also interpret $\bar{\phi}(x, t_3, t_5, \ldots; \lambda)$ as a characteristic value of the exponential of the vertex operator $X(x, t_3, \ldots; \lambda)$ acting on the space of τ functions.

Further, we see that if $\tau_{\text{old}} = 1$, then $\bar{\phi}$ is the new τ function and as such will satisfy the Hirota equations. Therefore, if one simply truncates the principal Painlevé series at the level of $\bar{\phi}^{-1}$ instead of the level $\bar{\phi}^0$, $\bar{\phi}$ will satisfy the Hirota equations. Thus the Hirota equations are a natural consequence of the WTC formalism.

The connection of the $\bar{\phi}$ function with the τ-function formulation of the integrable family carries over to those situations where the τ functions also depends on a discrete parameter. In the AKNS system, three functions $\tau(t, -1)$, $\tau(t, 0)$, and $\tau(t, 1)$ are particularly important. The function $\tau(t, 0)$ is $\bar{\phi}$ and the auxiliary ones are the leading coefficients in the truncation principal Painlevé expansions for the dependent variables.

Finally, we turn to the squared eigenfunction substitution (2.9). It turns out that this choice is entirely natural and not a fluke peculiar to the KdV equation. Observe that in the truncated expansion the coefficient u_1 of $\bar{\phi}^{-1}$ will satisfy the linearized equation, which, when linearized about the solution $u(x, t)$, is

$$u_{1t} = k_u u_1 , \tag{2.28}$$

where k_u is the variational derivative of the functional $k[u, u_x, \ldots]$ with respect to u. Therefore, u_1 embodies the symmetries of (2.4) and we know from previous work [24] how these symmetries relate to the "squared eigenfunctions" of the associated linear Lax equations. In particular the symmetries for the KdV hierarchy are generated by the x derivatives of the squared eigenfunctions, i.e., $u_1 \propto (\psi^2)_x$. But $u_1 = 2\phi_{xx}$ (for all members of the hierarchy) and hence we can make the identification $\phi_x = \psi^2$. Therefore we expect that in every exactly integrable situation, the coefficients of $\bar{\phi}^{-1}$ in the expansions for the dependent variables will be related to new functions, which are the "square" (in general they are appropriate quadratic combinations) of the functions which satisfy the Lax equations. We have shown explicitly [8] that this is indeed the case for not only the KdV hierarchy but also for the AKNS hierarchy and the Hénon–Heiles system. In fact it is true for all the other equations (Kadomtsev–Petviashvili, Boussinesq, sine-Gordon, etc.) which were tested in the original WTC paper [9].

In this subsection we have used the notation $\bar{\phi}$ to distinguish explicitly the singular manifold function in the truncated expansions from the ϕ used in the infinite Laurent series. In subsequent sections, however, the overbar is omitted since this distinction should now be self-evident.

2.4 Truncated Versus Non-truncated Expansions

Given that a series has the Laurent property, let us enumerate the various scenarios which can occur.

(i) The system has the Painlevé property and the truncated expansion yields a self-consistent overdetermined system of equations for which there is a one parameter family of $\phi(x, t; \lambda)$, i.e., (2.7a) is a general auto-Bäcklund transformation. The KdV and AKNS hierarchies are obvious examples of this case.

(ii) The system has the Painlevé property but the truncated expansion does not appear to yield a one parameter family of auto-Bäcklund transformations (e.g., the modified KdV and sine-Gordon equations [9]). This can indicate that the equation of interest is a reduction of some "larger" integrable system (e.g., the MKdV or sine-Gordon equations can be written as a system in the AKNS hierarchy) and whereas it is possible to introduce the free parameter, in general it will require some ingenuity. However, in some cases, e.g., the first Painlevé transcendent, although the equation has the Painlevé property, truncation at $O(\phi^0)$ does not yield any form of Bäcklund transformation because it is clear from the form of the equation that truncation at this level is inappropriate. Whether there exists a self-consistent truncation at a higher level has yet to be determined.

(iii) The system does not have the Painlevé property and does not even have the Laurent property for arbitrary ϕ. However, for certain choices of ϕ, the Laurent property can be restored. *Weiss* has called this the "conditional Painlevé property" [25]. In these cases we could have also attempted to find a self-consistent system of equations by truncating the Laurent expansion at $O(\phi^0)$ without regard to whether the system actually has the Laurent property. It turns out that solutions ϕ (to the truncated system) will be a subset of those ϕ's which guarantee the conditional Painlevé property of the full expansion. Using such techniques one can find interesting solutions to nonintegrable equations (e.g., double sine-Gordon [25] and Ginzburg–Landau equations [26]).

(iv) The incorrigibles. The system never satisfies the Painlevé or conditional Painlevé property for any choice of nontrivial ϕ. Even in these cases, however, analysis of the associated psi-series can still yield valuable insights [27].

Therefore, although we will only be discussing the WTC method in the context of classes (i) and (iv) in this paper, it is worth emphasizing that the power of the method goes well beyond integrable systems.

3. The Integrable Hénon–Heiles System: A New Result

The generalized Hénon–Heiles Hamiltonian

$$H = \tfrac{1}{2}(p_1^2 + p_2^2 + Aq_1^2 + Bq_2^2) + Dq_1q_2^2 + Cq_1^3$$

has been shown to pass the Painlevé test for ODEs for four special sets of parameters (A, B, C, D) [28]. One may again enquire as to whether the WTC approach can yield further information about the solutions of these systems. This has already been investigated by *Weiss* [29] who has derived certain specialized Bäcklund transformations for these cases (as well as an interesting formulation in terms of Schwarzian derivatives). For one of these integrable cases we are now able to show, using the ideas described in the introduction, how to derive the associated Lax pair, integrals of motion, and algebraic curve.

3.1 The Lax Pair

The system considered here is governed by the integrable Hamiltonian

$$H = \tfrac{1}{2}p_1^2 + p_2^2 + \tfrac{1}{2}q_1q_2^2 + q_1^3 \tag{3.1}$$

with the canonical symplectic form $\sum_{i=1}^2 dp_i \wedge dq_i$. (The quadratic potential energy terms of the general Hamiltonian may be dropped without loss of generality.) The equations of motion for q_1 and q_2 are

$$q_{1tt} = -3q_1^2 - \tfrac{1}{2}q_2^2 , \tag{3.2}$$

$$q_{2tt} = -q_1q_2 . \tag{3.3}$$

Expanding both q_1 and q_2 about the same singular manifold $\phi(t) = 0$, i.e.,

$$q_1(t) = \frac{1}{\phi^\alpha} \sum_{j=0}^\infty u_j(t)\phi^j , \quad q_2(t) = \frac{1}{\phi^\beta} \sum_{j=0}^\infty v_j(t)\phi^j , \tag{3.4}$$

it is easy to show that there are two possible balances. They are:

(i): $\alpha = -2, \quad \beta = -1,$
$u_0 = -2\phi_t^2,$
$v_0 = $ undetermined at leading order, with resonances at $-1, 1, 4, 6;$

(ii): $\alpha = -2, \quad \beta = -2,$
$u_0 = -6\phi_t^2, \ v_0 = \pm 12i\phi_t^2,$
with resonances at $-3, -1, 6, 8.$

Although much valuable information can be extracted from the singular branch, i.e., case (ii) (see the discussion in *Ercolani* and *Siggia* [20]), we shall work with the principal branch in order to obtain the Lax pair.

The truncated expansions

$$q_1 = \frac{u_0}{\phi^2} + \frac{u_1}{\phi} + u_2, \quad q_2 = \frac{v_0}{\phi} + v_1 \tag{3.5}$$

are substituted into the equations of motion (3.2, 3) yielding an over-determined system of equations for the five functions u_1, u_2, v_0, v_1, ϕ. We will now demonstrate that this system is self-consistent. The equations are

$$\phi^{-3} : 2v_0\phi_t^2 + u_0v_0 = 0 , \tag{3.6}$$

$$u_1 = 2\phi_{tt} , \tag{3.7}$$

$$\phi^{-2} : \ -2\phi_t v_{0t} + \phi_{tt}v_0 = 2v_1\phi_t^2 , \tag{3.8}$$

$$-8\phi_t\phi_{ttt} + 6\phi_{tt}^2 - 12u_2\phi_t^2 + \tfrac{1}{2}v_0^2 = 4y\phi\phi_t , \tag{3.9}$$

$$\phi^{-1} : \ v_{0tt} + v_0u_2 + 2\phi_{tt}v_1 = 0 , \tag{3.10}$$

$$2\phi_{tttt} + 12u_2\phi_{tt} + v_1v_0 = -4y\phi_t , \tag{3.11}$$

$$\phi : u_{2tt} = -3u_2^2 - \tfrac{1}{2}v_1^2 , \tag{3.12}$$

$$v_{1tt} = -v_1u_2 . \tag{3.13}$$

Note that in keeping with the ideas introduced earlier, we have added in a term $-4y\phi\phi_t$ to the coefficient at ϕ^{-2} and subtracted $-4y\phi_t$ in the coefficient at ϕ^{-1} which gives an extra free parameter y. The first equation in the sequence is automatically satisfied. Dividing (3.9) by $-\phi_t$, differentiating and subtracting (3.11) gives

$$6\left(\frac{\phi_{ttt}}{\phi_t} - \frac{1}{2}\frac{\phi_{tt}^2}{\phi_t^2} + 2u_2\right)_t = 0 \tag{3.14}$$

or

$$\frac{\phi_{ttt}}{\phi_t} - \frac{1}{2}\frac{\phi_{tt}^2}{\phi_t^2} + 2u_2 = -2\lambda .$$

This equation can be linearized by setting

$$\phi_t = \psi^2 , \tag{3.15}$$

whence

$$\psi_{tt} + (u_2 + \lambda)\psi = 0 . \tag{3.16}$$

Differentiating (3.15) three times, substituting into (3.11) gives

$$2\psi\psi_{ttt} + 6\psi_t\psi_{tt} + 6u_2\psi_{tt} + \tfrac{1}{2}v_0v_1 + 2y\psi^2 = 0 \tag{3.17}$$

which, using (3.16), can be rewritten as

$$4\psi\psi_{ttt} + 6u_2\psi\psi_t + 3\psi^2 u_{2t} + \tfrac{1}{4}v_0v_1 + y\psi^2 = 0 . \tag{3.18}$$

But, integrating (3.8) and using (3.15), gives

$$v_0 = -\psi \int^t v_1\psi \, dt ,$$

and therefore we find

$$y\psi = -4\psi_{ttt} - 6u_2\psi_t - 3u_{2t}\psi + \frac{1}{4}v_1 \int^t v_1\psi \, dt . \tag{3.19}$$

The integrability condition for (3.16, 19) is simply the equation set (3.12, 13). The Hirota equations for (3.2, 3) follow immediately by truncating the Laurent series at ϕ_{-1}. Setting $q_1 = 2d^2 \ln \phi / dt^2$, $q_2 = -\psi/\phi$ (remember $\int \psi v \, dt$ is a constant when $v = 0$) we obtain

$$D_t^4 \phi \cdot \phi = \phi\phi_{tttt} - 4\phi_t\phi_{ttt} + 3\phi_{tt} = \tfrac{1}{4}\psi^2 \quad \text{and} \quad D_t^2 \phi \cdot \psi = 0 . \tag{3.20}$$

In summary then, v_0 and u_1 are given by (3.19) and (3.7), and the three remaining functions u_2, v_1, and ψ (or ϕ) satisfy the four mutually consistent equations (3.12, 13, 16, 20), the first two of which contain the arbitrary parameters λ and y, respectively. The reader should check that the solvability for (3.16, 20) is the equation pair (3.12, 13). The operator appearing in (3.20), which takes the form (with $u_2 = u$, $v_1 = v$)

$$L = -4D^3 - 3(uD + Du) + \tfrac{1}{4}vD^{-1}v ,$$

appears to be associated with the loop algebra of $A_3^{(2)}$. It would be interesting to determine which family of partial differential equations contains (3.12, 13) as a stationary equation.

3.2 The Algebraic Curve and Integration of the Equations of Motion

Having identified the Bäcklund transformation

$$q_1 = 2\frac{\partial^2}{\partial t^2} \ln \phi + u , \quad q_2 = -\frac{\psi}{\phi} \int \psi v \, dt + v \tag{3.21}$$

(where we have set $u = u_2$ and $v = v_1$) which relates solutions (3.2, 3) and (3.12, 13) and having identified the Lax pair for these equations, we now turn to their explicit integration. It is convenient to rewrite (3.16, 19) in system form as

$$W_t = PW , \quad y\lambda W = QW , \tag{3.22}$$

with

$$W = \begin{pmatrix} \psi \\ \psi_t \end{pmatrix} , \quad P = \begin{pmatrix} 0 & 1 \\ -\lambda - u & 0 \end{pmatrix} ,$$

$$Q = \begin{pmatrix} \lambda u_t + \frac{1}{4}vv_t & 4\lambda^2 - 2\lambda u - \frac{1}{4}v^2 \\ -4\lambda^3 - 2\lambda^2 u - \lambda(u^2 + \frac{1}{4}v^2) + \frac{1}{4}v_t^2 & -\lambda u_t - \frac{1}{4}vv_t \end{pmatrix} . \tag{3.23}$$

The solvability condition of (3.22) is

$$Q_t = [P, Q] , \tag{3.24}$$

which, when written in component form, gives (3.12, 13). Further, if W is a fundamental solution matrix of (3.22), then the solution of (3.24) can be conveniently written as

$$Q = WQ_0W^{-1} , \tag{3.25}$$

with Q_0 independent of t. As a consequence of (3.25), the characteristic polynomial of Q:

$$\det(Q - y\lambda I) = 0 , \tag{3.26}$$

which is the condition that (3.22) has a nontrivial solution, is independent of t. In terms of y, λ, v and u, (3.26) is the algebraic curve

$$\lambda^2 y^2 = -16\lambda^5 + 2H\lambda^2 + G\lambda , \tag{3.27}$$

where

$$H = \frac{1}{2}(u_t^2 + v_t^2) + \frac{1}{2}uv^2 + u^3 \tag{3.28}$$

is the Hamiltonian (3.1) written in the canonically conjugate coordinate pairs u, $U = u_t$ and v, $V = v_t$ and

$$G = \frac{1}{4}v^2(u^2 + \frac{1}{4}v^2) + \frac{1}{2}V(vU - uV) . \tag{3.29}$$

Because (3.27) is independent of time, we can identify G as the second constant of the motion for the flow (3.12, 13) generated by the Hamiltonian H. G and H are in involution under the canonical Poisson bracket

$$\{G, H\} = \frac{\partial G}{\partial V}\frac{\partial G}{\partial v} + \frac{\partial G}{\partial U}\frac{\partial H}{\partial u} - \frac{\partial G}{\partial v}\frac{\partial H}{\partial V} - \frac{\partial G}{\partial u}\frac{\partial H}{\partial U} . \tag{3.30}$$

The choice of auxiliary variables which leads to the identification of the angle variables corresponding to the actions H and G is most conveniently made by using the matrix from the Lax equation (3.24). The new variables are μ_1, μ_2 given by

$$\mu_1 + \mu_2 = \tfrac{1}{2}u , \quad \mu_1\mu_2 = -\tfrac{1}{16}v^2 , \tag{3.31}$$

the zeros of the (1,2) element $e = 4\lambda^2 - 2\lambda u - v^2/4$ in Q [24]. The reason this choice separates the equations of motion can be seen by computing μ_{it}, $i = 1,2$ from the (1,2) element $e_t = -2h$ in (3.24) where $h = \lambda u_t + vv_t/4$. Estimating this equation at $\lambda = \mu_1$ gives

$$-4\mu_{1t}(\mu_1 - \mu_2) = -2h(\lambda = \mu_1) = -2\sqrt{-\det Q(\mu_1)} .$$

Hence

$$\frac{\mu_{1t}}{\sqrt{R(\mu_1)}} = \frac{1}{2}\frac{1}{\mu_1 - \mu_2} , \tag{3.32}$$

$$R(\lambda) = -16\lambda^5 + 2H\lambda^2 + G\lambda , \tag{3.33}$$

and similarly

$$\frac{\mu_{2t}}{\sqrt{R(\mu_2)}} = -\frac{1}{2}\frac{1}{\mu_1 - \mu_2} . \tag{3.34}$$

The reader can also verify that the choice of variables μ_1, μ_2 separates the Hamilton–Jacobi equation and gives (3.32, 34). We recommend strongly that the reader consult the work of *Ercolani* and *Siggia* [20] on these ideas. We can also directly calculate the equations for $\mu_{1\tau}, \mu_{2\tau}$ where τ is the time parameter of the flow generated by taking G as Hamiltonian. Concretely, by writing down the equation of motion, we find by using (3.30) that

$$V = \frac{4}{v}(\mu_{1\tau} + \mu_{2\tau}) , \quad U = -\frac{1}{\mu_1\mu_2}(\mu_{1\tau}\mu_1 + \mu_2\mu_{2\tau}) .$$

Substituting u, v, U and V into H and G and calculating $R(\mu_1) - R(\mu_2)$ gives $\mu_{1\tau}^2(\mu_1 - \mu_2)^2\mu_2^{-2} - \mu_{2\tau}^2(\mu_1 - \mu_2)^2\mu_1^{-2}$ which separates to give

$$\frac{\mu_{1\tau}}{\sqrt{R(\mu_1)}} = \frac{\mu_2}{\mu_1 - \mu_2} , \tag{3.35}$$

$$\frac{\mu_{2\tau}}{\sqrt{R(\mu_2)}} = -\frac{\mu_1}{\mu_1 - \mu_2} . \tag{3.36}$$

Hence

$$\frac{d\mu_1}{\sqrt{R(\mu_1)}} + \frac{d\mu_2}{\sqrt{R(\mu_2)}} = -d\tau \quad \text{and} \quad \frac{\mu_1 d\mu_1}{\sqrt{R(\mu_1)}} + \frac{\mu_2 d\mu_2}{\sqrt{R(\mu_2)}} = \frac{1}{2}dt ,$$

equations which can be directly integrated to give

$$\phi_1 = \int_{\infty}^{\mu_1} \frac{d\mu_1}{\sqrt{R(\mu_1)}} + \int_{\infty}^{\mu_2} \frac{d\mu_2}{\sqrt{R(\mu_2)}} = -\tau + \text{const} , \tag{3.37a}$$

$$\phi_2 = \int\limits_{\infty}^{\mu_1} \frac{\mu_1 d\mu_1}{\sqrt{R(\mu_1)}} + \int\limits_{\infty}^{\mu_2} \frac{\mu_2 d\mu_2}{\sqrt{R(\mu_2)}} = \frac{1}{2}t + \text{const} , \tag{3.37b}$$

The first equality sign in (3.37) provides a map Φ, called the Abel map, from the old coordinates to ϕ_1, ϕ_2, which live on its Jacobi variety. The map Φ takes C^2/P_2 (P_2 is the permutation group on two symbols; μ_1 and μ_2 are complex but it is only symmetric functions which count) onto $C^2/\wedge$ where $\wedge$ is the lattice in C^2 spanned by the vectors corresponding to the integration of the holomorphic differentials $d\mu/\sqrt{R(\mu)}$ and $\mu\, d\mu/\sqrt{R(\mu)}$ about the independent cycles associated with a Riemann surface of genus two. Namely, given μ_1, μ_2, the point ϕ_1, ϕ_2 is only determined up to its position within a lattice parallelogram. From the second half of the equations (3.37), we note that ϕ_1 and ϕ_2 are linear functions of the flow times t and τ of the two commuting flows generated by H and G, respectively. In [8], we have carried out the integration in detail and express the solutions in terms of the Riemann theta function. At this point, we want to address another matter, the form of the principal polar part of the Painlevé expansions for $u(t, \tau)$ and $v(t, \tau)$. Since we will be discussing poles of the solutions q_1, q_2 or u, v, we will be interested in points on the Riemann surface (3.33) where either μ_1, μ_2, or both, take on infinite values. Fixing one of the μ's at infinity corresponds to a surface (here a curve) of dimension one (genus minus one) on the Jacobi variety. Such a surface is called a theta divisor. On this divisor there may be a point at which two μ's, μ_1 and μ_2 become infinite. As we will see, near these multiple poles, the principal Painlevé expansion must be re-expanded because it will turn out that on the pole manifold $\phi(t, \tau) = 0$, ϕ_t is also zero and therefore the expansion is invalid. This re-expansion of the principal expansion in the neighborhood of a pole coalescence leads to the singular branch. Because the expansions correspond to a situation in which two of the μ's are fixed, we expect the resulting singular Painlevé expansion to have one less arbitrary constant.

3.3 The Role of the Rational Solutions in the Painlevé Expansions

We have already noted that the Painlevé expansions (3.4) for (3.2, 3) has two branches with leading order (ϕ^{-2}) behaviors

$$q_1 = -2\frac{\phi_t^2}{\phi^2} + \dots , \quad q_2 = 0 + \dots , \tag{3.38a}$$

$$q_1 = -6\frac{\phi_t^2}{\phi^2} + \dots , \quad q_2 = \pm\frac{12i\phi_t^2}{\phi^2} + \dots . \tag{3.38b}$$

In each case, we assume that we are expanding about a ϕ with the property that ϕ_t is nonzero on the pole manifold $\phi(t, \tau) = 0$. It may also be verified that solutions of (3.2, 3) rational in t are (up to translation in t)

$$q_1 = 2\frac{\partial^2}{\partial t^2}\log\phi , \quad \tfrac{1}{2}q_2^2 = q_{1tt} - 3q_1^2 , \tag{3.39}$$

with (a) $\phi = t$ and (b) $\phi = t^3 + 6\tau$. Specifically, in these cases,

$$q_1 = -\frac{2}{t^2}, \quad q_2 = 0, \tag{4.40a}$$

$$q_1 = -\frac{6t(t^3 - 12\tau)}{(t^3 + 6\tau)^2}, \quad q_2 = \pm\frac{12it}{t^3 + 6\tau}. \tag{3.40b}$$

Each of these rational solutions also satisfies the commuting flow generated by G. We have also learned that $(3.2, 3)$ are separated by the transformation (3.31).

$$q_1 = 2(\mu_1 + \mu_2), \quad q_2^2 = -16\mu_1\mu_2, \tag{3.40c}$$

and the equations for the μ's are given in (3.37). Let us study these two cases. In either case we need to know how to expand the integrals in (3.37) about $\mu_1 = \infty$. A little calculation shows that

$$\int_{\infty}^{\mu_1} \frac{\mu^n d\mu}{\sqrt{R(\mu)}} \sim \frac{1}{4i}\frac{\mu_1^{n-3/2}}{n - 3/2}. \tag{3.41}$$

From $(3.37b)$, we find that

$$\mu_1 \sim -\frac{1}{t^2}, \tag{3.42}$$

where we have incorporated into t (by translation) the constant term in $(3.37b)$ and the integral from ∞ to c, the value μ_2 attains as t tends to zero and μ_1 to infinity. The corresponding principal polar parts of q_1, q_2 are

$$q_1 = -\frac{2}{t^2}, \quad q_2 = \frac{4c}{t}. \tag{3.43}$$

Observe that the leading order behavior is the rational solution (3.39) with $\phi = t$.

Now let us turn to the second case where μ_1 and μ_2 both tend to infinity at the same point (t_0, τ_0) in (t, τ) space (equivalently at the same point on the theta divisor in the Jacobi variety). Repeating the former calculation, we find that (3.37) gives

$$\frac{i}{\mu_1^{3/2}} + \frac{i}{\mu_2^{3/2}} = 6(\tau - \tau_0), \quad \frac{i}{\mu_1^{1/2}} + \frac{i}{\mu_2^{1/2}} = t - t_0. \tag{3.44}$$

Calculation shows that the corresponding principal polar parts of q_1 and q_2 are (we translate t, τ by t_0, τ_0)

$$q_1 = -\frac{6t(t^3 - 12\tau)}{(t^3 + 6\tau)^2}, \quad q_2 = \pm\frac{12it}{t^3 + t\tau}. \tag{3.45}$$

Now in (3.45) we cannot simply re-expand q_1 and q_2 about each of the simple poles $t = (6\tau)^{1/3} \exp[(2s + 1)\pi i/3]$, $s = 0, 1, 2$ because both t and τ tend to zero as μ_1 and μ_2 approach infinity. If we did, we would simply regain the principal polar expansion. However as τ tends to zero, the three poles [or zeros of $\phi(t, \tau) = t^3 + 6\tau$] coalesce, and at $\tau = 0$ the principal polar parts of q_1 and q_2 are

$$q_1 = -\frac{6}{t^2}, \quad q_2 = \pm\frac{12i}{t^2}, \tag{3.46}$$

which is precisely the leading order behavior of the second (and, in this case, only) singular branch of the Painlevé expansion.

We learn from this illustration that the lower Painlevé branches (in this case branch) are simply a re-expansion of the principal branch about a location on the pole manifold $\phi = 0$ (in this case $\phi = t^3 + 6\tau$) at which ϕ_t is zero. What happens at these points is that the principal branch must be re-expanded in terms of another candidate for ϕ (in this case $\phi = t$) and this re-expansion gives rise to the singular branches. It turns out that the set of ϕ's which capture in a uniformly valid way the principal polar parts of the various Painlevé expansions are simply multiples of the rational solutions, the multiplying functions being analytic functions of the times which are nonzero on surfaces where the rational solutions have zeros.

4. A Mikhailov and Shabat Example

In the conference at Kiev some years ago, Mikhailov and Shabat challenged the audience to find the Lax (L, B) pair for the following system of partial differential equations:

$$\gamma_t = r_{xx} + (r + s)_{sx} - \tfrac{1}{6}(r + s)^3 \tag{4.1}$$

$$s_t = -s_{xx} + (r + s)v_x + \tfrac{1}{6}(r + s)^3, \tag{4.2}$$

or, with $r + s = p$, $r - s = q$,

$$p_t = q_{xx} + pp_x, \tag{4.3}$$

$$q_t = p_{xx} - pq_x - p^3/3. \tag{4.4}$$

These equations belong to the class for which Mikhailov and Shabat knew there existed an infinite number of conservation laws but for which they were unable to find the (L, B) pair. The Painlevé method immediately provides the answer. Let

$$p = \frac{v_0}{\varphi} + v, \quad q = \frac{u_0}{\varphi} + u, \tag{4.5}$$

substitute into (4.3, 4) and find that $u_0 = -6\varphi_x$, $v_0 = \beta\varphi_x$, $\beta^2 = -12$ and

$$\varphi_t = \tfrac{1}{2}\beta\varphi_{xx} + v\varphi_x + \mu\varphi, \tag{4.6}$$

$$4\beta\varphi_{xxx} + 12v\varphi_{xx} - \beta(u_x + v^2)\varphi_x + 6v_x\varphi_x = -6\mu\varphi_x . \tag{4.7}$$

The free parameter μ is added as follows. Add $-6\mu\varphi_x/\varphi + 6\mu\varphi_x\varphi/\varphi^2$ to (4.4) and add $-\mu\beta\varphi_x/\varphi + \mu\beta\varphi_x\varphi/\varphi^2$ to (4.3) after (4.5) is substituted. (See discussion in Sect. 2.2.) Defining

$$L = 4\beta D^3 + 12vD^2 - \beta(u_x + v^2)D + 6v_xD , \tag{4.8}$$

$$B = \frac{\beta D^2}{2} + vD , \quad D = \frac{\partial}{\partial x} , \tag{4.9}$$

whence (4.6, 7) read

$$\varphi_t = B\varphi + \mu\varphi , \quad L\varphi = -6\mu\varphi_x , \tag{4.10}$$

respectively. It is easy to show from (4.10) that

$$L_t = [B, L] \tag{4.11}$$

and that (4.11) gives equations (4.3, 4) with p replaced by v and q by u.

5. Some Comments on the KdV Hierarchy

In [8], we discuss the KdV and AKNS hierarchies. The principal results on the former are:

(1) A general formula for enumerating all the resonances in the Painlevé expansions is derived.

(2) It is proven that the principal Painlevé expansion for both the time-dependent and stationary general member of the KdV hierarchy can be consistently truncated to give the Lax pair and Bäcklund transformation.

(3) The appearance of the various branches of the Painlevé expansion is shown to be due to a confluence of poles. This leads to a situation in which the WTC ϕ function has a multiple zero, which necessitates a reordering of the principal expansion in which ϕ_x is nonzero when $\phi = 0$. Different numbers of coalescing poles will lead to different expansions. One of the key results is that near a confluence of $m(m+1)/2$ poles, the *uniformly valid* expression for the principal polar part of q is $2\,\partial^2 \ln R_m(x, t_3, \ldots, t_{2m-1})/\partial x^2$ where R_m is the mth member of the rational solution hierarchy. The re-expansion of q in terms of $\phi = R_1 = x$ about a point in $x, t_3, \ldots, t_{2m-1}$ space where R_m, as function of x, has an $m(m+1)/2$-fold zero, leads to the mth branch of the Painlevé expansion for the nth stationary equation in the KdV hierarchy. At these points, exactly m of the intermediate angle variables (called $\mu_1, \ldots, \mu_n$ and analogous to the μ's introduced in Sect. 3) also have poles.

(4) In the sense described in (3), the mth rational solution *unfolds* the singularity near the coalescence of $m(m+1)/2$ poles.

(5) The precise connection between the fluxes and the arbitrary coefficients of the most singular branches of the stationary flows is obtained.

Similar results are obtained for the AKNS hierarchy.

6. Connection with Symmetries and Algebraic Structure

As we have seen, the Painlevé property imposes algebraic structure on a system of differential equations. If there are parameters in the equations, the requirement that all singular solutions be Laurent series (or, more generally, series in ϕ^{-1}) restricts those parameters to just a few values. This is only understood on an "operational"level. It should be possible to relate those special parameter values to symmetry groups: one is reminded of the integrality conditions that permeate the theory of semi-simple Lie algebras, and indeed there are examples admitting just such an interpretation which we discuss below. There is also a geometric aspect to Painlevé analysis; one expects the various integers (resonances, for example) to reflect discreteness of topological invariants of the complex orbits of the system. After all, in most Painlevé examples that have been analyzed, the solutions live naturally on a compact subvariety of some complex projective space.

This idea is still quite vague, and at present the only sensible thing to do is to collect experimental evidence. We therefore study examples whose connections to symmetry groups and projective spaces are already understood, or at least lie near the surface, and we pretend to discover these features anew, starting from the Painlevé analysis.

The most obliging example in this respect is the *Toda* lattice [30]. The physical model of masses connected by springs with exponential restoring force is reduced by a change of variables to a polynomial system,

$$\dot{a}_i = a_i \sum_{j=1}^{l} N_{ij} b_j , \quad \dot{b}_i = \sum_{j=1}^{l} N_{ji} a_j . \tag{6.1}$$

For Toda's original lattice, the N_{ij} are certain fixed numbers, but let them be arbitrary for the moment, assume that $\det(N_{ij}) \neq 0$, and ask: when does (6.1) have the Painlevé property? According to [31,32], the necessary and sufficient condition is that the rows of (N_{ij}) be the simple roots of a crystallographic root system. In this one example, at least, one can follow the algebraic structure as it is created. When one solves recursively for the (vector) coefficients in the Laurent expansion $a(t) = t^{-2}(a^{(0)} + a^{(1)}t + \ldots)$, integral resonances arise only when a certain matrix has integral eigenvalues. This translates into the requirement that for any two distinct rows n_i, n_j of N_{ij}, the number $2n_i \cdot n_j / n_j \cdot n_j$ be a nonpositive integer. Fortunately, the classification of sets of vectors $n_1, \ldots, n_l$ with that property is exactly the concern of the theory of root systems, reflection groups, and semi-simple Lie algebras: if the system (6.1) has the Painlevé property, the Painlevé method identifies a semi-simple Lie algebra $\mathcal{G}$ whose structure determines the behavior of the solutions of that Toda system. As an aside, we should mention the pioneering work of *Bogoyavlenskii* [33]. He was the first to identify the integrable cases of (6.1) by establishing an entirely different and physically more concrete connection with reflection groups; the relation between Painlevé analysis and Bogoyavlenskii's compactification of the phase space of (6.1) is still under investigation.

The success of Painlevé methods in identifying integrable generalized Toda systems (6.1) suggest the following, so far open, problem: Starting with a system of PDEs

$$\frac{\partial u_i}{\partial t} = K_i(u_1, \ldots, u_l), \quad i = 1, \ldots, l$$

(of suitably restricted form), use the WTC analysis to rediscover Drinfel'd and Sokolov's generalized KdV equations associated with affine Lie algebras. For the practical person, this would only be the 17th method of identifying a known class of equations, but one might still learn a bit more about the WTC method, and understand how it imposes algebraic constraints.

Let us return now to (6.1). Painlevé implies that the rows n_i, $i = 1, \ldots, l$, of (N_{ij}) are the simple roots of a semi-simple complex Lie algebra $\mathcal{G}$. We fix $\mathcal{G}$, and correspondingly, a matrix N_{ij} (there is a lot of freedom in the choice of basis). Painlevé analysis provides more data than we may want: to deduce that the rows n_i must form a root system, one need only ask that there exist Laurent solutions with $2l - 1$ free parameters allowing any chosen a_i, b_i to have a pole. These are the principal balances. There are still all the lower balances, however, and, as in the KdV example summarized above, they present a bewildering variety of arithmetic information. The complete list of balances and resonances has been computed by *Flaschka* and *Zeng* [34]; the calculation draws on general properties of roots, weights, and exponents of semi-simple lie algebras, and even provides new identities for root systems.

In most of the examples that have been worked out in some detail, the balances and resonances have not been related to any familiar mathematical structure. Toda lattices are different; all conceivable symmetries are built in at the start and make themselves known at the slightest provocation. The balances of (6.1) are indexed by subsets $\Theta \subset \{1, \ldots, l\}$. The Laurent series solution corresponding to Θ has the form

$$a_i(t) = a_i^{(0)} t^{-2} + a_i^{(1)} t^{-1} + \ldots$$

with $a_i^{(0)} \neq 0 \Leftrightarrow i \in \Theta$, and there are $2l - |\Theta|$ arbitrary constants in this series. The collection of subsets $\{n_{i_1}, \ldots n_{i_k}\}$ of a root system is a geometric object called a *Coxeter complex* and one finds that the balances of an integrable Toda lattice are indexed by such a Coxeter complex; moreover, they merge into each other according to inclusion relations among subsets of $\{1, \ldots, l\}$.

It is now time to switch from algebraic considerations to phase-space geometry. A "balance" is, after all, just a collection of points at infinity in phase space; one of the dependent variables, at least, is infinite and the parameters in the Laurent series coordinatize those singular points. Again, the Toda lattice serves as a useful illustration.

Pick an $i_0 \in \{1, \ldots, l\}$. There exist Laurent series solutions

$$a(t) = t^{-2} \left\{ \begin{pmatrix} 0 \\ \vdots \\ 1 \\ \vdots \\ 0 \end{pmatrix} + O(t^2) \right\} \tag{6.2}$$

with $2l - 1$ arbitrary constants. We think of each of those series as defining a point at infinity, attached to the phase space $= \{(a_1, \ldots, a_l, b_1, \ldots, b_l)\}$. There are sets $D_{i_0}, i_0 = 1, \ldots, l$, to be added to C^{2l}: in D_{i_0}, the component a_{i_0} alone becomes infinite, and there are $2l - 1$ parameters to describe exactly how this happens. The closures of the ideal sets D_{i_0} will intersect. For $l = 2$, for example, we add 2 three-dimensional manifolds at infinity which in a two-dimensional drawing might be viewed as a straight line which touches a parabola, the intersection of these two sets (the point of contact in a drawing) is a two-dimensional object. It represents singular solutions

$$a(t) = t^{-2} \left\{ \begin{pmatrix} 2 \\ 2 \end{pmatrix} + O(t^2) \right\}$$

in which both a_1 and a_2 blow up; there are only 2 parameters (rather than 3) in the Laurent series. This is a *lower balance* (in fact, the *lowest* balance in this example).

For the general semi-simple Toda lattice, one adds l noncompact manifolds $D_1, \ldots, D_l$ of dimension $2l - 1$ at infinity; those represent the principal balances. At their boundaries, there are the manifolds D_{i_j} of dimension $2l - 2$, where the closures of D_i and D_j intersect. Then come $\bar{D}_i \cap \bar{D}_j \cap \bar{D}_k$ of dimension $2l - 3$ and so forth, down to the lowest balance $\bar{D}_1 \cap \ldots \cap \bar{D}_l$ of dimension l. The *geometric problem* of Painlevé analysis is to piece together those additional phase-space parts at infinity in order to create a partial compactification of the original finite phase space C^{2l}. One wants to produce a complex manifold on which all solutions exist for all time.

For the Toda systems where the balances are indexed by subsets $\Theta \subset \{1, \ldots, l\}$, one augments the phase space by adding manifolds D_Θ of dimension $2l - |\Theta|$. The closures $\bar{D}_\Theta$, $\bar{D}_{\Theta'}$ intersect along $D_{\Theta \cap \Theta'}$, and the manner of intersection (transverse, tangent, higher-order ...) is determined by the Dynkin diagram of the associated Lie algebra. The geometric relations that arise here are a special case of the geometry of the flag manifolds and Schubert (or, more precisely, Birkhoff) cells. (For more details see [35].)

Of course, "adding points at infinity" is not accomplished just by listing the extra points. One must define coordinate patches with transition functions, so that everything fits together into a complex manifold M containing the original phase space as dense open subset.

Example: The $sl(2)$ Toda equations are

$$\dot{a} = ab, \quad \dot{b} = a$$

the function $a - b^2/2$ is a constant of motion. Set

$$u = a - \frac{1}{2}b^2, \quad v = \frac{1}{b}. \tag{6.3}$$

Then

$$\dot{u} = 0, \quad \dot{v} = -uv^2 - \tfrac{1}{2}. \tag{6.4}$$

The partial compactification of the phase space C^2 is the complex manifold M defined by two coordinate patches: $U_1 = \{(a, b) \in C^2\}$, $U_2 = \{(u, v) \in C^2\}$; the intersection is $U_1 \cap U_2 = \{(a, b) \in C^2 | b \neq 0\} = \{(u, v) \in C^2 | v \neq 0\}$ and the coordinate change is given by (6.3).

One may picture this particular M as $C \times CP^1$. To each point $c \in C$, attach the compactification of the level set $\{(a, b) | a - b^2/2 = c\}$; this is the Riemann sphere or, equivalently, the one-dimensional complex projective space CP^1. A better description can be gotten if one knows that $CP^1 \times CP^1$ is a conic in three-dimensional projective space CP^3. The manifold $M = C \times CP^1$ is that conic with $\{\infty\} \times CP^1$ removed, i.e., a conic less one (projective) line. M is not compact; the missing piece is the set $a - b^2/2 = \infty$ of infinite energy phase points.

While other examples of this type have been worked out, the only algorithm of any generality is due to *Ercolani* and *Siggia* [20]. For each balance of a system generated by a polynomial Hamiltonian $H(q_1, \ldots, q_n, p_1, \ldots, p_n) = H(q, p)$, they find an asymptotic solution of the Hamiltonian–Jacobi equation $H(\partial S/\partial q, q) = E$ and use the first few terms of that solution to define a symplectic coordinate change from C^{2n} to a neighborhood of points at infinity. This produces what they call a *minimal augmentation* of phase space, i.e., the smallest complex manifold M on which solutions of the Hamiltonian system exist for all time.

For $\ddot{q} = 6q^2$, for example, the augmentation M is obtained by adding one point to each energy level curve $p^2 - 4q^3 - c = 0$, thus producing a family of elliptic curves indexed by $c \in C$; M is a dense open subset of an *elliptic surface*. In most cases, however, M can only be described implicitly, by coordinate charts, and in order to arrive at a general understanding one must continue to study various nontrivial examples.

For systems that have principal *and* lower balances, such as stationary KdV, AKNS, the Toda lattice, etc., the augmentation of phase space has a characteristic feature: the original phase space is some Euclidean space C^n, and the balances to be added at infinity, being coordinatized by the parameters in Laurent series, are lower-dimensional affine varieties. Sometimes, as in the Toda lattice, they are Euclidean spaces of lower dimensions. They are then called *cells* in topology, and the augmented phase space M comes with a natural *cell decomposition*. This would give powerful geometric information if M were known to be a compact projective variety; Painlevé analysis, unfortunately, overlooks the infinite-energy points, which may change the topology. Too few examples have been worked out. One can see, however, in some of the most familiar systems, that the cell decomposition of M arises from classical properties of Grassmannians. This is hoped to be a guide to a more comprehensive theory. We conclude this section with a sketch of the KdV geometry, which reinforces some of the general points.

First, recall that the solution to the KdV hierarchy is determined by a single tau function, $\tau(t_1, t_3, t_5, \ldots)$,

$$u(t_1, t_3, \ldots) = 2\frac{\partial^2}{\partial t_1^2} \log \tau(t_1, t_3, \ldots).$$

The solution acquires a pole whenever $\tau = 0$. According to *Sato* [36] and *Segal and Wilson* [37], the tau function and its zero-set are governed by the geometry of an infinite-dimensional Grassmann manifold. It would take too much space to explain the setup, but a short description of a finite-dimensional analogue might be useful. The relevance to the KdV equation will be outlined later.

The Grassmannian $G(k, n)$ is the set of all k-dimensional subspaces of C^n. Every such subspace is obtained as a linear transformation of the reference k-plane Π_0 spanned by

$$v_1 = \begin{pmatrix} 1 \\ 0 \\ 0 \\ \vdots \\ 0 \end{pmatrix}, \quad v_2 = \begin{pmatrix} 0 \\ 1 \\ 0 \\ \vdots \\ 0 \end{pmatrix}, \ldots, v_k = \begin{pmatrix} 0 \\ \vdots \\ 0 \\ 1 \\ 0 \\ \vdots \\ 0 \end{pmatrix} - k\text{th entry}.$$

So, if g is any invertible complex $n \times n$ matrix, i.e., an element in the group $GL(n, C)$, the k-plane Π_g spanned by $gv_1, \ldots, gv_k$ is another point of $G(k, n)$. All of $G(k, n)$ is obtained that way. The representation is not unique, however, $\Pi_{g_1} = \Pi_{g_2}$ if $g_2^{-1}g_1$ leaves Π_0 fixed. The subgroup P_k of $GL(n, C)$ fixing Π_0 consists of all matrices of the form

$$\begin{matrix} k \\ n-k \end{matrix} \left\{ \begin{matrix} \\ \\ \end{matrix} \begin{pmatrix} * & * \\ 0 & * \end{pmatrix} \right. $$
$$\underbrace{\quad\quad}_{k} \underbrace{\quad\quad}_{n-k}$$

and $G(k, n)$ may be identified with the quotient space $GL(n, C)/P_k$. It is useful to describe each $\Pi \in G(k, n)$ by listing a basis determined according to some algorithm. The model $GL(n,C)/P_k$ suggests such an algorithm. Try to write each $g \in GL(n,C)$ in a unique way as a product $g = bp$, where $p \in P_k$, the distinguished basis for Π_g will then be $bv_1, \ldots, bv_k$. The matrix b should be lower-triangular and have a special form.

Example: If $n = 3$ and $k = 2$, one looks for a factorization

$$g = \begin{pmatrix} 1 & 0 & 0 \\ 0 & 1 & 0 \\ r & s & 1 \end{pmatrix} \begin{pmatrix} u & v & w \\ x & y & z \\ 0 & 0 & \delta \end{pmatrix} = bp \, . \tag{6.5}$$

Since p must be invertible, the minor $\begin{vmatrix} u & v \\ x & y \end{vmatrix}$ cannot vanish. Thus, only g's with nonzero principal 2×2 minor determinant can be factored in this way. When that minor determinant vanishes, one needs a different factorization. If

$$g = \begin{pmatrix} u & v & w \\ x & y & z \\ \alpha & \beta & \gamma \end{pmatrix}$$

and $\begin{vmatrix} u & v \\ x & y \end{vmatrix} = 0$, then since g is invertible, eight $\begin{vmatrix} u & v \\ \alpha & \beta \end{vmatrix} \neq 0$ or $\begin{vmatrix} x & y \\ \alpha & \beta \end{vmatrix} \neq 0$.

Suppose that $\begin{vmatrix} u & v \\ \alpha & \beta \end{vmatrix} \neq 0$. The matrix g has the form

$$\begin{pmatrix} u & v & w \\ hu & hv & z \\ \alpha & \beta & \gamma \end{pmatrix}$$

and can be factored

$$\begin{pmatrix} 1 & 0 & 0 \\ h & 1 & 0 \\ 0 & 0 & 1 \end{pmatrix} \begin{pmatrix} 1 & 0 & 0 \\ 0 & 0 & 1 \\ 0 & 1 & 0 \end{pmatrix} \begin{pmatrix} u & v & w \\ \alpha & \beta & \gamma \\ 0 & 0 & \delta \end{pmatrix} \, . \tag{6.6}$$

If $\begin{vmatrix} u & v \\ x & y \end{vmatrix} = 0 = \begin{vmatrix} u & v \\ \alpha & \beta \end{vmatrix}$, but $\begin{vmatrix} x & y \\ \alpha & \beta \end{vmatrix} \neq 0$, then $u = v = 0$. In that case,

$$g = \begin{pmatrix} 0 & 0 & w \\ x & y & z \\ \alpha & \beta & \gamma \end{pmatrix} \, ,$$

and this has the factorization

$$\begin{pmatrix} 1 & 0 & 0 \\ 0 & 1 & 0 \\ 0 & 0 & 1 \end{pmatrix} \begin{pmatrix} 0 & 0 & 1 \\ 1 & 0 & 0 \\ 0 & 1 & 0 \end{pmatrix} \begin{pmatrix} x & y & z \\ \alpha & \beta & \gamma \\ 0 & 0 & w \end{pmatrix} \, . \tag{6.7}$$

Therefore, the two-planes in C^3 can be assigned coordinates as follows:

(i) most of them are parametrized by a pair $(r, s) \in C^2$ as in (6.5);

(ii) a smaller set is parametrized by $h \in C$ as in (6.6);

(iii) one exceptional plane is given by (6.7).

These are the *Birkhoff cells* mentioned earlier. We have exhibited $G(2,3)$ as a disjoint union

$$C^2 \cup C^1 \cup C^0 , \tag{6.8}$$

which is the same as the *cell decomposition* of the two-dimensional projective plane: C^2 is the finite part, C^2 is the line at infinity, and the last point C^0 is the point at infinity of the line at infinity. The relevance of all this to integrable systems is as follows. We define a function $\tau(\Pi_g)$ on $G(2,3)$ to be the principal 2×2 minor determinant of g. (Actually, τ is not uniquely defined, but its zero set *is* determined.) On the *big cell C^2*, $\tau \neq 0$. On the next-largest cell C^1 (6.6), τ has – roughly speaking – a single zero, and on the smallest cell C^0 (6.7), τ has – again, roughly speaking – a double zero.

Of course, there is no integrable system in sight – that is an extra structure which makes the situation more complicated. For the standard two-particle Toda lattice (6.1), there are two tau functions (one for each particle) and the balances in which the position coordinate of the second particle blows up have precisely the cell structure (6.8). This is explained in [35].

In each cell (6.5–7), there is a distinguished point, corresponding, respectively, to $r = s = 0$, $h = 0$, and to the smallest cell C^0 which is already a single point (6.7). We call these points the *centers* of the cells, and they are represented by the permutation matrices

$$\begin{pmatrix} 1 & & \\ & 1 & \\ & & 1 \end{pmatrix} , \quad \begin{pmatrix} 1 & & \\ & & 1 \\ & 1 & \end{pmatrix} , \quad \begin{pmatrix} & & 1 \\ 1 & & \\ & 1 & \end{pmatrix}$$

in (6.5–7). In the Toda system, they turn out to characterize certain rational solutions, and subgroup of the permutation matrices may be identified with the group of Bäcklund transformations that send a given rational solution to any other rational solution.

Now, this very same picture, when extended to infinite dimensions, provides the setting for the KdV tau function. Instead of C^n, we have the Hilbert space $H = L^2(S^2, C)$ of square-integrable functions on the unit circle. The Grassmann manifold $G(k, n)$ is replaced by a certain collection $Gr^{(2)}$ of infinite-dimensional subspaces, obtained from the reference subspace

$$H^+ = \text{span of } 1 , \quad e^{2\pi i\Theta} , \quad e^{4\pi i\Theta} , \quad \dots$$

by linear transformations from a big group G. The KdV tau function is a determinant: if it vanishes, a certain factorization cannot be performed. There is a cell decomposition of the infinite-dimensional Grassmannian $Gr^{(2)}$, and the *centers* of the cells correspond to the rational solutions of KdV, whose role in the Painlevé analysis was described in [8] and summarized in Sect. 5 above.

The details may be found in [36, 37] and the application to the Painlevé analysis of KdV will be published elsewhere. For present purposes, the important point is this: of all the "algorithmic" procedures that test integrability and produce Lax pairs, Hirota equations and the like, only the Painlevé analysis manages to identify the geometric structure of the (possible infinite-dimensional) phase space. The "balances" are the "cells" – that much is obvious. Where the geometric theory is already understood, as it is with the Toda lattice or the KdV equation, one can easily reconstruct the Grassmannians. The major open problem is to find the analogous geometric setting for all the other Hamiltonian systems that have the Painlevé property: Hénon–Heiles, various spinning-top equations, geodesic flows on Lie groups, and so forth. Since the other characteristic features of integrable systems fit naturally into the Grassmannian framework (Hirota equations are Plücker relations, Lax pairs are reductions of linear equations on the cotangent space of the group that acts on the Grassmannian, special Bäcklund transformations permute the centers of the cells) one expects to find most of these other properties as consequences of a geometric Painlevé analysis.

7. Integrating the Nonintegrable

In the preceding sections we have examined the ways in which the local, singular manifold expansions can be used to integrate – in the sense of finding Lax pairs and Bäcklund transformations – integrable differential equations. We now turn to an investigation of the information contained in the local expansion of nonintegrable systems for which the movable singularities are no longer simple poles. That nonintegrable systems exhibit pathological distributions of movable singularities was demonstrated by *Chang* et al [38] in a study of the Hénon–Heiles system. Here the singularities were found to cluster recursively to form self-similar natural boundaries. Subsequent analysis by *Chang* et al [39] was able to provide an accurate analytical description of this structure and, indeed, to demonstrate that such structures are typical for systems whose movable singularities have complex order. Here we will concentrate on systems with movable logarithms, branch points such as the Duffing oscillators and Lorenz equations. These singularities must be locally represented by logarithmic psi-series. Remarkably, a renormalization type procedure can be used to resum these series in such a way as to provide explicitly analytical representations of the solution in the neighborhood of a singularity. In addition, these series can then be used to identify and construct integrals of motion that may still exist in nonintegrable parameter regimes. It is this collection of ideas that suggests the notion of "integrating the nonintegrable".

We first consider the Duffing oscillator

$$\ddot{x} + \mu\dot{x} + \tfrac{1}{2}x^3 = \varepsilon F(t) \tag{7.1}$$

which for nonzero damping and periodic driving can exhibit chaotic behavior and even a strange attractor. (Nondissipative chaos is also possible for zero damping, provided that the driving force is sufficiently strong.) We note that in the limit of $\mu = \varepsilon = 0$ (7.1) reduces to the integrable Hamiltonian system

$$\ddot{x} + \tfrac{1}{2}x^3 = 0 \tag{7.2}$$

whose solution can be represented in terms of the Lemniscate elliptic function, which exhibits a square lattice of poles in the complex t-plane. This "underlying" integrable part of (7.1) will play an important role in our subsequent analysis.

It is an easy matter to show that about an arbitrary movable singularity t_0 in the complex t-plane, the solution to (7.2) may be locally represented as a simple Laurent series of the form

$$x(t) = \sum_{j=0}^{\infty} a_j(t - t_0)^{j-1} . \tag{7.3}$$

Direct substitution of (7.3) into the equations of motion leads to the recursion relations for the a_j

$$a_j(j + 1)(j - 4) = -\tfrac{1}{2} \sum_k \sum_j a_{j-k-1} a_k a_j , \tag{7.4}$$

where $a_0 = 2i$, $a_1 = a_2 = a_3 = 0$, $a_4 = $ arbitrary coefficient, etc. The arbitrary pole position t_0 and coefficient a_4 constitute the two pieces of arbitrary data consistent with a local representation of the general solution to the second order equation (7.2). The first integral of (7.2) is, of course, the quantity

$$\tfrac{1}{2}\dot{x}^2 + \tfrac{1}{8}x^4 = I_1 . \tag{7.5}$$

By Liouville's Theorem the left-hand side of (7.5) must be an entire function. Thus if the Laurent series for $x(t)$ and $\dot{x}(t)$ are substituted into (7.5) the singular parts must cancel identically. In this way one can find the relationship between the arbitrary coefficient a_4 and the value of the integral I_1, namely, $a_4 = iI_1/5$. This cancellation of singular terms will also play an important role in our analysis.

Introduction of dissipation or driving into (7.2) leads to the break down of the Laurent series (7.3) since it becomes no longer possible to introduce an arbitrary coefficient at $j = 4$. It is a standard matter to rectify this problem by adding logarithmic terms to the expansion thereby obtaining the psi-series

$$x(t) = \sum_{j=0}^{\infty} \sum_{k=0}^{\infty} a_{jk}(t - t_0)^{j-1}[(t - t_0)^4 \ln(t - t_0)]^k . \tag{7.6}$$

Computation of the recursion relations for the a_{jk} is tedious but straightforward and is described in *Fourier* et al [40]. Following the technique developed in [41] we now look for a closed set of recursion relations among the a_{jk}. These are the set a_{0k}, $k = 0, 1, 2, \ldots$ which satisfy

$$4k(k-1)a_{0k} + ka_{0k} + \tfrac{1}{2}a_{0k} = -\tfrac{1}{8} \sum_{s}\sum_{q} a_{0,\,k-s}a_{0,\,s-q}a_{0q} \,. \tag{7.7}$$

Introduction of the generating function

$$\Theta(z) = \sum_{k=0}^{\infty} a_{0k} z^k \,, \tag{7.8}$$

where z is some yet unspecified independent variable, the following differential equation for $\Theta(z)$ is obtained:

$$16z^2 \Theta''(z) + 4z\Theta'(z) + 2\Theta(z) + \tfrac{1}{2}\Theta^3(z) = 0 \,, \tag{7.9}$$

where prime denotes differentiation with respect to z.

The differential equation (7.9) may be obtained via a different, more direct, route by using the procedure described in *Chang* et al [38]. In the limit $t \to t_0$ we concentrate on the terms in the psi-series (7.6) involving powers of $t^4 \ln t$ and therefore make the substitution

$$x(t) = \frac{1}{t - t_0} \Theta_0(z) \,, \tag{7.10}$$

where

$$z = (t - t_0)^4 \ln(t - t_0) \,, \tag{7.11}$$

into (7.1). In the limit $t \to t_0$ it is easy to show formally that $\Theta_0(z)$ again satisfies (7.9), provided that there is an ordering in which $|t - t_0| \ll |z|$. Due to the infinite multivaluedness of the logarithm this is indeed perfectly possible for large (absolute) value of the argument of $(t - t_0)$. We point out that this approach can be thought of as a type of "renormalization" in that the differential equation (7.9) can be regarded as the original equation of motion "rescaled" in the neighborhood of a given singularity. However, this equation is now found to have the Painlevé property and, furthermore, by making the substitution

$$\Theta_0(z) = z^{1/4} f\left(z^{1/4}\right) \,, \tag{7.12}$$

the equation transforms to

$$f''(y) + \tfrac{1}{2}f^3(y) = 0 \,, \tag{7.13}$$

where prime denotes differentiation with respect to the variable $y = z^{1/4} = (t - t_0)(\ln(t - t_0))^{1/4}$. In keeping with the ideas of renormalization we again stress that through the two-step transformation (7.10, 12) the general Duffing equation (7.1) has been locally mapped onto the integrable case (7.2).

From these results the following picture of singularity clustering emerges. To each singularity t_0 in the complex t-domain one can "attach" an associated y-plane. The lattice of singularities exhibited by (7.13) in the y-plane is then mapped back into the t-space according to the multivalued transformation $z = t^4 \ln t$ which results in a t-plane pattern of four-armed "stars" of singularities. An immensely complicated, multisheeted structure in the t-space results with the degree of clustering (about a given t_0) determined by both the degree of "scaling" and "rotating" of the lattice in the y-plane and the intricacies of the mapping $z \to t$. Furthermore, any one of these singularities in the t-plane can have its own clustered, i.e., four-armed star, substructure. The recursive nature of this clustering can clearly lead to a singularity structure of pathological complexity. Numerical results, which strikingly confirm this picture, are described in [40]. Despite this complexity there is a rather remarkable feature of the above analysis. Combining equations (7.10, 12, 13) leads to an explicit local expression for $x(t)$ which accurately predicts the location and nature of neighboring singularities. Coupled with the higher order terms, described below, this therefore provides, if effect, an integration of motion traditionally regarded as "nonintegrable" and, for some parameter values, even known to exhibit chaos.

The substitution (7.10) can be thought of as just the first term in a more general expansion of the form

$$x(t) = \sum_{k=0}^{\infty} \Theta_k(z) t^{k-1} , \tag{7.14}$$

where $z = t^4 \ln t$ and for notational convenience we have set $t_0 = 0$. Substitution of (7.14) into (7.1) and taking the limit $t \to 0$ leads to a hierarchy of coupled differential equations for the Θ_k of the general form

$$16 z^2 \ddot{\Theta}_k + 4(2k+1) z \dot{\Theta}_k + [(k-1)(k-2) + \tfrac{3}{2}\Theta_0^2]\Theta_k$$
$$= R_k(\Theta_j : j < k) . \tag{7.15}$$

All the differential equations for Θ_k, $k \geq 1$, are linear inhomogeneous equations. The general homogeneous counterpart is just

$$16 z^2 \ddot{\Theta}_k + 4(2k+1) z \dot{\Theta}_k + [\tfrac{3}{2}\Theta_0^2 + (k-1)(k-2)]\Theta_k = 0 , \quad k \geq 1 . \tag{7.16}$$

Remarkably, this equation can be solved by making the substitution

$$\Theta_k(z) = (z^{1/4})^{1-k} \psi_k(z^{1/4}) , \quad k \geq 0 , \tag{7.17}$$

which, with a little manipulation, reduces (7.16) to

$$\ddot{\psi}_k + \tfrac{3}{2}\psi_0^2 \psi_k = 0 . \tag{7.18}$$

The function ψ_0 is (to within a scaling) the lemniscate elliptic function solution of (7.13). Thus, (7.18) is readily identified as a form of the Lamé equation. Thus, in principle, the solutions to all the inhomogeneous equations (7.15) may be obtained explicitly.

A significant way of regarding the expansion (7.14) is to compare it directly with the psi-series expansion (7.6) and recognize that each Θ_j is the generating function for the set of coefficients a_{jk}, i.e.,

$$\Theta_j(z) = \sum_{k=0}^{\infty} a_{jk} z^k , \tag{7.19}$$

where $z = t^4 \ln t$. For the case $j = 0$, the recursion relations for the a_{0k} are closed whereas for all other cases they are coupled to preceding coefficients. From this point of view we may regard (7.14) as a resummation of the psi-series (7.6).

These ideas can be used to good effect in studying the Lorenz equations

$$\dot{X} = \sigma(Y - X); \tag{7.20a}$$

$$\dot{Y} = -XZ + RX - Y , \tag{7.20b}$$

$$\dot{Z} = XY - BZ , \tag{7.20c}$$

where σ, B, and R are the adjustable system parameters. It is a straightforward matter to show that the system has the Painlevé property for three special parameter sets ($\sigma = \frac{1}{2}$, $B = 1$, $R = 0$; $\sigma = 1$, $B = 2$, $R = \frac{1}{9}$ and $\sigma = \frac{1}{3}$, $B = 1$, R arbitrary) [42, 43]. In each of these cases two time dependent integrals can be identified and the solution expressed in terms of standard functions. More interestingly, *Segur* [42] noticed that there are two cases, namely

(i) $B = 1$, $R = 0$, σ arbitrary and
(ii) $B = 2\sigma$, R arbitrary

for which the system has just one time dependent integral but does not posses the Painlevé property. For example, for case (ii) the integral is the quantity

$$I = (x^2 - 2\sigma Z) e^{2\sigma t} . \tag{7.21}$$

A subsequent investigation by *Kus* [44] revealed several other such cases, namely

(iii) $\sigma = 1$, $B = 1$, R arbitrary,
(iv) $\sigma = 1$, $B = 4$, R arbitrary,
 (v) $B = 6\sigma - 2$, $R = 2\sigma - 1$.

Since these single integrals must still be entire functions, the singular parts of the local expansions of the associated dependent variables must cancel at any movable singularity position. However, since the singularities are no longer simple poles – in fact, they are logarithmic branch points – this cancellation is much more delicate than before.

Typically, the variables must be expanded as psi-series of the form

$$X = \sum_{j=0}^{\infty} \sum_{k=0}^{\infty} a_{jk} \tau^{j-1} (\tau^2 \ln \tau)^k , \tag{7.22a}$$

$$Y = \sum_{j=0}^{\infty} \sum_{k=0}^{\infty} b_{jk} \tau^{j-2} (\tau^2 \ln \tau)^k , \tag{7.22b}$$

$$Z = \sum_{j=0}^{\infty} \sum_{k=0}^{\infty} c_{jk} \tau^{j-2} (\tau^2 \ln \tau)^k \ , \tag{7.22c}$$

where $\tau = t - t_0$. For certain parameter values the logarithmic corrections enter as powers of $\tau^4 \ln \tau$. In the same spirit as the Duffing equation analysis it is again possible to extract closed recursion relations for the coefficient set a_{0k}, b_{0k}, c_{0k}. Similarly, the associated generating functions can also be obtained by means of the rescaling transformation (in the limit $\tau \to 0$)

$$X = \frac{1}{\tau} \Theta_0(z) \ , \tag{7.23a}$$

$$Y = \frac{1}{\tau^2} \Phi_0(z) \ , \tag{7.23b}$$

$$Z = \frac{1}{\tau^2} \Psi_0(z) \ , \tag{7.23c}$$

which leads to the system of equations

$$2z\Theta_0'(z) - \Theta_0(z) - \sigma \Phi_0(z) = 0 \ , \tag{7.24a}$$

$$2z\Phi_0'(z) - 2\Phi_0(z) + \Theta_0(z)\Psi_0(z) = 0 \ , \tag{7.24b}$$

$$2z\Psi'(z) - 2\Psi_0(z) - \Theta_0(z)\Phi_0(z) = 0 \ , \tag{7.24c}$$

where $t = \tau^2 \ln \tau$. This system of equations admits the (z-dependent) integral

$$\Theta_0^2 - 2\sigma \Psi_0(z) = \gamma z \ , \tag{7.25}$$

where γ is a certain constant, which can then be used to reduce the system (7.24) to

$$4z^2 \Theta_0'' - 2z\Theta_0' - 3i\lambda \sigma z \Theta_0 + \tfrac{1}{2} \Theta_0(\Theta_0^2 + 4) = 0 \ , \tag{7.26}$$

where λ is a certain function of B and σ. The transformation

$$\Theta_0(z) = z^{1/2} f(z^{1/2}) \tag{7.27}$$

reduces (7.26) to the Jacobi elliptic equation

$$f''(y) - 3i\lambda \sigma f(y) + \tfrac{1}{2} f^3(y) = 0 \ , \tag{7.28}$$

where prime denotes differentiation with respect to the variable $y = z^{1/2} = \tau(\ln \tau)^{1/2}$. Equation (7.28) exhibits a regular lattice of poles in the y-plane. These map back to the t-plane, through the transformation $z = \tau^2 \ln \tau$ yielding a recursive singularity structure consisting of families of two-armed stars of singularities. As with the Duffing equation, this type of analysis leads to a very accurate picture of the singularity distribution in the t-plane. Notice, however, hat the shape of the pole lattice of the elliptic function $f(y)$ can now depend on the value of λ and the value of the integral

$$I = (f')^2 - 3i\lambda\sigma f^2 + \tfrac{1}{4}f^4 . \tag{7.29}$$

Furthermore, the functions Θ_0, Φ_0, Ψ_0 are (again) just the leading terms in the resummed psi-series

$$X = \sum_{j=0}^{\infty} \tau^{j-1}\Theta_j(z) , \tag{7.30}$$

$$Y = \sum_{j=0}^{\infty} \tau^{j-2}\Phi_j(z) , \tag{7.31}$$

$$Z = \sum_{j=0}^{\infty} \tau^{j-2}\Psi_j(z) , \tag{7.32}$$

for which a hierarchy of linear equations for the Θ_j, Φ_j, $\Psi_j (j > 1)$ can also be obtained.

We conclude by summarizing the recent work of *Levine* and *Tabor* [43] which demonstrates that the special parameter values at which the single integral appears [cases (i) to (v) given above] are associated with subtle changes in the singularity clustering. Using (7.28, 29) one may show that the elliptic function reduces to a circular function for the parameter values

$$B = 6\sigma - 2 \quad \text{and} \quad B = \sigma + 1 \tag{7.33}$$

and to a hyperbolic function for the values

$$B = 1 \quad \text{and} \quad B = 3\sigma - 1 . \tag{7.34}$$

These include, apart from the R–dependence, the special parameter sets (i) to (v). At these special parameter values the lattice of poles in the y-plane collapses to a single line [along the imaginary axis for (7.33)] with the result that the singularity clustering in the t-plane, as governed by the transformation $y = \tau(\log \tau)^{1/2}$, must be drastically simplified.

The "local" integral (7.29) can be written as

$$\sigma\Phi_0^2 + \Theta_0^2\Psi_0 - \frac{1}{4\sigma}\Theta_0^4 = \frac{1}{\sigma}Iz^2 \tag{7.35}$$

and remarkable, for the special parameter values (7.33, 34) it becomes z-independent. Specifically for set (7.33) it takes the form

$$\sigma\Phi_0^2 + \Theta_0^2\Psi_0 - \frac{1}{4\sigma}\Theta_0^4 = 0 \tag{7.36}$$

and for set (7.34)

$$\Phi_0^2 + \Psi_0^2 = 0 . \tag{7.37}$$

This z-independence suggests that these local integrals may now have global properties. Recalling the resummed psi-series (7.30–32) we can rewrite (7.36, 37) in terms of the original variables as

$$\sigma Y^2 + X^2 Z - \frac{1}{4\sigma} X^4 \tag{7.38}$$

and

$$Y^2 + Z^2 , \tag{7.39}$$

respectively. These expressions are not, themselves, integrals of the Lorenz equations since, although their most singular parts [at $O(t - t_0)^{-4}$] cancel identically, one finds that they still have singular parts at order $(t - t_0)^{-3}$ and lower. However, by careful use of the resummed psi-series one can systematically find the various combinations of X, Y, Z (this involves some nice symmetry properties) at special R values that when added to (7.38, 39) result in entire functions of t! In this way one is explicitly able to construct all the special case integrals of motion. This procedure is analogous to the cancellation of singular terms that we mentioned above in the case of the integral of the integrable limit of the Duffing equation. In that case, since the local expansions were Laurent, the cancellation was just that of *constants*. In this case the resummed psi-series can be thought of as a pseudo-Laurent series and the cancellation of singular terms is now due to the cancellation of various combinations of the *functions* $4\Theta_j$, Φ_j and Ψ_j – which can only occur when the singularity clustering is sufficiently simple.

References

1   H. Flaschka, A. C. Newell: Commun. Math. Phys. **76**, 65–116 (1980); Physica **3** D. 203–222 (1981); *Math Studies* Vol. **61**, ed. by A. Bishop, D. Campbell, B. Nicolaenko (North-Holland, Amsterdam 1982) 65–91

2   M. Sato, T. Miwa, M. Jimbo: Physica **1** D, 80 (1980); in *Mathematical Problems in Theoretical Physics*, ed. by K. Osterwalder, Lect. Notes Phys. Vol. 116 (Springer, New York 1980) 126; in *Nuclear Spectroscopy*, ed. by G. F. Bertsch, D. Kurath, Lect. Notes Phys. Vol. 119 (Springer, New York 1980)

3   C. S. Gardner, J. M. Greene, M. D. Kruskal, R. M. Miura: Phys. Rev. Lett. **19**, 1095 (1967); Comm. Pure Appl. Math. **27**, 97 (1974)

4   V. E. Zakharov, L. D. Faddeev: Anal. Appl. **5**, 280 (1971)

5   C. S. Gardner: J. Math. Phys. **12**, 1548 (1971)

6   M. D. Kruskal, N. J. Zabusky: Phys. Rev. Lett. **15**, 240 (1965)

7   S. Kovalevskaya: Acta Math. **12**, 177 (1889); **14**, 81 (1890)

8   A. C. Newell, M. Tabor, Y. Zeng: Physica, **12** D, 1 (1987)

9   J. Weiss, M. Tabor, G. Carnevale: J. Math. Phys. **24**, 522–526 (1983)

10   J. Weiss: J. Math. Phys. **24**, 1405–1413 (1983)

11   J. Weiss: J. Math. Phys. **25**, 13–24 (1984)

12   J. Weiss: J. Math. Phys. **26**, 258–269 (1985)

13   J. Weiss: J. Math. Phys. **26**, 2174–2180 (1985)

14   D. V. Chudnovsky, G. V. Chudnovsky, M. Tabor: Phys. Lett. A **97**, 268–274 (1983)

15   J.D. Gibbon, P. Radmore, M. Tabor, D. Wood: Stud. Appl. Math. **72**, 39–63 (1985)
16   J.G. Gibbon, M. Tabor: J. Math. Phys. **26**, 1956–1960 (1985)
17   M. Adler, P. van Moerbeke: Invent. Math. **67**, 297–331 (1982)
18   M. Adler, P. van Moerbeke: Commun. Math. Phys. **83**, 83–106 (1982)
19   L. Haine: Commun. Math. Phys. **94**, 271 (1986)
20   N. Ercolani, E. Siggia: *"The Painlevé Property and Integrability"*, Phys. Lett. **A 119**, 112 (1986)
21   J.D. Gibbon, A.C. Newell, M. Tabor, Y.B. Zeng: *"Lax Pairs, Bäcklund Transformations and Special Solutions for Ordinary Differential Equations"*, Nonlinearity **1**, 481 (1988)
22   R. Hirota: *"Direct Methods in Soliton Theory"*, in *Solitons*, ed. by R.K. Bullough, P.J. Caudrey, Topics Curr. Phys. Vol. 17 (Springer, New York 1980)
23   E. Date, M. Jimbo, M. Kashiwara, T. Miwa: *"Transformation Groups for Soliton Equations"*, in *Proc. RIMS Symp. Nonlinear Integrable Systems – Classical and Quantum Theory*, ed. by M. Jimbo, T. Miwa (World Scientific, Singapore 1983)
24   A.C. Newell: *"Solitons in Mathematics and Physics"*, Conf. Bd. Math. Soc. **43**, Soc. Indust. Appl. Maths. (1985)
25   J. Weiss: J. Math. Phys. **25**, 2226–2235 (1984)
26   F. Cariello, M. Tabor: *"Painlevé Expansions for Nonintegrable Evolution Equations"* Physica **39 D**, 77 (1989)
27   J.D. Fourier, G. Levine, M. Tabor: *"Singularity Clustering in the Duffing Oscillator"*, J. Phys **A 21**, 33 (1988)
28   Y.T. Chang, M. Tabor, J. Weiss: J. Math. Phys. **23**, 531–536 (1982)
29   J. Weiss: Phys. Lett. **A 105**, 387 (1984)
30   M. Toda: *Theory of Nonlinear Lattices*, Springer Ser. Solid-State Sci. Vol. 20 (Springer, Berlin–Heildelberg 1981)
31   M. Adler, P. van Moerbeke: Commun. Math. Phys. **83**, 83 (1982)
32   H. Yoshida: in *Nonlinear Integrable Systems – Classical Theory and Quantum Theory*, ed. by M. Jimbo, T. Miwa (World-Scientific, Singapor 1983) 273
33   O.I. Bogoyavlenskii: Commun. Math. Phys. **51**, 201 (1976)
34   H. Flaschka, Y. Zeng: in preparation
35   H. Flaschka: *"The Toda Lattice in the Complex Domain"*, preprint, Univ. of Arizona (1987)
36   M. Sato: RIMS Kokyuroku **439**, 30 (1981)
37   G. Segal, G. Wilson: Publ. Math. Inst. Haute Etudes Scientifique **61**, 5 (1985)
38   Y.F. Chang, M. Tabor, J. Weiss: J. Math. Phys. **23**, 531 (1982)
39   Y.F. Chang, J.M. Greene, M. Tabor, J. Weiss: Physica **8 D**, 183 (1983)
40   J.D. Fourier, G. Levine, M. Tabor: J. Phys. **A 21**, 33 (1988)
41   M. Tabor, J. Weiss: Phys. Rev. **A 24**, 2157 (1981)
42   H. Segur: *"Solitons and the Inverse Scattering Transform"*, in *Lectures given at the International School of Physics, "Enrico Fermi"*, Varenna, Italy 7–9 July, 1980
43   G. Levine, M. Tabor: *"Integrating the Nonintegrable: Analytic Structure of the Lorenz System Revisited"*, Physica **33 D**, 189 (1988)
44   M. Kus: J. Phys. **A**, L689 (1983)

The Symmetry Approach to Classification of Integrable Equations

A.V. Mikhailov, A.B. Shabat, and V.V. Sokolov

Introduction

In this volume each of the contributors proposes his own test to recognize integrable PDEs. We believe that, independently from the basic definition of integrability, the test must satisfy some general requirements. Namely, it has to be

- effective (in other words, if an equation has passed through the test, then there are almost no doubts about its integrability);
- sufficiently algorithmical, yet able to admit a proper realization in a symbolic computer language (like Reduce, Formac, Macsyma, MuMath, AMP, etc.);
- applicable to a large class of PDEs.

Probably, the best way to judge the value of the test is by attempting to obtain a complete list of integrable equations from a certain simple class. After that one could see what equations have been missed and which ones passed through the test that are not, in fact, integrable. Another way to compare different tests is through a competition. Such jousts were popular in the 16th century (Cardano, Tartaglia, Viete, ..., etc) but in our age of glasnost the participants should not hide their tools.

The well-known notion of higher symmetry was a starting point of our study. The Sophus Lie classical theory of the contact transformations and his concept of locality has been serving us as a solid base to develop a symmetry approach. The inverse scattering transform has been another origin for us. The mutual influence of these theories has led us to a fundamental abstract concept which we call a *formal symmetry* that is more basic than symmetry. It has been proved that one can come to this concept by beginning with higher conservation laws, the Bäcklund transformations or L-A pair representation. In this sense a formal symmetry is a universal object.

Mainly, in this paper we shall discuss nonlinear evolution equations of a comparatively simple form:

$$u_t = F(x, u, u_x, u_{xx}) \,,$$
$$u_t = A(x, u, u_x, u_{xx})u_{xxx} + B(x, u, u_x, u_{xx}) \,,$$
$$u_t = u_{xxxxx} + F(u, u_x, u_{xx}, u_{xxx}, u_{xxxx}) \,,$$
$$u_t = A(u)u_{xx} + F(u, u_x) \,, \quad u = (u^1, u^2) \,, \quad \det A(u) \neq 0 \,.$$

Exhaustive lists of these equations[1] that have passed through our integrability test are given in the last section. Our test of integrability is based on a few necessary conditions of existence of a formal symmetry.[2] It is a miracle that all the equations from these lists prove to be integrable (C-integrable or S-integrable in the terminology of Calogero).

Sometimes we are asked why a certain integrable equation is absent from the list. A lot of equations which seem to be different at first glance are in fact related via simple transformations. Each list is supplied with comments which describe classes of transformations that our classification is based upon. If the equation possesses local higher symmetries or conservation laws then these transformations enable one to put it into the list. If these local properties are violated, then we propose an algorithm to attack the problem and to restore the locality as a rule (see example in Sect. 2.1.5). The true value of transformations is becoming quite clear in the classification problems. On this account we pay attention to the contact transformations and their generalizations which are closely related to classical symmetries and conservation laws.

Essentially, our test is quite simple and algorithmical. At each stage one has to verify that a certain recursively determined function (*a canonical density*) is a density of a local conservation law. Sometimes this process involves enormously tedious computations. We prefer to perform some of these with the help of a computer. Moreover, a PC-program that has allowed one to answer the question of whether a given second order evolution equation is integrable or not is available now.

This article is largely based on joint work with S. I. Svinolupov and R. I. Yamilov. The results presented here are the most advanced part of a scientific program dedicated to a classification of integrable equations. One of the objectives is a classification of integrable chains and a construction of the finite-dimensional models of PDEs. There is an intriguing problem to include the theory of multi-dimensional integrable equations into the frame of the symmetry approach.

1. Basic Definitions and Notations

1.1 Classical and Higher Symmetries

1.1.1 General Notion of Symmetry. In the classical theory of Sophus Lie the concept of symmetries is connected with one parameter groups of transformations like Lorentz. Galileian, scaling, etc. If a partial differential equation admits such

---

[1] No assumptions (like a polynomiality) on the form of the functions in the lhs of equations have been imposed.

[2] These conditions are formulated in terms of the so-called canonical conservation laws.

a group then the group action on a given solution generates a one parameter family of solutions. For example, the Galileian transformations

$$x' = x - 6\tau t , \quad t' = t , \quad u' = u + \tau , \tag{1.1.1}$$

do not change the form of the KdV equation

$$u_t = u_{xxx} + 6uu_x \tag{1.1.2}$$

and the action of group (1.1.1) on the solution $u(x, t)$ gives us a one parameter family of solutions

$$u(x, t, \tau) = u(x + 6\tau t, t) + \tau . \tag{1.1.3}$$

It follows from (1.1.3) that

$$u_\tau = 1 + 6tu_x . \tag{1.1.4}$$

Since (1.1.2) holds identically in τ one obtains that the partial derivative u_τ satisfies the linear equation

$$\left(\frac{\partial}{\partial t} - \left(\frac{\partial}{\partial x} \right)^3 - 6u\frac{\partial}{\partial x} - 6u_x \right) (u_\tau) = 0 , \tag{1.1.5}$$

i.e., the function $f = 1 + 6tu_x$ satisfies (1.1.5).

Any function f of

$$x, t, u, u_x, u_{xx}, u_{xxx}, \ldots \tag{1.1.6}$$

we shall call *a symmetry* of the KdV equation if it satisfies (1.1.5). Here we mean that when substituting $f(x, t, u, u_x, \ldots)$ instead of u_τ into (1.1.5) one has to express the derivatives $u_t, u_{xt}, u_{tt}, \ldots$ in terms of (1.1.6) via (1.1.2). After that (1.1.5) has to be fulfilled identically [the variables (1.1.6) are treated as independent ones]. It is well-known that the KdV equation possesses symmetries that are unusual from the classical viewpoint. The simplest example of such symmetry is

$$f = u_{xxxxx} + 10uu_{xxx} + 20u_x u_{xx} + 30u^2 u_x .$$

It is easy to generalize the above definition on any partial differential equation

$$H(x, t, u, u_x, u_t, u_{xx}, u_{xt}, u_{tt}, \ldots) = 0 . \tag{1.1.7}$$

A function $f(x, t, u, u_x, u_t, \ldots)$ is called symmetry of PDE (1.1.7) if it satisfies the "linearization" of (1.1.7) [c. f. (1.1.5)]

$$\left(\frac{\partial H}{\partial u} + \frac{\partial H}{\partial u_x} \frac{\partial}{\partial x} + \frac{\partial H}{\partial u_t} \frac{\partial}{\partial t} + \frac{\partial H}{\partial u_{xx}} \left(\frac{\partial}{\partial x} \right)^2 \right.$$

$$\left. + \frac{\partial H}{\partial u_{xt}} \frac{\partial}{\partial t} \frac{\partial}{\partial x} + \frac{\partial H}{\partial u_{tt}} \left(\frac{\partial}{\partial t} \right)^2 + \ldots \right) (f) = 0 . \tag{1.1.8}$$

For example, in the case of equation[3]

$$u_{tx} = F(u) \tag{1.1.9}$$

(1.1.8) has the form

$$\frac{\partial}{\partial t}\frac{\partial}{\partial x}f = F'(u)f .$$

The last equation must be fulfilled identically in the variables u, t, u, u_x, u_t, u_{xx}, u_{tt}, u_{xxx}, u_{ttt} .... All mixed derivatives are excluded via (1.1.9) (i.e. $u_{xt} = F(u)$, $u_{xxt} = F'(u)u_x$, ... etc.). In the general case (1.1.7), a particular choice of independent variables is defined by the form of H. For evolution equations $u_t = F(x, t, u, u_x, u_{xx}, ...)$ we shall use independent variables (1.1.6).

Symmetries of the form $f = f(x, t, u, p, q)$, where $p = u_x$, $q = u_t$ we shall call *classical*. In this case the symmetry generates a one parameter group of transformations

$$t' = (\tau, x, t, u, p, q) , \quad x' = \phi(\tau, x, t, u, p, q) ,$$
$$u' = \Psi(\tau, x, t, u, p, q) , \quad p' = \chi(\tau, x, t, u, p, q) , \tag{1.1.10}$$
$$q' = \omega(\tau, x, t, u, p, q)$$

that can be reconstructed by means of the following dynamical system

$$\frac{dt}{d\tau} = -\frac{\partial f}{\partial q} , \quad \frac{dx}{d\tau} = -\frac{\partial f}{\partial p} ,$$
$$\frac{du}{d\tau} = -\frac{\partial f}{\partial p}p - \frac{\partial f}{\partial q}q + f , \tag{1.1.11}$$
$$\frac{dp}{d\tau} = \frac{\partial f}{\partial x} + \frac{\partial f}{\partial u}p , \quad \frac{dq}{d\tau} = \frac{\partial f}{\partial t} + \frac{\partial f}{\partial u}q .$$

New variables t', x', u', p', q' are obtained from t, x, u, p, q as a result of a shift in τ along the trajectories of (1.1.11). In other words, the functions $t'(\tau)$, $x'(\tau)$, $u'(\tau)$, $p'(\tau)$, $q'(\tau)$ represent a solution of the initial value problem $t'(0) = t$, $x'(0) = x$, $u'(0) = u$, $p'(0) = p$, $q'(0) = q$ for (1.1.11).

The transformation (1.1.10) turns a function $u(x, t)$ into a new one

$$u'(\tau, x', t') = \Psi(\tau, x, t, u(x, t), u_x(x, t), u_t(x, t)) \tag{1.1.12}$$

where the relation of x, t to x', t' is determined by

$$x' = \phi(\tau, x, t, u(x, t), u_x(x, t), u_t(x, t)) , \tag{1.1.13}$$

$$t' = \vartheta(\tau, x, t, u(x, t), u_x(x, t), u_t(x, t)) . \tag{1.1.14}$$

If $u(x, t)$ is a solution of (1.1.7), then the condition (1.1.8) provides that $u'(\tau, x', t')$ satisfies the same equation for any τ.

---

[3] All equations of this type possessing higher symmetries have been classified in [1].

For instance, in the case (1.1.4) $f = 1 + 6tp$, (1.1.11) have the form

$$\frac{dt}{d\tau} = 0 , \quad \frac{dx}{d\tau} = -6t , \quad \frac{du}{d\tau} = 1 , \quad \frac{dp}{d\tau} = 0 , \quad \frac{dq}{d\tau} = 6p ,$$

and

$$t'(\tau) = t , \quad x'(\tau) = x - 6\tau t , \quad u'(\tau) = u + \tau .$$

This gives us the one parameter family of solutions (1.1.3) of the KdV equation. It is not a difficult problem to obtain all symmetries of the form $f(x, t, u, u_x, u_t)$ of the KdV equation. The answer is

$$f = c_1 u_x + c_2 u_t + c_3(2u + x u_x + 3 t u_t) + c_4(1 + 6 t u_x) , \tag{1.1.15}$$

where $c_j \in \mathbb{C}$.

Equations (1.1.11) are just the well-known characteristic equations for the PDE of the first order

$$u_\tau = f(x, t, u, u_x, u_t) . \tag{1.1.16}$$

The condition (1.1.8) ensures the compatibility of (1.1.16) and (1.1.7) and the function $u(\tau, x, t)$ constructed above is a solution of these equations.

Symmetries that are not of the form of (1.1.16) are usually called the *higher symmetries*. Now we consider one more well-known example of higher symmetry for the Burgers equation

$$u_t = u_{xx} + 2 u u_x . \tag{1.1.17}$$

The evolution equation corresponding to this symmetry is

$$u_\tau = u_{xxx} + 3 u u_{xx} + 3 u_x^2 + 3 u^2 u_x . \tag{1.1.18}$$

The relationship (1.1.8) is just a compatibility condition of (1.1.17, 18). It is interesting to note that the common solution of these equations $u(\tau, x, t)$ satisfies

$$3 u_{tt} - 4 u_{\tau x} + u_{xxxx} + 12 u_x u_{xx} = 0$$

and obviously gives a solution ($v = u_x$) of the famous KP equation

$$3 v_{tt} - (4 v_\tau + v_{xxx} + 12 v v_x)_x = 0 .$$

1.1.2 The ∗-Operation.

Here we consider vector evolution equations of the form

$$u_t = F(x, t, u, u_x, \ldots, \underbrace{u_{xx \ldots x}}_{m-\text{times}}) , \tag{1.1.19}$$

where $u = (u^1, u^2, \ldots, u^N)$. To underline the independence of the variables $u, u_x, u_{xx}, \ldots$ we shall use the following notations

$$u_x \to u_1 , \quad u_{xx} \to u_2 , \quad u_{xxx} \to u_3 , \dots \tag{1.1.20}$$

and sometimes call these variables the dynamical ones. The definition of symmetry given in the previous section can be generalized to the vector case with little effort. Namely, the vector function $G(x, t, u, u_1, u_2, \dots, u_n)$ is called a *symmetry* of (1.1.19) if it satisfies

$$\frac{d}{dt} G = \sum_{i=0}^{m} \frac{\partial F}{\partial u_i} D^i(G) . \tag{1.1.21}$$

Here $\partial F / \partial u_i$ denotes the Jacobi matrix,

$$D = \frac{\partial}{\partial x} + \sum_{i=0}^{\infty} u_{i+1} \frac{\partial}{\partial u_i} , \tag{1.1.22}$$

$$\frac{d}{dt} = \frac{\partial}{\partial t} + \sum_{i=0}^{\infty} D^i(F) \frac{\partial}{\partial u_i} . \tag{1.1.23}$$

The operators (1.1.22, 23) of total differentiation with respect to x and t act on the functions of variables $t, x, u, u_1, u_2, \dots$.

It is convenient to introduce a special notation for the operation of linearization ($*$-operation)

$$S_*(v) = \frac{d}{d\varepsilon} S(x, t, (u + \varepsilon v), \dots, D^n(u + \varepsilon v)) \big|_{\varepsilon=0} \tag{1.1.24}$$

that maps any function $S(x, t, u, u_1, \dots, u_k)$ to the linear differential operator

$$S_* = \sum_{i=0}^{k} (\partial S / \partial u_i) D^i . \tag{1.1.25}$$

Now (1.1.21) can be written as

$$\frac{d}{dt} G = F_*(G) . \tag{1.1.26}$$

It follows from the definition (1.1.24) that for any scalar functions f, g the relations

$$(fg)_* = f g_* + g f_* ,$$
$$(D(f))_* = D(f_*) + f_* \circ D ,$$
$$\left(\frac{d}{dt} f \right)_* = \frac{d}{dt}(f_*) + f_* \circ F_*$$

hold.

Applying the $*$-operation to (1.1.26) one can obtain the following operator equation

$$\frac{d}{dt}(G_*) - \frac{d}{d\tau}(F_*) + [F_*, G_*] = 0 , \tag{1.1.27}$$

where

$$\frac{d}{d\tau} = \sum_{i=0}^{\infty} D^i(G) \frac{\partial}{\partial u_i} . \tag{1.1.28}$$

The relationship (1.1.27) can be interpreted as a compatibility condition of equations

$$\frac{d}{dt}(v) = F_*(v) , \qquad \frac{d}{d\tau}(v) = G_*(v) .$$

By the *order of symmetry* ord(G) we mean the degree of the differential operator G_*.

1.2 Local Conservation Laws

The notion of first integrals, in contrast to symmetries, is not so important in the case of PDEs. Instead of them, the concept of conservation laws arises. The well-known links among the classical symmetries and conservation laws has been formulated in the famous theorem by E. Noether. For instance, the invariance of the nonlinear Schrödinger equation

$$i\psi_t = \psi_{xx} + |\psi|^2 \psi$$

under the phase transformation $\psi \to \psi e^{i\tau}$ leads to the conservation law

$$(|\psi|^2)_t = (i\psi_x^* \psi - i\psi^* \psi_x)_x$$

with the conserved density $|\psi|^2$.

Usually the correspondence of higher symmetries and conservation laws associates with the Hamiltonian structure [2]. However, if we have at least one higher symmetry, then there exists an unexpected but straightforward and algorithmic way to construct not one but many conservation laws (Sect. 3). These conservation laws contain a lot of important knowledge about the equation under consideration.

A function $\varrho = \varrho(x, t, u, u_1, u_2, \ldots, u_n)$ is called a *density* of a conservation law of (1.1.19) if there exists a local function[4] σ such that

$$\frac{d}{dt}(\varrho) = D(\sigma) . \tag{1.2.1}$$

Like the symmetry equation (1.1.26) the relation (1.2.1) has to be fulfilled identically in independent variables $t, x, u, u_1, \ldots$.

---

[4] Here by "local" we mean a function of a finite number of variables $x, t, u, u_1, u_2, \ldots$.

The relation (1.2.1) is obviously satisfied if $\varrho = D(h)$ for any h. In this case $\sigma = h_t$. Such a conservation law is called a *trivial* one. Two conserved densities ϱ_1, ϱ_2 are considered to be equivalent ($\varrho_1 \simeq p_2$) if the difference $\varrho = \varrho_1 - \varrho_2$ is a trivial density. By the *order of a conserved density* ϱ we shall mean the order of the differential operator

$$R = \left(\frac{\delta\varrho}{\delta u}\right)_* , \tag{1.2.2}$$

where the variational derivative is defined as

$$\frac{\delta}{\delta u} = \sum_{i=0}^{\infty}(-D)^i \frac{\partial}{\partial u_i} . \tag{1.2.3}$$

One can verify the following useful properties of the variational derivative

$$\frac{\delta}{\delta u}(Df) = 0 , \tag{1.2.4}$$

$$\left(\frac{\delta f}{\delta u}\right)_* = \left[\left(\frac{\delta f}{\delta u}\right)\right]_*^\dagger , \tag{1.2.5}$$

$$\frac{d}{dt}\frac{\delta f}{\delta u} = \frac{\delta}{\delta u}\frac{df}{dt} - F_*^\dagger \frac{\delta f}{\delta u} , \tag{1.2.6}$$

where f any scalar function, $\delta f/\delta u$ is a row-vector, $(\delta f/\delta u)_*$ is a differential operator with matrix coefficients and $\dagger$ denotes the conjugation

$$A = \sum a_i D^i \Rightarrow A^\dagger = \sum (-D)^i \circ a_i^T \tag{1.2.7}$$

(a_i^T is a transposed matrix). It follows from (1.2.4) that equivalent densities have the same order.

The variational derivative of a conserved density satisfy the well-known equation (see, for instance, [3])

$$\frac{d}{d\tau}\frac{\delta\varrho}{\delta u} + F_*^\dagger \frac{\delta\varrho}{\delta u} = 0 , \tag{1.2.8}$$

which follows immediately from (1.2.1, 4, 6). It is interesting to note that (1.2.8) is dual with equation (1.1.26) which symmetries satisfy to. Applying the $*$-operation to (1.2.8) one can obtain the following operator equation [cf. (1.1.26, 27)]

$$R_t + R \circ F_* + F_*^\dagger \circ R = Q , \tag{1.2.9}$$

where

$$Q = F_*^\dagger \circ \left(\frac{\delta\varrho}{\delta u}\right)_* - \left(F_*^\dagger \left(\frac{\delta\varrho}{\delta u}\right)\right)_* = \sum_{j,k,n} \frac{\partial\Phi_j}{\partial u_k^n} D^j \left(\frac{\delta\varrho}{\delta u^n}\right) D^k .$$

1.3 PDEs and Infinite-Dimensional Dynamical Systems

In the above defining symmetries and conservation laws we use, following S. Lie, a standard set of independent variables $t, x, u, u_1, \ldots$ [cf. (1.1.20)]. Any function of these variables is called local. However for shift-invariant equations

$$u_t = F(u, u_1, \ldots, u_m) \tag{1.3.1}$$

we shall often restrict ourselves to a reduced set of variables $u, u_1, u_2 \ldots$. In this case, by local functions we mean the shift-invariant ones. The choice of the set of basic variables depends on the particular problem. For instance, to find exact solutions the so-called chain variables (see below) seem to be very convenient.

Any partial differential equation (1.3.1) can be represented as a pair of compatible infinite-dimensional dynamical systems

$$(u_m)_x = u_{m+1} \ . \tag{1.3.2}$$

$$(u_m)_t = F_m(u_0, u_1, \ldots, u_{s_m}) \ . \tag{1.3.3}$$

Here $m = 0, 1, \ldots$; $u_0 = u$; $F_m = D^m(F)$. The vector fields corresponding to dynamical systems (1.3.2, 3) are just

$$D = \sum_{i=0}^{\infty} u_{i+1} \frac{\partial}{\partial u_i} \ , \qquad \frac{d}{dt} = \sum_{i=0}^{\infty} D^i(F) \frac{\partial}{\partial u_i} \ .$$

The representation of a PDE by means of a pair of compatible infinite-dimensional dynamical systems we regard as a guiding principle for constructing the finite-dimensional models of integrable equations.

There exist nonstandard ways to represent (1.3.1) by a pair of dynamical systems. Now we give an example of a chain representation of the well-known integrable equation [4]

$$\begin{aligned} u_t &= u_{xx} + (2uv + u^2)_x \\ -v_t &= v_{xx} - (2uv + v^2)_x \ . \end{aligned} \tag{1.3.4}$$

Consider two compatible chain equations[5]

$$(q_n)_x = q_n(q_{n+1} - q_{n-1}) \ , \quad n \in \mathbb{Z} \tag{1.3.5}$$

$$(q_n)_t = q_n q_{n+1}(q_{n+2} + q_{n+1} + q_n) + q_n q_{n-1}(q_{n-2} + q_{n-1} + q_n) \ . \tag{1.3.6}$$

We choose and fix an integer n and define new variables $u = q_n$, $v = q_{n-1}$. Equation (1.3.5) enables one to express easily the variables $q_{n-2}, q_{n+1}, q_{n+2}$ in terms of $u, v, u_x, v_x, u_{xx}, v_{xx}$:

$$q_{n-2} = u - v_x/v \ , \quad q_{n+1} = v + u_x/u \ , \quad q_{n+2} = (\log(v + u_x/u))_x + u \ . \tag{1.3.7}$$

---

[5] In other words, the chain (1.3.6) is a symmetry for (1.3.5).

One can check that after being rewritten in the variables u, v, (1.3.6) coincides with (1.3.4). Solutions of PDEs which correspond to different values of n are related to each other. Namely, these relationships are just the Bäcklund transformations for (1.3.4).

Any chain possessing a symmetry enables one to construct the associated PDE in a similar way [5]. The fact that the Bäcklund transformations of the nonlinear Schrödinger equation and the Toda lattice are closely related has been discussed in [6].

1.4 Transformations

In the next sections dealing with certain classes of equations, we shall choose an appropriate class of transformations. Sometimes we have to go beyond the framework of the classical theory.

1.4.1 The Lie Algebra of Classical Symmetries.

If $G_i(t, x, u, u_i, \ldots, u_{n_i})$, $i = 1, 2$ are symmetries of (1.1.19), then one can construct a new symmetry

$$G_3 = G_{1_*}(G_2) - G_{2_*}(G_1) \, . \tag{1.4.1}$$

It is easy to check that the corresponding vector fields [see (1.1.28)]

$$\frac{d}{d\tau_i} = \sum_{k=0}^{\infty} D^k(G_i) \, \partial/\partial u_k$$

satisfy the usual relation

$$\frac{d}{d\tau_3} = \left[\frac{d}{d\tau_2}, \frac{d}{d\tau_1} \right] \, .$$

Thus, the set of all local symmetries constitutes a Lie algebra with the bracket (1.4.1). The classical symmetries connected with one parameter groups of transformations define a subalgebra of this Lie algebra. In other words, the commutator (1.4.1) of classical symmetries is a classical symmetry as well.

Below, in Sect. 4 we shall classify scalar integrable equations of the form

$$u_t = F(x, u, u_1, \ldots, u_m) \, , \quad m \geq 2 \, . \tag{1.4.2}$$

The subalgebra of the classical symmetries of the form

$$G = G(x, u, u_1) \tag{1.4.3}$$

will be widely used in this classification (Sect. 2.3). Quite often integrable equations possess a nontrivial algebra of symmetries (1.4.3) that enables one to extend a module of transformations and relate equations with essentially different commutation representations.

Now we present a few equations possessing extremely rich algebras S of classical symmetries (1.4.3). It can be proved that the dimension $\dim_{\mathbb{C}} S \leq m+3$, where m denotes the order of (1.4.2) [7].

Consider the equations

$$u_t = u_1(u_3 u_1^{-1} - 3u_2^2 u_1^{-2}/2)^{-1/2} \tag{1.4.4}$$

$$u_t = u_2^3(u_2^2 u_5 - 5u_2 u_3 u_4 + 40u_3^3/9)^{-2/3} \tag{1.4.5}$$

$$u_t = u_3^4(u_3^2 u_7 - 7u_3^2 u_4 u_6 - 49u_3^2 u_5^2/10 + 28u_3 u_4^2 u_5 - 35u_4^4/2)^{-3/4} . \tag{1.4.6}$$

Equation (1.4.4) is invariant under the six parameter group of point-transformations

$$t' = t , \quad x' = (ax + b)/(cx + d) , \quad ad - bc = 1 , \quad a, b, c, d \in \mathbb{C} ,$$
$$u' = (\alpha u + \beta)/(\gamma u + \delta) , \quad \alpha\delta = \beta\gamma = 1 , \quad \alpha, \beta, \gamma, \delta \in \mathbb{C} .$$

The corresponding Lie algebra S of classical symmetries is isomorphic to $so(4, \mathbb{C})$ and generated by u_1, xu_1, $x^2 u_1$, 1, u, u^2.

The algebra S of (1.4.5) is isomorphic to $sc(3, \mathbb{C})$ and generated by 1, x, u_1, xu_1, u, uu_1, $xu - x^2 u_1$, $u^2 - xu_1 u$. The corresponding one parameter groups of point-transformations are

$$\begin{cases} x' & = x \\ u' & = u + \tau_1 \end{cases} \quad \begin{cases} x' & = x \\ u' & = u + \tau_2 x \end{cases} \quad \begin{cases} x' & = x + \tau_3 \\ u' & = u \end{cases}$$

$$\begin{cases} x' & = x \exp(\tau_4) \\ u' & = u \end{cases} \quad \begin{cases} x' & = x \\ u' & = u \exp(\tau_5) \end{cases} \quad \begin{cases} x' & = x + \tau_6 u \\ u' & = u \end{cases}$$

$$\begin{cases} x' & = x/(1 + \tau_7 x) \\ u' & = u/(1 + \tau_7 x) \end{cases} \quad \begin{cases} x' & = x/(1 + \tau_8 u) \\ u' & = u/(1 + \tau_8 u) \end{cases}$$

Classical symmetries of (1.4.6) are 1, x, x^2, u_1, xu_1, u, $xu - x^2 u_1/2$, u_1^2, $u_1 u - xu_1^2/2$, $u^2 - xu_1 u + x^2 u_1^2/4$. They generate the Lie algebra isomorphic to $sp(4, \mathbb{C})$. The first seven symmetries correspond to the group of point-transformations like in the previous case. The one parameter groups generated by the last three symmetries are the ones of contact transformations. They can be found from the dynamical system (1.1.11).

1.4.2 Generalized Contact Transformations.

For (1.4.2) whose rhs do not depend on t explicitly we shall restrict ourselves with the contact transformations of the form[6]

$$x' = \phi(x, u, u_1) , \quad u' = \psi(x, u, u_1) , \quad t' = t . \tag{1.4.7}$$

The functions ϕ, ψ are constrained with the relation

---

[6] The well-known example of such a transformation is the Legendre transformation $x' = u_1$, $u' = xu_1 - u$, $u_1' = x$, $u_2' = 1/u_2, \ldots$.

$$D(\phi)\frac{\partial\psi}{\partial u_1} = D(\psi)\frac{\partial\phi}{\partial u_1} \quad \text{and} \qquad (1.4.8)$$

$$u'_k = \left(\frac{1}{D(\phi)}D\right)^k \psi , \quad k = 1, 2, \ldots . \qquad (1.4.9)$$

Moreover, one can verify easily that

$$(u')_{t'} = \left(\frac{\partial\psi}{\partial u} - \frac{D(\psi)}{D(\phi)}\frac{\partial\psi}{\partial u}\right)(u)_t . \qquad (1.4.10)$$

If $\partial\psi/\partial u_1 = \partial\phi/\partial u_1 = 0$, then the condition (1.4.8) holds and the transformation (1.4.7) becomes the point one. In the general case condition (1.4.8) means that the function $u'_1 = D(\psi)/D(\phi)$ does not depend on u_2. This transformation is invertible in variables x, u, u_1. Formulas (1.4.7, 9, 10) enable one to reconstruct the local conservation laws and symmetries of the resulting equation. It is clear that the variable t can be excluded from a set of dynamical variables in the case of (1.4.2).

For shift-invariant equations (1.3.1) we shall often restrict ourselves to a subset of dynamical variables

$$u, u_1, u_2 \ldots . \qquad (1.4.11)$$

In this case, by local functions we mean the shift-invariant ones. Correspondingly, the notion of a local conservation law will change. Conservation law (1.2.1) we call local in the set of dynamical variables (1.4.11) if both functions ρ and σ are shift-invariant. So the conservation law $u_t = (u_1 + x)_x$ for equation $u_t = u_2 + 1$ becomes nonlocal in the set (1.4.11). Usually we shall write down a conservation law in the form $\rho_t \in \mathrm{Im}\{D\}$. One should take into account the fact that the space $\mathrm{Im}\{D\}$ expands or contracts depending on the choice of the set of dynamical variables.

The exclusion of x and t from the set of dynamical variables leads us to the problem of the description of invertible transformations in the subset (1.4.11) [8]. In the scalar case

$$x' = x + \phi(u, u_1) , \quad u' = \psi(u, u_1) , \quad t' = t \qquad (1.4.12)$$

is an example of such a contact transformation. It follows from (1.4.9) that $u'_k = \psi_k(u, u_1, u_2, \ldots)$, and so (1.4.12), generates an invertible map of the set (1.4.11) into itself. In general, transformations under consideration are not necessary local with respect to x, t. This makes it possible to generalize (1.4.12) and introduce very useful transformations applicable both to scalar equations and to vector ones. These transformations are defined by the formulas[7]

$$dx' = \alpha(u, u_1) dx + \beta(u, u_1, u_2, \ldots) dt , \quad t' = t , \qquad (1.4.13)$$

---

[7] In the case $\alpha = 1 + D(\phi(u))$ the transformation (1.4.13, 14) is a point-transformation of the form (1.4.12).

$$u' = \psi(u) , \quad u'_k = \left(\frac{1}{\alpha}D\right)^k \psi , \quad k = 1, 2, \dots . \tag{1.4.14}$$

The functions α, β have to satisfy relations

$$\frac{d}{dt}(\alpha) = D(\beta) , \quad \alpha \notin \mathrm{Im}\{D\} , \tag{1.4.15}$$

For a given solution $u(x, t)$ of (1.3.1) the differential form (1.4.13) is closed by virtue of the first condition (1.4.15) and the dependence of x' on x, t is defined by potentiation. It can be checked that

$$(u')_{t'} = F'(u, u_1, u_2, \dots, u_m) \equiv \frac{\partial \psi}{\partial u}(u_t - \alpha^{-1}\beta u_1) . \tag{1.4.16}$$

For example the equation

$$u_t = u^2 u_2 \tag{1.4.17}$$

possesses the conservation law (1.4.15) where $\alpha = u^{-1}$, $\beta = -u_1$ [9]. The transformation

$$dx' = u^{-1}dx - u_1 dt , \quad u' = u$$

which is generated by this conservation law reduces (1.4.17) to the linear heat equation $u'_t = u'_2$.

The condition $\alpha \notin \mathrm{Im}\{D\}$ guarantees invertibility of the transformation (1.4.14) of variables $u, u_1, u_2, \dots$. Namely, this condition implies that the matrix $\partial(u', u'_1)/\partial(u, u_1)$ is nondegenerated [8]. For the inverse transformation

$$dx = \alpha'(u', u'_1) dx' + \beta'(u', u'_1, u'_2, \dots) dt' ,$$

where

$$\alpha' = \alpha^{-1} , \quad \beta' = -\beta\alpha^{-1} . \tag{1.4.18}$$

The relation $d(\alpha')/dt' = D'(\beta')$ follows from a general formula

$$\varrho = \alpha\varrho' , \quad \sigma = \sigma + \beta\varrho' \tag{1.4.19}$$

which relates the conservation laws $\varrho_t = \sigma_x$ and $\varrho'_{t'} = \sigma'_{x'}$ of (1.3.1) and (1.4.16), respectively.

The fact that u^{-1} is conserved density for (1.4.17) reflects the following property of integrable equations (Sect. 3): the function $\alpha = (\partial F/\partial u_m)^{-1/m}$ is a conserved density for any scalar integrable equation (1.4.2). If such an equation has the form

$$u_t = A(u, u_1)u_m + B(u, u_1, \dots, u_{m-1}) \tag{1.4.20}$$

and $\alpha = A^{-1/m} \notin \mathrm{Im}\{D\}$ then the transformation (1.4.13, 14) gives

$$u'_t = u'_m + B'(u', u'_1, \dots, u'_{m-1}) . \tag{1.4.21}$$

In the case $\alpha \in \text{Im}\{D\}$, equations (1.4.20, 21) are connected via the contact transformation (1.4.12).

For vector equations of the form

$$u_t = A(u)u_2 + B(u, u_1) , \quad u = (u^1, u^2) , \quad \det A \neq 0 \tag{1.4.22}$$

the integrability conditions are (Sect. 3)

$$\text{Tr}\{A\} = 0 , \quad \frac{d}{dt}[(-\det A)^{-1/4}] \in \text{Im}\{D\} .$$

The transformation (1.4.13, 14) with $\alpha = (-\det A)^{-1/4}$ reduces (1.4.22) to [10]

$$\begin{aligned} u_t &= u_2 + f(u, v, u_1, v_1) \\ -v_t &= v_2 + g(u, v, u_1, v_1) \end{aligned} \tag{1.4.23}$$

In conclusion, we present an interesting example of the application of such transformations in the case of hyperbolic equations [11]

$$\begin{aligned} &a(u, u_x, u_y)u_{xx} + b(u, u_x, u_y)u_{xy} + c(u, u_x, u_y)u_{yy} \\ &+ d(u, u_x, u_y) = 0 . \end{aligned} \tag{1.4.24}$$

A straightforward generalization of (1.4.13) is

$$dx' = \alpha\, dx + \beta\, dy , \quad dy' = \gamma\, dx + \delta\, dy , \quad u' = u \tag{1.4.25}$$

where $\alpha, \beta, \gamma, \delta$ are functions of u, u_x, u_y which satisfy the constraints

$$\alpha_y = \beta_x , \quad \gamma_y = \delta_x , \quad \alpha\delta - \beta\gamma \neq 0 . \tag{1.4.26}$$

If the local conservation laws (1.4.26) are nontrivial then the transformation (1.4.25) is invertible and the resulting equation is of the same form as (1.4.24).

This transformation allows one to turn the nonlinear Klein-Gordon equation

$$u_{xy} = dh(u)/du \tag{1.4.27}$$

into equation of the form of (1.4.24) that corresponds to the Lagrangian with the density

$$L = h^{-1}\left(1 + \sqrt{1 - 2hu_x' u_y'}\right) .$$

For $\alpha, \beta, \gamma, \delta$ one has to choose the components of the energy-momentum tensor

$$\alpha = u_x^2/2 , \quad \beta = \gamma = h(u) , \quad \delta = u_y^2/2 .$$

2. The Burgers Type Equations

The linearizable equations in the theory of integrable equations are like a fruit fly by Zakharov's colorful expression. The best known example is the Burgers equation which has been linearized by the Cole–Hopf substitution [12, 13]. A few scalar and vector generalizations of this equation are known [9, 14–18]. In this section we develop a regular approach which enables one to recognize and linearize such equations. It serves as a good illustration of the general theory to which this paper is devoted.

2.1 Classification in the Scalar Case

2.1.1 Integrable Equations of Second Order. Any linear equation

$$u_t = u_2 + q(x)u \tag{2.1.1}$$

possesses an infinite sequence of local symmetries of the form

$$u_\tau = (D^2 + q)^k(u), \quad k \in \mathbb{N}.$$

If there exists a differential operator M of odd order that commutes with the operator $L = D^2 + q$, then (2.1.1) has additional symmetries $u_\tau = M^k(u)$, $k \in \mathbb{N}$. The question arises whether there exist nonlinear equations of the form

$$u_t = F(x, u, u_1, u_2) \tag{2.1.2}$$

with higher symmetries, which can not be linearized via the contact transformations. It is well-known that the Burgers equation

$$u_t = u_2 + 2uu_1 \tag{2.1.3}$$

provides such an example. According to [16] the exhaustive list of nonlinear integrable equations of the form (2.1.2) can be written as follows:

$$u_t = u_2 + 2uu_1 + h(x) \tag{2.1.4}$$

$$u_t = D(u_1 u^{-2} + \alpha x u + \beta u) \tag{2.1.5}$$

$$u_t = D(u_1 u^{-2} - 2x). \tag{2.1.6}$$

To decide whether a given equation of the form of (2.1.2) can be reduce via the contact transformation to one from the above list we propose the following algorithm. First, one must check the condition

$$(\partial F / \partial u_2)^{-1/2} = A(x, u, u_1)u_2 + B(x, u, u_1). \tag{2.1.7}$$

Violation of (2.1.7) means that the equation under consideration does not possess higher symmetries and cannot be reduced to the basic equations (2.1.1, 4–6). If

$$(\partial F/\partial u_2)^{-1/2} = D(\alpha(x, u, u_1)) ,$$

then (2.1.2) is reduced to an equation of the form

$$u_t = u_2 + f(x, u, u_1) \tag{2.1.8}$$

by means of the contact transformation (1.4.7–9), where $\phi = \alpha(x, u, u_1)$ and ψ is any solution of (1.4.8). Function f is to have the form

$$f = \alpha(x, u)u_1^2 + \beta(x, u)u_1 + \gamma(x, u) , \tag{2.1.9}$$

otherwise (2.1.2) is not integrable. Now the point-transformation $x' = x$, $u' = \psi(x, u)$, where $\partial^2\psi/\partial u^2 = \alpha(x, u)(\partial\psi/\partial u)^2$, reduces (2.1.9) to

$$u_t = u_2 + a(x, u)u_1 + b(x, u) . \tag{2.1.10}$$

The integrable (2.1.10) must coincide (up to the linear transformations $x' = x$, $u' = p(x)u + q(x)$) with (2.1.1) or (2.1.4).

If

$$\left(\frac{\partial F}{\partial u_2}\right)^{-1/2} = D(\alpha(x, u, u_1)) + b(x, u) , \quad \frac{\partial b}{\partial u} \neq 0 \tag{2.1.11}$$

or

$$\left(\frac{\partial F}{\partial u_2}\right)^{-1/2} = D(\alpha(x, u, u_1)) + b(x, u, u_1) , \quad \frac{\partial^2 b}{\partial u_1^2} \neq 0 \tag{2.1.12}$$

the integrable (2.1.2) can be reduced to the form

$$u_t = u_2 u^{-2} + g(x, u, u_1) . \tag{2.1.13}$$

First consider (2.1.11). If (2.1.2) is integrable, then $\partial\alpha/\partial u_1 = 0$. The point-transformation $x' = \phi(x, u)$, $u' = \psi(x, u)$, where $(\partial\phi/\partial x)\psi = \partial\alpha/\partial x + b$, $(\partial\phi/\partial u)\psi = \partial\alpha/\partial u$ reduces (2.1.2) to (2.1.13). In the case of (2.1.12), equation (2.1.2) is turned into (2.1.13) via the contact-transformation (1.4.7–9), where

$$\frac{\partial^2\alpha}{\partial u_1^2} = \frac{\partial\phi}{\partial u}\frac{\partial\psi}{\partial u_1} - \frac{\partial\psi}{\partial u}\frac{\partial\phi}{\partial u_1} ,$$

$$\frac{\partial^3\alpha}{\partial u_1^2\partial u} + \frac{\partial^2\alpha}{\partial u_1\partial x} - \frac{\partial\alpha}{\partial u} = \frac{\partial\phi}{\partial u}\frac{\partial\psi}{\partial x} - \frac{\partial\psi}{\partial u}\frac{\partial\phi}{\partial x} .$$

If (2.1.13) is integrable then it must be of the form

$$u_t = D(u_1 u^{-2} + \alpha(x)u + \beta(x) + \gamma(x)u^{-1}) . \tag{2.1.14}$$

The following change of variable $\bar{x} = y(x)$, $\bar{u} = u/y'(x)$, where $y'' + \gamma(x)y = 0$, kills the coefficient $\gamma(x)$ and gives an equation of the form

$$u_t = D(u_1 u^{-2} + a(x)u + b(x)) . \tag{2.1.15}$$

Up to the obvious rescaling the obtained equation must coincide with (2.1.5) or (2.1.6) if it is integrable.

2.1.2 The First Integrability Condition. The classification of the integrable equation (2.1.2) is based on a few necessary conditions for the existence of higher symmetries. To derive the first of these conditions we substitute

$$G_* = l_n D^n + l_{n-1} D^{n-1} + \ldots + l_0$$

into (1.1.27), where

$$F_* = F_2 D^2 + F_1 D + F_0 \,,$$
$$F_2 = \frac{\partial F}{\partial u_2} \,, \quad F_1 = \frac{\partial F}{\partial u_1} \,, \quad F_0 = \frac{\partial F}{\partial u} \,.$$

Collecting terms at D^{m+1} in (1.1.27) we obtain

$$2F_2 D(l_m) - m l_m D(F_2) = 0 \,. \tag{2.1.16}$$

Therefore

$$l_m = (F_2)^{m/2} \,. \tag{2.1.17}$$

At D^m we have the relation

$$\frac{dl_m}{dt} = F_2 D^2(l_m) - \frac{m(m-1)}{2} l_m D^2(F_2) + 2F_2 D(l_{m-1})$$
$$\qquad - (m-1) l_{m-1} D(F_2) + F_1 D(l_m) - m l_m D(F_1)$$

which can be rewritten in the form of the conservation law

$$\frac{dF_2^{-1/2}}{dt} = D\left((m-2)D(F_2^{1/2}) + F_1 F_2^{-1/2} \right.$$
$$\qquad \left. - \left(\frac{2}{m}\right) l_{m-1} F_2^{(1-m)/2} \right) \,. \tag{2.1.18}$$

Hence the existence of the third or higher order symmetry implies that

$$\frac{d\varrho_{-1}}{dt} \in \mathrm{Im}\{D\} \,, \quad \varrho_{-1} = F_2^{-1/2} \,, \tag{2.1.19}$$

i.e., the function ϱ_{-1} must be a conserved density.

Note that we have already used the function ϱ_{-1} in the above algorithm [see (2.1.7)]. If this conservation law is trivial ($\varrho_{-1} \in \mathrm{Im}\{D\}$), then, as we have noticed above, there exists a contact transformation that reduces (2.1.2) to (2.1.8). The fact that (2.1.2) possesses the nontrivial conservation law drastically restricts its form.

Lemma 2.1. *Any nontrivial conserved density ϱ of* (2.1.2) *has order less or equal to* 2.

i) *If* ord(ϱ) = 0, *then* (2.1.1) *is quasilinear*

$$u_t = A(x, u, u_1)u_2 + B(x, u, u_1) . \qquad (2.1.20)$$

ii) *If* ord(ϱ) = 2, *then* (2.1.2) *can be rewritten*

$$u_t = (A(x, u, u_1)u_2 + B(x, u, u_1))^{-1} + C(x, u, u_1) . \qquad (2.1.21)$$

Proof. Let ord(ϱ) = $2m$, $m \geq 2$. Without loss of generality we can set $\varrho = \varrho(x, u, \ldots, u_m)$, $\varrho_{mm} = \partial^2\varrho/\partial u_m^2 \neq 0$. Evaluating a derivative with respect to time one obtains

$$\varrho_t = \frac{\partial\varrho}{\partial u_m}D^m F + \ldots \frac{\partial\varrho}{\partial u}F \simeq \frac{\partial\varrho^2}{\partial u_m^2}F_2 u_{m+1}^2 + H(x, u, \ldots u_m) \in \mathrm{Im}\{D\} .$$

But this is impossible if $m \geq 2$.

i) Let $\varrho = \varrho(x, u)$, then $\varrho_t = (\partial\varrho/\partial u)F \in \mathrm{Im}\{D\}$ and, consequently, the function F has to be linear with respect to u_2, as indicated in (2.1.20).

ii) Let $\varrho = \varrho(x, u, u_1)$, $\partial^2\varrho/\partial u_1^2 \neq 0$. Then

$$\varrho_t\frac{\partial\varrho}{\partial u_1}DF + \frac{\partial\varrho}{\partial u}F \simeq \left(\frac{\partial\varrho}{\partial u} - \frac{\partial^2\varrho}{\partial u\,\partial u_1}u_1 - \frac{\partial^2\varrho}{\partial u_1^2}u_2\right)F \in \mathrm{Im}\{D\} .$$

Therefore

$$\left(\frac{\partial\varrho}{\partial u} - \frac{\partial^2\varrho}{\partial u\,\partial u_1}u_1 - \frac{\partial^2\varrho}{\partial u_1^2}u_2\right)F = D(g(x, u, u_1)) .$$

Hence

$$F = \left(\frac{\partial g}{\partial u}u_1 + \frac{\partial g}{\partial u_1}u_2\right)\Big/\left(\frac{\partial\varrho}{\partial u} - \frac{\partial^2\varrho}{\partial u\,\partial u_1}u_1 - \frac{\partial^2\varrho}{\partial u_1^2}u_2\right) ,$$

and (2.1.2) can be rewritten as (2.1.21). ∎

It follows from the Lemma and (2.1.19) that relation (2.1.7) used in the algorithm must hold if (2.1.2) is integrable. Moreover, in the case of (2.1.11) the considered equation has to be quasilinear and therefore $\partial\alpha/\partial u_1 = 0$ holds.

2.1.3 Sketch of the Classification. If the first integrability condition (2.1.19) is fulfilled, then the contact transformations indicated in Sect. 2.1.1 reduce the general case of (2.1.2) to quasilinear subcases (2.1.10) or (2.1.13). Consider, for example, the problem of classification of integrable equations of the form of (2.1.13). Equation (2.1.10) can be handled in a similar way.

It follows from (2.1.13, 19) that $\varrho_{-1} = u$ is a conserved density. Therefore (2.1.13) has to be of the form

$$u_t = D(u_1 u^{-2} + g(x, u)) . \qquad (2.1.22)$$

To describe all integrable cases of (2.1.22) we use the two next integrability conditions ϱ_{0t}, $\varrho_{1t} \in \text{Im}\{D\}$ (3.1.25, 26), the derivation of which we defer to Sect. 3. In the particular case of (2.1.13) the canonical conserved density ϱ_0 can be represented in the form:

$$\varrho_0 \simeq ug - u^2 \partial g / \partial u , \quad (\varrho_{0t} \in \text{Im}\{D\}) . \qquad (2.1.23)$$

For any conserved density $\varrho = \varrho(x, u)$ we have

$$\varrho_t \simeq -\frac{\partial^2 \varrho}{\partial u^2} u^{-2} u_1^2 + H(x, u) \in \text{Im}\{D\} \Rightarrow \frac{\partial^2 \varrho}{\partial u^2} = 0 .$$

Thus $ug - u^2 \partial g / \partial u = b(x)u + c(x)$, and therefore [cf. (2.1.14)]

$$g = a(x)u + b(x) + \frac{c(x)}{(2u)} .$$

The change of variables $\bar{x} = y(x)$, $\bar{u} = u/y'(x)$, where $2y'' + c(x)y = 0$, kills the coefficient $c(x)$. Now we have

$$u_t = D(u_1 u^{-2} + A(x)u + B(x)) , \quad \varrho_0 = B(x)u , \qquad (2.1.24)$$

and

$$(\varrho_0)_t \simeq u^{-1} B''(x) + A(x)B'(x)u \in \text{Im}\{D\} \Leftrightarrow B'' = 0 , \quad AB' = 0$$

When $A = 0$ we have $B = c_1 x + c_2$, and (2.1.24) coincides with (2.1.6) up to the rescaling. If $B = 0$, then $\varrho_1 \simeq -A'(x)u$, and one can check easily that $A'' = 0$, i.e., the equation coincides with (2.1.5).

2.1.4 Linearization of the Burgers Type Equations.

According to the Sophus Lie theory, no invertible transformations except (1.4.7–9) that are admissible to the dynamical system (1.3.2) exist [see, for instance, Refs. 8, 19]. Here we extend a little a class of contact transformations by adding the so-called potentiation

$$v_1 = u, \ldots, v_{k+1} = u_k, \ldots . \qquad (2.1.25)$$

Essentially we add a new dynamical variable v such that $D(v) = u$ and spoil the invertibility from the classical viewpoint. We can use this transformation to equations of the form

$$u_t = D(A(x, u, u_1)) . \qquad (2.1.26)$$

As a result we obtain

$$v_t = A(x, v_1, v_2) . \qquad (2.1.27)$$

The Burgers equation (2.1.3) cannot be linearized by means of the contact transformation, but after the potentiation (2.1.25) it reduces to

$$v_t = v_2 + v_1^2 \ .$$

(2.1.28)

After the point-transformation $\tilde{u} = \exp v$, equation (2.1.28) becomes the linear heat equation

$$\tilde{u}_t = \tilde{u}_2 \ .$$

(2.1.29)

The Cole-Hopf substitution $u = \tilde{u}_1/\tilde{u}$ which is the composition of these above transformations enables one to find the general solution of the Burgers equation.

All equations (2.1.4–6) can be linearized by means of composition of the potentiations and point-transformations. Indeed the composition of (2.1.25) and $\tilde{x} = v, \tilde{u} = x$

$$x = \tilde{u} \ , \quad u = \frac{1}{\tilde{u}_1}$$

(2.1.30)

maps (2.1.5) into the linear

$$\tilde{u}_t = \tilde{u}_2 - \alpha\tilde{u} - \beta$$

(2.1.31)

and (2.1.6) into the Burgers equation (2.1.3). Transformations such as the Cole-Hopf transformation allow one to reconstruct a solution of the pointed nonlinear equations based on a solution of the respective linear equations in the explicit form. Finally we note that the higher symmetries of (2.1.4–6) can be reconstructed from the symmetries of (2.1.1) as well.

2.1.5 Extension of the Set of Dynamical Variables.

For the layperson, an equation reducible by means of some transformation to a linear form is naturally considered as integrable. Of course the list (2.1.4–6) is not complete from this viewpoint since some sophisticated transformation may spoil the local structure of symmetries. Nevertheless, if the symmetries of (2.1.2) become local after a finite extension of the set of dynamical variables, our approach can also be applicable [19].

The simplest way to extend the set of dynamical variables is by potentiation (2.1.25). More generally potentiation can be considered as adding a new dynamical variable v, such that

$$v_1 = \varrho(x, u) \ .$$

(2.1.32)

and ϱ is a conserved density of (2.1.2). It follows from $\varrho_t = D(\sigma(x, u, u_1))$ that

$$v_t = \sigma(x, u, u_1) \ .$$

(2.1.33)

Expressing u in terms of x, v_1 from (2.1.32) and substituting $u = U(x, v_1)$, $u_1 = D(U)$ in (2.1.33) one obtains an equation of the form of (2.1.27). It is clear that the transformation (2.1.32) is a composition of the point-transformation $u' = \varrho(x, u)$ and the potentiation $v_1 = u'$ (2.1.25). In the general case $\varrho = \varrho(x, u, u_1)$, the potentiation $v_1 = \varrho$ can be represented as a composition of a contact transformation and (2.1.25) [20].

Extending of the set of dynamical variables enlarges the linear space $\text{Im}\{D\}$. For instance, potentiation adds the variable u to $\text{Im}\{D\}$. The above classification of (2.1.2) is based on the conditions of the form $(\varrho_k)_t \in \text{Im}\{D\}$, $k = -1, 0, 1$. If these conditions are violated we ought to enlarge the space $\text{Im}\{D\}$ in order to fulfill them.

Example 2.2. Let us consider the equation

$$u_t = \frac{u_2 u^2}{x^2} - \frac{4 u_1 u^2}{u^3} + \frac{6 u^3}{x^4} + \lambda u_1 , \quad \lambda \in \mathbb{C} . \tag{2.1.34}$$

The function $\varrho_1 = F_2^{-1/2}$ is equal to x/u. One can see easily that

$$\frac{d\varrho_{-1}}{dt} = D\left(\frac{-u_1}{x} + \frac{3u}{x^2} + \frac{\lambda x}{u}\right) - \frac{\lambda}{u} .$$

If $\lambda \neq 0$, the first integrability condition (2.1.19) is violated and hence there are no local higher symmetries. If we want to satisfy this condition for any λ we must add a new variable v to the set of the dynamical variables such that $v_1 = 1/u$. It is just possible since the function $1/u$ is a conserved density! There is no surprise that the equation

$$v_t = D\left(-\frac{1}{v_1 x^2} + \lambda v\right) \tag{2.1.35}$$

satisfies the condition (2.1.19). This equation is of the type (2.1.11), and the change of variables $x = \lambda v/2$, $u = 2x/\lambda$ reduces the equation to the form of (2.1.6) (see the algorithm).

In the example, we have used the first integrability condition (2.1.19). The advanced linearization scheme of equations (2.1.2) is a straightforward generalization of the above procedure. It is based on the extension of the set of dynamical variables and involves three integrability conditions $(\varrho_k)_t \in \text{Im}\{D\}$, $k = -1, 0, 1$ (2.1.19), (3.1.25, 26). This method has allowed us to find the linearization transformation for all linearizable equations of the form (2.1.2) which we know. This is clear since the described procedure of extension of the set of dynamical variables is quite general. It can be applied to equations of higher order with almost local symmetries.

2.2 Systems of Burgers Type Equations

2.2.1 The Standard Form.
Here, following [18], we consider the vector equations

$$\boldsymbol{u}_t = \boldsymbol{u}_2 + \boldsymbol{F}(\boldsymbol{u}, \boldsymbol{u}_1) \tag{2.2.1}$$

where

$$\boldsymbol{u} = (u^1, \ldots, u^N), \quad \boldsymbol{F} = (F^1, \ldots, F^N), \quad \boldsymbol{u}_k = \partial^k u/\partial x^k$$

which have higher symmetries of the form

$$u_\tau = u_n + H(u, \ldots, u_{n-1}) \,.$$

As in the previous section the integrability conditions can be derived from (1.1.27)

$$\partial_t(L) - [D^2 + F_1 D + F_0, L] = F_{1\tau} D + F_{0\tau} \,. \tag{2.2.2}$$

Here

$$L = D^n + \sum_{i=0}^{n-1} l_i D^i \,, \quad l_i = l_i(u, \ldots, u_{k_i}) \,,$$

l_i, F_1, F_0 are the $N \times N$-matrixes,

$$(F_1)_j^i = \frac{\partial F^i}{\partial u_1^j} \,, \quad (F_0)_j^i = \frac{\partial F^i}{\partial u^j} \,, \quad \partial_t(L) = \sum_{i=0}^{n-1} \partial_t(l_i) D^i \,.$$

Collecting in (2.2.2) the coefficients at D^{n+1}, D^n, D^{n-1} one obtains

$$D(P) + \frac{[F_1, P]}{2} = 0 \,, \tag{2.2.3}$$

$$D(Q) + \frac{[F_1, Q]}{2} = (F_1 + P)_t - D^2(P) + PD(F_1) - F_1 D(P) - [F_0, P] \,, \tag{2.2.4}$$

$$D(R) + \frac{[F_1, R]}{2} = (2F_0 + Q)_t - D^2(Q) - F_1 D(Q) + 2PD(F_0) - [F_0, Q] \,, \tag{2.2.5}$$

where unknowns P, Q, R are related to $l_{n-1}, l_{n-2}, l_{n-3}$ by the formulas[8]

$$l_{n-1} = \frac{n}{2}(F_1 + P) \,,$$

$$l_{n-2} = \frac{n(n-2)}{4} D(F_1) + \frac{n}{2} F_0 + \frac{n(n-2)}{8} F_1^2$$
$$+ \frac{n(n-2)}{4} PF_1 + \frac{n(n-2)}{8} P^2 + \frac{n}{4} Q \,,$$

$$l_{n-3} = \frac{1}{2} D(l_{n-2}) + \frac{n(n-1)(n-2)}{12} D^2(F_1) + \frac{n(n-1)}{4} D(F_0)$$
$$+ \frac{n(n-2)}{8}(P + F_1)_t \frac{n(n-2)}{8} QF_1$$
$$+ \frac{n(n-2)(n-4)}{12} F_1 D(F_1) + \frac{n(n-1)(n-2)}{12} D(F_1)F_1$$
$$+ \frac{n(n-2)}{16} PQ + \frac{n(n-2)}{16} QP + \frac{n(n-2)}{8} F_1 F_0$$
$$+ \frac{n(n-1)}{8} F_0 F_1 + \frac{n(n-1)}{4} PF_0 + \frac{n}{8} R \,.$$

---

[8] As a matter of fact, P, Q, R are just the coefficients of the formal series $L^{1/n} = D + P + QD^{-1} + RD^{-2} + \ldots$ (Sect. 3).

The matrix equations (2.2.3–5) have the form

$$D(X) + [A, X] = B ,\qquad(2.2.6)$$

where $A = A(u, \ldots, u_m)$, $B = B(u, \ldots, u_k)$ are given matrices. If $k > m$ and the matrix $X(u, \ldots, u_n)$ satisfies (2.2.6), then $n = k - 1$ and the relations

$$\frac{\partial^2 B}{\partial u_k^i \, \partial u_k^j} = 0 ,\qquad(2.2.7)$$

$$\frac{\partial^2 B}{\partial u_k^i \, \partial u_{k-1}^j} - \frac{\partial^2 B}{\partial u_k^j \, \partial u_{k-1}^i} = 0 ,\qquad(2.2.8)$$

$i, j = 1, \ldots, N$ hold. The conditions (2.2.7, 8) mean that

$$B = D(C(u, \ldots, u_{k-1})) + E(u, \ldots, u_{k-1}) .$$

The function

$$\tilde{X}(u, \ldots, u_{k-1}) = X - C(u, \ldots, u_{k-1})$$

obeys the equation

$$D(\tilde{X}) + [A, \tilde{X}] = \tilde{B} ,\quad \tilde{B}(u, \ldots, u_{k-1}) = B - [A, C] ,$$

where the function $\tilde{B}$ depends on a smaller number of variables. If $k - 1 > m$, relations (2.2.7, 8) for $\tilde{B}$ are also true and so on. Now we use conditions of the type (2.2.7, 8) in order to prove the theorem.

Theorem 2.3. *Let equation (2.2.1) possess the symmetry* $u_\tau = u_n + H(u, \ldots, u_{n-1})$ *of order* $n \geq 4$. *Then the equation can be reduced to the form*

$$u_t^i = u_2^i + A_{jk}^i u^j u_1^k + B_{jkm}^i u^j u^k u^m + C_{jk}^i u^j u^k$$
$$+ E_j^i u_1^j + M_j^i u^j + N^i ,\quad i = 1, \ldots, N\qquad(2.2.9)$$

by means of the invertible point transformation

$$u^i = \varphi^i(v) .\qquad(2.2.10)$$

Proof. It follows from (2.2.3) that $P = P(u)$. Taking this into account one can easily check that (2.2.4) has the form

$$D(Q) + \frac{[F_1, Q]}{2} = \frac{\partial F_1}{\partial u_1^j} u_3^j + \left(\frac{\partial F_1}{\partial u_1^j} \frac{\partial F^j}{\partial u_1^k} + \frac{\partial F_1}{\partial u_1^k} + P \frac{\partial F_1}{\partial u_1^k} \right) u_2^k + \kappa(u, u_1) .$$

The function

$$\tilde{Q}(u, u_1) = Q - \frac{\partial F_1}{\partial u_1^j} u_2^j$$

satisfies (2.2.6), where $A = F_1/2$,

$$B = -\frac{\partial^2 F_1}{\partial u_1^k \, \partial u_1^j} u_2^k u_2^j$$
$$+ \left(\frac{\partial F_1}{\partial u^j} \frac{\partial F^j}{\partial u^k} + \frac{\partial F_1}{\partial u^k} + P \frac{\partial F_1}{\partial u^k} - \frac{\partial^2 F_1}{\partial u^j \partial u_1^k} u_1^j + \frac{1}{2} \left[\frac{\partial F_1}{\partial u_1^k}, F_1 \right] \right)$$
$$\times u_2^k + \kappa(u, u_1) \, .$$

Conditions (2.2.7) are equivalent to the conditions

$$\frac{\partial^2 F_1}{\partial u_1^k \, \partial u_1^j} = 0 \, , \quad k, j = 1, \ldots, N \, .$$

Therefore (2.2.1) has the form

$$u_t^i = u_2^i + \Gamma_{jk}^i(u) u_1^j u_1^k + M_j^i(u) u_1^j + N^i(u) \, . \tag{2.2.11}$$

Using (2.2.8) we obtain

$$\frac{\partial \Gamma(j)}{\partial u^i} - \frac{\partial \Gamma(i)}{\partial u^j} = [\Gamma(i), \Gamma(j)] \, , \quad i, j = 1, \ldots, N \, . \tag{2.2.12}$$

Here $\Gamma(j)$ is the $N \times N$-matrix with the elements $\Gamma(j)_k^i = \Gamma_{jk}^i(u)$.

It is easy to see that any transformation (2.2.10) maps (2.2.11) into another one of the same type

$$v_t^i = v_2^i + \tilde{\Gamma}_{jk}^i(v) v_1^j v_1^k + \tilde{M}_j^i(v) v_1^j + \tilde{N}^i(v) \, ,$$

where

$$\frac{\partial \phi^i}{\partial v^m} \tilde{\Gamma}_{jk}^m(v) = \frac{\partial^2 \phi^i}{\partial v^j \partial v^k} + \Gamma_{ml}^i(\phi) \frac{\partial \phi^m}{\partial v^j} \frac{\partial \phi^l}{\partial v^k} \, . \tag{2.2.13}$$

It follows from (2.2.12, 13) that the functions $\Gamma_{jk}^i(u)$ are just the Christoffel symbols of some Euclidean affine connection. The zero-curvature conditions (2.2.12) guarantee existence of transformation (2.2.10) that reduces (2.2.11) to the form

$$u_t^i = u_2^i + M_j^i(u) u_1^j + N^i(u) \, . \tag{2.2.14}$$

The functions $\varphi^i(v)$ can be found from the system of equations

$$\frac{\partial^2 \phi^i}{\partial v^j \partial v^k} + \Gamma_{ml}^i(\phi) \frac{\partial \phi^m}{\partial v^j} \frac{\partial \phi^l}{\partial v^k} = 0 \, . \tag{2.2.15}$$

The relations (2.2.12) are compatibility conditions of (2.2.15).

For (2.2.14) it follows from (2.2.3) that P is a constant matrix. Equation (2.2.4) can be written as

$$D(Q) + \frac{[M,Q]}{2} = \frac{\partial M}{\partial u^i} u_2^i + \kappa_i(u)u_1^i + \gamma(u) .$$

The function $\tilde{Q}(u) = Q - (\partial M/\partial u^i)u_1^i$ satisfies (2.2.6) where

$$A = \frac{1}{2} M , \quad B = -\frac{\partial^2 M}{\partial u^i \partial u^j} u_1^k u_1^j + \tilde{\kappa}_i(u)u_1^i + \gamma(u) .$$

Consequently, the matrix function $M(u)$ is linear. In other words, (2.2.14) has a form

$$u_t^i = u_2^i + A_{jk}^i u^j u_1^k + E_j^i u_1^j + N^i(u) , \quad i = 1, \dots, N . \tag{2.2.16}$$

The structure of the functions $N^i(u)$ is determined by (2.2.5).   ■

2.2.2 Linearization in the Homogeneous Case.

We set $C_{jk}^i = 0$, $E_j^i = 0$, $M_j^i = 0$, $N^i = 0$ in (2.2.9) and restrict ourselves to the so-called homogeneous equations

$$u_t^i = u_2^i + A_{jk}^i u^j u_1^k + B_{jkm}^i u^j u^k u^m , \quad i = 1, \dots, N . \tag{2.2.17}$$

Following the investigation of the solvability conditions of (2.2.3–5) one can obtain that the constant coefficients B_{jkm}^i are expressed in terms of A_{kl}^i:

$$B_{jkm}^i = \frac{S_{jl}^i T_{km}^l + S_{kl}^i T_{jm}^l + S_{ml}^i T_{jk}^l}{24} , \tag{2.2.18}$$

where

$$S_{kl}^i = A_{kl}^i - A_{lk}^i , \quad T_{kl}^i = A_{kl}^i + A_{lk}^i . \tag{2.2.19}$$

The coefficients A_{kl}^i satisfy the following relations[9]

$$[A(j), A(k)] = \left(A_{jk}^i - A_{kj}^i \right) A(i) \tag{2.2.20}$$

where $A(j)$ is a matrix with the elements $(A(j))_k^i = A_{jk}^i$. The *necessary* and *sufficient* integrability conditions of (2.2.17) are (2.2.18–20). Namely, these conditions provide existence of the linearizing substitution

$$u^i = \phi_j^i(v)v_1^j \tag{2.2.21}$$

that links (2.2.17) with

$$v_t^i = v_2^i . \tag{2.2.22}$$

The functions ϕ_j^i have to be found from the overdetermined system of equations

---

[9] The matrices $A(i)$ define a representation of the Lie algebra that looks alike but differs from the adjoint representation.

$$\frac{\partial \phi^i_j}{\partial v^k} = -\frac{1}{2} A^i_{l s} \phi^l_k \phi^s_j .$$

(2.2.23)

As a matter of fact, (2.2.23) represents a set of dynamical systems $\partial \hat{\phi} \partial v^k = \hat{F}_k(\hat{\phi})$, $k = 1, \ldots, N$. The compatibility conditions of these systems coincide with (2.2.20).

A linear change of variables

$$\overline{v}^i = J^i_j v^j$$

(2.2.24)

preserves the form of (2.2.17). The corresponding transformation of the matrices $A(j)$ is

$$A(j) = J^{-1} J^i_j \tilde{\tilde{A}}(i) J$$

(2.2.25)

(J is a matrix with elements J^i_j). Two equations which are related by (2.2.24) we consider to be equivalent.

In the first nontrivial case of $N = 2$ a complete list of equations (2.2.17) satisfying the conditions (2.2.18–20) can be obtained. It can be represented in the following form

$$u_t = u_2 + c_1 u u_1 ,$$
$$v_t = v_2 + c_2 u u_1 + c_3 u v_1 + \tfrac{1}{4} c_2 c_3 u^3 + \tfrac{1}{4} c_3 (c_3 - c_1) u^2 v ;$$

(2.2.26)

$$u_t = u_2 + u u_1 ,$$
$$v_t = v_2 + v v_1 ;$$

(2.2.27)

$$u_t = u_2 + u u_1 ,$$
$$v_t = v_2 + u v_1 + v u_1 ;$$

(2.2.28)

$$u_t = u_2 + u u_1 ,$$
$$v_t = v_2 + v u_1 + u u_1 - u^3/4 ;$$

(2.2.29)

$$u_t = u_2 + u u_1 ,$$
$$v_t = v_2 + v u_1 + u v_1 + c(c - 1) v u^2 /4 ;$$

(2.2.30)

$$u_t = u_2 - 2 v u_1 - u^3 - u v^2 ,$$
$$v_t = v_2 - 2 u u_1 - 4 v v_1 ;$$

(2.2.31)

where c, c_i are arbitrary constants.

In the case of (2.2.26), the matrices $A(1)$, $A(2)$ form the Lie algebra G [see (2.2.20)] such that $\dim(G) \leq 1$. Equations (2.2.27, 28) correspond to the commutative two-dimensional algebra G. In the last three cases (2.2.29–31) the Lie algebra G is solvable but not commutative.

In cases (2.2.26–30) the first equation of the system is split. It is impossible to reduce (2.2.31) to the split form by the linear change of variables (2.2.24). The linearizing substitution (2.2.21) one can obtain easily in all cases since (2.2.23) are integrable in the explicit form. In the most intricate case of (2.2.31) the substitution is

$$u = (V - U^2)^{-1/2} U_1 , \quad v = (2UU_1 - V_1)(V - U^2)^{-1} . \tag{2.2.32}$$

2.2.3 Concluding Remarks. The conditions (2.2.18–20) must be valid for the integrable equations of the general form (2.2.9) as well. The above scheme enables one to obtain other algebraic constraints on the coefficients of (2.2.9). In the particular interesting case of

$$u_t^i = u_2^i + C_{jk}^i u^j u^k , \tag{2.2.33}$$

the integrability conditions have the form

$$C(j)C(k) = 0 , \quad \text{for any} \quad j, k \tag{2.2.34}$$

where $C(j)_k^i = C_{jk}^i$. The substitution

$$u^i = v_1^i - \frac{C_{jk}^i v^j v^k}{2}$$

reduces (2.2.33, 34) to $v_t^i = v_2^i$. Unfortunately, in the integrable cases, (2.2.33) can be split by means of the linear change of variables (2.2.24). Namely, it can be represented in the form

$$v_t^i = v_2^i , \quad i = 1, \ldots, M$$

$$u_t^i = u_2^i + H_{jk}^i v^j v^k \quad i = 1, \ldots, N - M .$$

For the general case of (2.2.9) (with $N \geq 3$) a straightforward analysis of algebraic constraints seems to be a difficult task even with a homogeneous (2.2.17).

An interesting example of the Burgers type equation has been found [15]:

$$u_t = \lambda_1 u_2 + 2\lambda_1 (u + v) u_1 + (\lambda_1 - \lambda_2) uv_1 + (\lambda_1 - \lambda_2)(u + v) uv$$
$$v_t = \lambda_2 v_2 + 2\lambda_2 (u + v) v_1 + (\lambda_2 - \lambda_1) vu_1 + (\lambda_2 - \lambda_1)(u + v) uv . \tag{2.2.35}$$

The corresponding linearizing substitution is

$$u = \frac{U_1}{(U + V)} \quad v = \frac{V_1}{(U + V)} , \tag{2.2.36}$$

If $\lambda_1 \neq \lambda_2$ this equation cannot be reduced to (2.2.1). The integrability conditions for equations

$$\boldsymbol{u}_t = A\boldsymbol{u}_2 + F(\boldsymbol{u}, \boldsymbol{u}_1) , \quad A = \text{diag}(\lambda_1, \ldots, \lambda_N) , \quad \lambda_i \neq \lambda_j$$

will be indicated in the next section.

2.3 Lie Symmetries and Differential Substitutions

It has been shown in Sect. 2.1.2 that equations (2.1.2), (2.2.1) with higher symmetries can be reduced to linear equations. It is possible to construct these linearizable equations starting from the linear ones.

2.3.1 The Scalar Case. The Cole-Hopf transformation is closely related to a symmetry group of the linear equation (2.1.29). Namely, the heat equation is invariant under the following group of transformations

$$x \to x , \quad \tilde{u}_i \to \exp(\tau)\tilde{u}_i , \quad \tau \in \mathbb{C} . \tag{2.3.1}$$

It is clear that

$$x , \quad u = \frac{\tilde{u}_1}{\tilde{u}}, \dots, u_k = D^k \frac{\tilde{u}_1}{\tilde{u}} \tag{2.3.2}$$

are invariant functions with respect to (2.3.1). Moreover, any invariant function $\tilde{f}(x, \tilde{u}, \dots, \tilde{u}_k)$ can be expressed through invariants (2.3.2): $\tilde{f} = f(x, u, \dots, u_{k-1})$. Since the heat equation admits the group (2.3.1), the respective differentiation d/dt maps an invariant function into invariant ones. In particular du/dt is invariant, so it can be represented as $du/dt = F(x, u, u_1, \dots)$. The explicit form of the function F can be found by a simple computation. Indeed

$$\frac{d}{dt} u = \frac{d}{dt} \frac{\tilde{u}_1}{\tilde{u}} = \frac{\tilde{u}_3}{\tilde{u}} - \frac{\tilde{u}_1 \tilde{u}_2}{\tilde{u}^2} = D^2 \frac{\tilde{u}_1}{\tilde{u}} + \frac{2\tilde{u}_1}{\tilde{u}} D \frac{\tilde{u}_1}{\tilde{u}} = u_2 + 2uu_1 .$$

Thus the Cole-Hopf substitution can be interpreted as a restriction of the heat equation on the invariants of the group (2.3.1). It is clear that one can perform such a restriction for any equation (not necessarily a linear one) which admits this group.

The above trick has been based on the premise that a set I_G of all invariants of the group G is generated by two invariants ϕ and ψ (namely $\phi = x$, $\psi = \tilde{u}_1/\tilde{u}$)[10] in the following sense: any invariant taken from I_G is a function of

$$y = \phi , \quad u = \psi , u_1 = D(\psi)/D(\phi), \dots ,$$

$$u_k = \left(\frac{1}{D(\phi)} D \right)^k (\psi), \dots . \tag{2.3.3}$$

In new variables y, u we come to an evolution equation

$$\frac{d}{dt} u = \left(\frac{d}{dt} - \frac{1}{D(\phi)} \frac{d\phi}{dt} D \right) \psi = F(y, u, u_1, \dots) . \tag{2.3.4}$$

Example 2.4. Consider another group of transformations

---

[10] An existence of such basic invariants for any finite parameter group of contact transformations (1.4.7–9) has been proved in [21].

$$x \rightarrow x + \tau , \quad \tilde{u}_i \rightarrow \tilde{u}_i$$

that the heat equation admits. Functions $\phi = \tilde{u}$, $\psi = \tilde{u}_1$ may be chosen as basic invariants. Simple computations show that in this case (2.3.4) is

$$\frac{d}{dt} u = u^2 u_2 . \tag{2.3.5}$$

Thus our scheme gives a regular way to obtain the well-known linearizing substitution $y = \tilde{u}$, $u = \tilde{u}_1$ for (2.3.5) [9, 22]. Note that another choice of the basic invariants leads to an equation reducible to (2.3.5) by a contact transformation. ∎

One may use a multi parameter group G, too. For example, the heat equation is invariant under the following two parameter group of transformations

$$x \rightarrow x + \tau_1 , \quad \tilde{u}_i \rightarrow \exp(\tau_2)\tilde{u}_i , \quad \tau_1, \tau_2 \in \mathbb{C} . \tag{2.3.6}$$

The set I_G is generated with $\phi = \tilde{u}_1/\tilde{u}$ and $\psi = (D(\tilde{u}_1/\tilde{u}))^{-1}$. The corresponding equation (2.3.4) is just (2.1.6).

Note that in all above discussed cases the group G has been solvable. The solvability of the group implies that the corresponding substitution can be represented as a composition of invertible transformations and substitutions opposite to potentiation (2.1.25) [21]. The next example shows that in the case of a simple symmetry group such a decomposition contains the Miura type transformation.

Example 2.5. Consider the equation

$$\frac{d}{dt} \tilde{u} = \tilde{u}_3 - 3\tilde{u}_2^2 \tilde{u}_1^{-1}/2 \tag{2.3.7}$$

which is invariant under the following transformations

$$x \rightarrow x , \quad \tilde{u} \rightarrow \frac{\alpha\tilde{u} + \beta}{\gamma\tilde{u} + \delta} , \quad \alpha\delta - \beta\gamma \neq 0 , \quad \alpha, \beta, \gamma, \delta \in \mathbb{C} .$$

The set I_G is generated by $\phi = x$, $\psi = \tilde{u}_3\tilde{u}_1^{-1} - 3\tilde{u}_2^2\tilde{u}_1^{-2}/2$. One can check that the corresponding (2.3.4) is the KdV equation

$$\frac{d}{dt} u = u_3 + 3u_1 u . \tag{2.3.8}$$

In this case the substitution (2.3.3) can be decomposed into

$$u = D(w) - \frac{w^2}{2} , \tag{2.3.9}$$

$$w = D(\log(v)) , \quad v = D(\tilde{u}) . \tag{2.3.10}$$

Here (2.3.9) is the well-known Miura transformation. Transformations that are inverse to substitutions (2.3.10), are the potentiation (2.1.25). ∎

What symmetries correspond to a general substitution

$$y = \phi(x, \tilde{u}, \tilde{u}_1), \quad u = \psi(x, \tilde{u}, \tilde{u}_1) \tag{2.3.11}$$

that relates a scalar evolution equation

$$\tilde{u}_t = \tilde{F}(x, \tilde{u}, \tilde{u}_1, \dots, \tilde{u}_m) \tag{2.3.12}$$

with another one

$$u_t = F(y, u, u_1, \dots, u_m) \tag{2.3.13}$$

of the same type?[11] The answer to this question is the following [24]. Equation (2.3.12) has to possess a nonlocal symmetry[12]

$$g = \exp\left(\int \Omega(x, \tilde{u}, \tilde{u}_1, \tilde{u}_2)\, dx\right). \tag{2.3.14}$$

Note that by substituting (2.3.14) into (1.1.26) which a symmetry has to satisfy, one obtains the local conservation law

$$\Omega_t = \tilde{D}\left(\sum_{i=0}^{m} \frac{\partial \tilde{F}}{\partial \tilde{u}_i}(\tilde{D} + \Omega)^i(1)\right). \tag{2.3.15}$$

In the case $\partial^2 \Omega / \partial \tilde{u}_2^2 = 0$, the functions (2.3.11) can be found as solutions of the first-order PDE

$$-\frac{\partial \Omega}{\partial \tilde{u}_2}\tilde{D}(\chi) + \Omega\frac{\partial \chi}{\partial \tilde{u}_1} + \frac{\partial \chi}{\partial \tilde{u}} = 0. \tag{2.3.16}$$

In the case $\partial^2 \Omega / \partial \tilde{u}_2^2 \neq 0$, the corresponding substitution has a slightly more general form than (2.3.11).

The classical example of the substitution (2.3.11) gives the Miura transformation (2.3.9) which relates the mKdV and KdV equations. In this case $\tilde{F} = \tilde{u}_3 - 3\tilde{u}^2\tilde{u}_1$, $\Omega = 2\tilde{u}$ and from

$$2\tilde{u}\frac{\partial \chi}{\partial \tilde{u}_1} + \frac{\partial \chi}{\partial \tilde{u}} = 0$$

one obtains $\phi = x$, $\psi = \tilde{u}_1 - \tilde{u}^2$. Any other choice of the solution of (2.3.16) corresponds to a point-transformation of the KdV equation.

The differential substitutions (2.3.11) play an important role in Sect. 4 in the problem of classification of scalar integrable evolution equations of the third and fifth order.

2.3.2 The Vector Case. A linear vector equation

---

[11] The problem of describing differential substitutions (2.3.11) into a given equation (2.3.13) has been discussed in [23].

[12] The substitutions which we have discussed correspond to the case $\Omega \in \text{Im}\{\tilde{D}\}$.

$$u_t = Au_2 \quad A = \text{diag}(\lambda_1, \ldots, \lambda_N), \quad \lambda_i \neq \lambda_j \tag{2.3.17}$$

is invariant under the following group of transformations:

$$u^i \rightarrow a^i u^i + b^i + c^i x, \quad x \rightarrow x + \lambda, \quad a^i, b^i, c^i, \lambda \in \mathbb{C}. \tag{2.3.18}$$

One can choose a subgroup G of this group and restrict (2.3.17) on the invariants of G. The obtained equation is related to (2.3.17) via a differential substitution.

Example 2.6. If $N = 2$ and G is a group of transformations of the form

$$x \rightarrow x, \quad u \rightarrow \exp(\tau)u + c, \quad v \rightarrow \exp(\tau_2)v - c,$$

the set I_G is generated by the functions $\phi = x$, $\psi^1 = u_1/(u+v)$, $\psi^2 = v_1/(u+v)$. The respective PDE is (2.2.35).

Example 2.7. Let G be a group of transformations

$$x \rightarrow x + \tau, \quad u \rightarrow u + c, \quad v \rightarrow v - c.$$

Then I_G is generated by the functions $\phi = u+v$, $\psi^1 = u_1$, $\psi^2 = v_1$. The respective PDE has the form

$$u_t = \lambda_1(u + v)^2 u_2 + (\lambda_1 - \lambda_2)(u + v)u_1 v_1,$$
$$v_t = \lambda_2(u + v)^2 v_2 + (\lambda_2 - \lambda_1)(u + v)u_1 v_1.$$

The equation

$$u_t = u_2 \tag{2.3.19}$$

admits a larger group than (2.3.17). In particular, (2.3.19) is invariant under the affine transformations

$$u^i \rightarrow a^i_j u^j + b^i, \quad x \rightarrow x, \quad \det(a^i_j) \neq 0, \quad a^i_j, b^i, \in \mathbb{C}. \tag{2.3.20}$$

Any transitive N-parameter subgroup of the affine group of $\mathbb{C}^N$ defines a substitution which relates (2.3.17) to the equation of the form (2.2.17) [21]. Namely, one can verify that besides $\phi = x$ there exist N basis invariants $\psi^1, \ldots, \psi^N$ of the form

$$\psi^i = \phi^i_j(u)u^j_1$$

which generate the set I_G. Moreover, any equation of the form of (2.2.17–20) can be obtained by means of an appropriate subgroup of the affine group. For instance, (2.2.31) arises as a restriction of the linear equation on the basis invariants $\phi = x$, $\psi^1 = (v - u^2)^{-1/2}u_1$, $\psi^2 = (2uu_1 - v_1)(v - u^2)^{-1}$ of the following subgroup:

$$x \rightarrow x, \quad u \rightarrow \exp(\tau)u + \frac{c}{2}, \quad v \rightarrow \exp(2\tau)v + c\exp(\tau)u + \frac{c^2}{4}.$$

Let us consider the case of $N = n^2$. The vectors of $\mathbb{C}^N$ we represent by $n \times n$ matrices (a matrix which corresponds to vector u we denote by $\hat{u}$). It is obvious that equation $\hat{u}_t = \hat{u}_2$ is invariant under the transformations

$$\hat{u} \to \hat{a}\hat{u} , \quad \hat{a} \in GL(n) \tag{2.3.21}$$

which form a subgroup of affine group of $\mathbb{C}^N$. As basis invariants we choose $\hat{v} = \hat{u}^{-1}\hat{u}_1$. One can check easily that $\hat{v}$ satisfies the well-known matrix equation [14]

$$\hat{v}_t = \hat{v}_2 + 2\hat{v}\hat{v}_1 .$$

3. Canonical Conservation Laws

The main result of this section is an algorithm for constructing a sequence of *canonical* conservation laws. This algorithm is found on a concept of *formal symmetry* upon which the symmetry approach is based.

3.1 Formal Symmetries

3.1.1 Definitions. In this section we consider scalar equations of the form

$$u_t = F(u, \ldots, u_m) , \quad m \geq 2 . \tag{3.1.1}$$

The differential operator

$$L = l_n D^n + l_{n-1} D^{n-1} + \ldots + l_0 , \quad l_n \neq 0 \tag{3.1.2}$$

we call a formal symmetry of order N of (3.1.1) if

$$\deg(L_t - [F_*, L]) \leq \deg(F_*) + \deg(L) - N . \tag{3.1.3}$$

For instance, the differential operator $L = F_*$ is a formal symmetry of order $m = \deg(F_*)$. If $G(x, u, \ldots, u_n)$ is a higher symmetry of (3.1.1), then it follows from (1.1.25) and (3.1.3) that $L = G_*$ is a formal symmetry of order n. Moreover, the following theorem holds [8, 25].

Theorem 3.1. *If (3.1.1) possesses a local symmetry of order N or two conserved densities of orders*[13] $N_1 > N_2 \geq N + m$, *then it possesses a formal symmetry of order N.*

We can now prove the following useful proposition:

---

[13] We recall that the order of the conserved density ϱ has been defined as $\mathrm{ord}(\varrho) = \deg(R_*)$, where $R = \delta\varrho/\delta u$.

Proposition 3.1. *Let L_1, L_2 be formal symmetries of orders N_1, N_2, respectively. Then $L = L_1 \circ L_2$ is a formal symmetry of order $N = \min(N_1, N_2)$.*

Proof. It follows from

$$L_{it} - [F_*, L_i] = Q_i , \quad i = 1, 2$$

that

$$L_t - [F_*, L] = L_1 Q_2 + Q_1 L_2 ,$$

so

$$\deg(L_1 Q_2 + Q_1 L_2) \leq \max(\deg(L_1) + \deg(Q_2), \ \deg(Q_1) + \deg(L_2))$$
$$\leq \max(\deg(F_*) + \deg(L) - N_2, \ \deg(F_*) + \deg(L) - N_1)$$
$$= \deg(F_*) + \deg(L) - \min(N_1, N_2) . \qquad \blacksquare$$

Thus if L is a formal symmetry, then L^k, $k \in \mathbb{N}$ is the formal symmetry of the same order. Below we shall show that $L^{1/n}$ is the formal symmetry as well if $n = \deg(L)$. For this purpose we move from differential operators to formal series[14] of the form

$$A = a_n D^n + a_{n-1} D^{n-1} + \ldots + a_0 + a_{-1} D^{-1} + \ldots , \quad a_n \neq 0 \qquad (3.1.4)$$

whose coefficients a_k are functions of the dynamical variables. Now $\deg(A) = n \in \mathbb{Z}$. The product of two series is defined by

$$aD^k \circ bD^m = a(bD^{m+k} + \tbinom{k}{1} D(b) D^{m+k-1} + \tbinom{k}{2} D^2(b) D^{m+k-2} + \ldots) ,$$

where $k, m \in \mathbb{Z}$ and

$$\tbinom{k}{j} = \frac{k(k-1) \cdot \ldots \cdot (k - j + 1)}{j!} .$$

The associativity of the product can be verified by a direct calculation.

For any A of the form (3.1.4) one can define the series $B = v_1 D + b_0 + \ldots$ such that $B^n = A$. The coefficients of B can be determined via recurrent relations

$$b_1^n = a_n , \quad n b_1^{n-1} b_0 = a_{n-1} - \tbinom{n}{1} b_1^{n-1} D(b_1), \ \ldots ,$$

$$n b_1^{n-1} b_{1-k} = a_{n-k} + \theta_k(a_n, a_{n-1}, \ldots, a_{n-k+1}, b_1, b_0, \ldots, b_{2-k}, \ldots) ,$$

where θ_k are some differential polynomials. The series $A^{1/n}$ is unique up to a constant factor ω ($\omega^n = 1$). In the same way one can obtain the formal series A^{-1}.

Let L be a formal symmetry (3.1.2) of order N. Then the series $L^{1/n}$ satisfies

$$L_t - [F_*, L] = \sum_{k=0}^{n} L^{k/n} (L_t^{1/n} - [F_*, L^{1/n}]) L^{(n-k-1)/n} .$$

---

[14] A comprehensive study of properties of such formal series can be found in [8, 26–28].

That implies [cf. (3.1.3)]

$$\deg(L_t^{1/n} - [F_*, L^{1/n}]) \le \deg(F_*) + 1 - N . \tag{3.1.5}$$

Definition 3.3. The series

$$L = l_n D^n + l_{n-1} D^{n-1} + \ldots + l_0 + l_{-1} D^{-1} + \ldots, \quad l_n \ne 0 \tag{3.1.6}$$

is said to be *a formal symmetry of order* N of (3.1.1) if it satisfies (3.1.3).

It follows from (3.1.5) that by extracting the n-th root one does not change the order of a formal symmetry. By the Definition 3.3 the coefficients l_k, where $k \le n - N + 1$, are nonessential since their changing does not violate (3.1.3). Two formal symmetries L_1, L_2 of order N of the form (3.1.6) are called equivalent ($L_1 \simeq L_2$) if they coincide up to the nonessential coefficients.

Proposition 3.4. *Let series L of the form* (3.1.6) *be a formal symmetry of order N. Then any formal symmetry $\tilde{L}$ of the same order is equivalent to*

$$\tilde{L} \simeq \sum_{k=l-N+2}^{l} c_k L^{k/n}$$

where $l = \deg(\tilde{L})$. The constant coefficients c_k can be found uniquely.

Proof. It follows from (3.1.3) that in $[F_*, L]$ the coefficient at D^{n+m-1} vanishes. Thus (2.1.17)

$$l_n = \alpha \left(\frac{\partial F}{\partial u_m} \right)^{n/m} , \quad \alpha \in \mathbb{C} . \tag{3.1.7}$$

Similarly the leading coefficient of $\tilde{L}$ is $\beta(\partial F/\partial u_m)^{l/m}$. Therefore $\tilde{L} - c_l L^{l/n}$, where $c_l = \beta \alpha^{-l/n}$ is a formal symmetry of order less than N. The induction of N proves this proposition. ∎

3.1.2. Canonical Densities. The coefficient a_{-1} of the formal series (3.1.4) is called a *residue* and denoted $\mathrm{res}(A) = a_{-1}$. A *logarithmic residue* we define by

$$\mathrm{res}\log(A) = \frac{a_{n-1}}{a_n} .$$

A logarithmic residue satisfies the following identities

$$\mathrm{res}\log(A \circ B) = \mathrm{res}\log(A) + \mathrm{res}\log(B) + nD(\log(b_m)) \tag{3.1.8}$$

$$\partial(\mathrm{res}\log(A)) = \mathrm{res}((\partial A) \circ A^{-1}) , \tag{3.1.9}$$

where $n = \deg(A)$, $m = \deg(B)$, (i.e., b_m is a leading coefficient of (B) and ∂ is any differentiation. It follows from (3.1.8) that

$$\mathrm{res}\log(A \circ B \circ A^{-1} \circ B^{-1}) = D(S(A, B)) \tag{3.1.10}$$

where

$$S(A, B) = n \log(b_m) - m \log(a_n) .$$

In the previous section we have shown that the function $\varrho_{-1} = (\partial F/\partial u_2)^{-1/2}$ must be a conserved density if (3.1.1) ($m = 2$) possesses higher symmetries. We will note here that this density can be represented as $\varrho_{-1} = \mathrm{res}(F_*^{-1/2})$. We shall prove

Theorem 3.5. *Let L be a formal symmetry* (3.1.6) *of order $N > m$ of* (3.1.1). *Then* (3.1.1) *possesses $N - m$ conserved densities*[15]

$$\varrho_k = \begin{cases} \mathrm{res}(L^{k/n}), & k \neq 0 \\ \mathrm{res} \log(L), & k = 0 \end{cases} \tag{3.1.11}$$

where $k = -1, 0, \ldots, N - m - 2$.

Proof. It follows from

$$\deg(L_t^{k/n} - [F_*, L^{k/n}]) \leq \deg(F_*) + k - N \tag{3.1.12}$$

that

$$\frac{d}{dt}\mathrm{res}(L^{k/n}) = \mathrm{res}([F_*, L^{k/n}]) \tag{3.1.13}$$

if $-1 \leq k \leq N - m - 2$. If $N > m + 1$ then (3.1.9) and inequality (3.1.3) imply

$$\frac{d}{dt}\mathrm{res} \log(L) = \mathrm{res}([F_*, L^{-1}, L]) . \tag{3.1.14}$$

According to *Adler*'s formula [29] [see (3.1.16) below] the residue of the commutator of any two formal series is a total derivative. Thus (3.1.13, 14) have the form of a conservation law. ∎

It follows from (3.1.7) that for any formal symmetry L ($\mathrm{ord}(L) \geq 2$) we have

$$\varrho_{-1} = \gamma \left(\frac{\partial F}{\partial u_m} \right)^{-1/m} , \quad \gamma \in \mathbb{C} . \tag{3.1.15}$$

If (3.1.1) possesses a formal symmetry of order higher than m then the theorem means that ϱ_{-1} is a density of a local conservation law. This general result has proved to be very useful in the theory of integrable equations. For instance, it has allowed us in Scct. 2 to reduce the classification problem of the second order equations to a quasilinear case. It has been noted there that a complete solution of a classification problem is required to derive the next two integrability conditions. These conditions are to be found in a regular way below.

Theorem 3.5 indicates the regular way to construct a set of conservation laws of (3.1.1)

---

[15] It might happen that some of these local densities are trivial.

$$(\varrho_k)_t = D(\sigma_k) \quad k = -1, 0, 1, \ldots$$

by the known formal symmetry L. The densities ϱ_k are determined by (3.1.11), and the functions σ_k are defined by the Adler formula which holds for any formal series A, B

$$\mathrm{res}([A, B]) = D(\sigma(A, B)), \tag{3.1.16}$$

where

$$\sigma(A, B) = \sum_{p \le \deg A, \, q \le \deg B}^{p+q+1>0} \binom{q}{p+q+1} \sum_{s=0}^{p+q} (-1)^s D^s(a_q) D^{p+q-s}(b_p). \tag{3.1.17}$$

Namely

$$\sigma_k = \sigma(F_*, L^{k/m}), \quad k = -1, 1, 2, \ldots \tag{3.1.18}$$

$$\sigma_0 = \sigma(F_* \circ L^{-1}, L). \tag{3.1.19}$$

Now we describe the basic *algorithm* which has recently been realized on the symbolic computer languages Reduce and Formac by Bakirov and Tarnopolskii. The algorithm is applicable to (3.1.1) which are not assumed to be integrable a priori.

First we calculate differential polynomials σ_k of variables $l_{-1}, \ldots, l_k$ by (3.1.17–19). We choose a series L in the form[16]

$$L = F_* + l_{-1} D + l_0 + l_1 D^{-1} + \ldots. \tag{3.1.20}$$

It follows from (3.1.17) that

$$\sigma_{-1} = \sigma(F_*, L^{-1/m}) = -F_m^{-1/m} l_{-1} + \theta_{-1}$$

where θ_{-1} depends on F_* (namely θ_{-1} is a differential polynomial of the variables $F_m^{-1/m} F_m^{1/m}$, $F_{m-1}, \ldots, F_0$). For example if $m = 2$

$$\sigma_{-1} = -F_2^{-1/2} l_{-1}. \tag{3.1.21}$$

In a general case we have

$$\sigma_0 = m l_0 + \theta_0(l_{-1}),$$
$$\sigma_k = k F_m^{k/m} l_k + \theta_k(l_{-1}, l_0, \ldots, l_{k-1}), \quad k \ne 0 \tag{3.1.22}$$

where θ_k is a differential polynomial with the coefficients depending on F_*.

Second, by resolving the relations (3.1.22) recurrently we get l_k as a differential polynomial

$$l_k = l_k(\sigma_{-1}, \ldots, \sigma_k). \tag{3.1.23}$$

---

[16] If there exists a formal symmetry of order N then it follows from Proposition 3.4 that there exists one of the form (3.20).

Substituting (3.1.23) into (3.1.11) we obtain the differential polynomials

$$\varrho_k = \varrho_k(\sigma_{-1}, \sigma_0, \ldots, \sigma_{k-m+1}) \quad k = -1, 0, 1, \ldots \qquad (3.1.24)$$

which we shall call *canonical densities*. The first $m-1$ densities ϱ_k are explicitly expressed in terms of the coefficients F_* and do not depend on variables σ_i. For instance, if $m > 2$ the density ϱ_0 is

$$\varrho_0 = \mathrm{res}\,\log(L) = \mathrm{res}\,\log(F_*) = \frac{\partial F/\partial u_{m-1}}{\partial F/\partial u_m} \; ;$$

the density ϱ_{-1} has been obtained above [see (3.1.15)].

By applying this algorithm to the second order equations we get

$$\varrho_0 = F_2^{-1}l_{-1} + F_2^{-1}F_1 = F_2^{-1}F_1 - F_2^{-1/2}\sigma_{-1} \qquad (3.1.25)$$

and at the next step

$$\sigma_0 = D(l_{-1}) + 2l_0 - l_{-1}F_2^{-1}(l_{-1} + F_1)$$

$$\varrho_1 = F_2^{-1/2}(\tfrac{1}{4}\sigma_0 + \tfrac{1}{2}F_0 + \tfrac{1}{8}\sigma_{-1}^2 - \tfrac{1}{32}F_2^{-1}(2F_1 - D(F_2)^2) \,. \qquad (3.1.26)$$

More stringent then Theorem 3.5 is

Theorem 3.6. *A formal symmetry of order N of (3.1.1) exists if and only if the canonical densities ϱ_k, $k = -1, \ldots, N-m-2$ are densities of local conservation laws, i.e., the following system of equations*

$$D(\sigma_k) = \frac{d}{dt}\varrho_k(\sigma_{-1}, \sigma_0, \ldots, \sigma_{k-m+1}) \qquad (3.1.27)$$

is solvable in the class of local functions[17] *[8, 27].*

It is a crucial point that the above theorem can be applied to (3.1.1) where integrability properties are under question. The algorithm of calculation of canonical densities (3.1.24) given above provides us with a criterion of existence of a formal symmetry of any fixed order. For instance, we have used the canonical densities ϱ_{-1}, ϱ_0, ϱ_1 to classify the second order equations (2.1.2). Theorem 3.6 means that (2.1.2) can be reduced to one of (2.1.1, 4–6) via an invertible transformation if and only if it possesses a formal symmetry of the fifth order. At the end of Sect. 2.1.4 we have given an example of application of Theorem 3.6 to find a linearizing substitution when symmetries are almost local.

---

[17] Recall that by the local functions we mean the functions depending on a finite number of dynamical variables.

3.2 The Case of a Vector Equation

In this section we generalize the algorithm of calculation of canonical densities in the case of vector equations. We also indicate some additional properties of the canonical densities for equations with higher conservation laws.

3.2.1 Formal Diagonalization. The formal symmetries of a vector equation

$$u_t = F(x, u, u_1, \ldots, u_m), \quad u = (u^1, \ldots, u^M) \tag{3.2.1}$$

are defined exactly as in the scalar case (Definition 3.3), but the coefficients of F_* and L are now $M \times M$ matrices. One can check that for any formal series A, B with matrix coefficients [cf. (3.16)]

$$\operatorname{res} \operatorname{Tr}([A, B]) \in \operatorname{Im}\{D\} .$$

As above, this enables one to produce local conservation laws connected with a given formal symmetry.

We assume that eigenvalues $\lambda_1, \ldots, \lambda_M$ of the matrix F_m

$$(F_m)^i_j = \frac{\partial F^i}{\partial u^j_m} \tag{3.2.2}$$

are all distinct ($\lambda_i \neq \lambda_j$ if $i \neq j$). It is known that in this case the "operator" equation

$$\left[\frac{d}{dt} - F_*, L\right] = 0$$

can be diagonalized by a formal gauge transformation

$$F_* \to \hat{F}_* = T \circ F_* \circ T^{-1} + T_t \circ T^{-1}, \quad L \to \hat{L} = T \circ L \circ T^{-1}, \tag{3.2.3}$$

where T is a formal series.

Proposition 3.7. *Let the matrix F_m with different eigenvalues $\lambda_1, \ldots, \lambda_M$, ($\lambda_i \neq \lambda_j$ if $i \neq j$) be diagonalized by the similarity transformation T_0*

$$F_m = T_0^{-1} \Lambda T_0 \quad \Lambda = \operatorname{diag}(\lambda_1, \ldots, \lambda_M)$$

Then there exists the unique formal series

$$T = T_0(I + T_{-1}D^{-1} + T_{-2}D^{-2} \ldots) \tag{3.2.4}$$

such that $\operatorname{diag}(T_k) = 0$ *($k = -1, -2, \ldots$) and all coefficients of the series*

$$\Phi = T \circ F_* \circ T^{-1} + T_1 \circ T^{-1} \tag{3.2.5}$$

are diagonal. Moreover, if L is a formal symmetry of order N, then the first N coefficients of

$$\hat{L} = T \circ L \circ T^{-1} \tag{3.2.6}$$

are diagonal.

Proof. Substituting $\Phi = \Lambda D^m + \Phi_{m-1} D^{m-1} + \dots$ in (3.2.4) and collecting coefficients at $D^{m-k}(k = 1, 2, \dots)$ we obtain the following chain of equations

$$[\Lambda, T_{-1}] + \Phi_{m-i} = \theta_i ,$$

where θ_i can be expressed through $T_k(k > -i)$, $\Phi_j(j > m - i)$. Taking into account that $\mathrm{diag}(T_k) = 0$, we find inductively

$$\Phi_{m-i} = \mathrm{diag}(\theta_i) , \quad T_{-i} = ad_\Lambda^{-1}(\theta_i - \Phi_{m-i}) .$$

Let $\hat{L} = \hat{l}_n D^n + \hat{l}_{n-1} D^{n-1} + \dots$. It follows from (3.2.3) that

$$\deg(\hat{L}_t - [\Phi, \hat{L}]) \leq \deg(\Phi) + \deg(\hat{L}) - N .$$

Collecting and equating to zero the coefficients of $\hat{L}_t - [\Phi, \hat{L}]$ at $D^{m+n-k}(k = 0, 1, \dots, N-1)$, we obtain a chain of relationships. The first of them, $[\Lambda, \hat{l}_n] = 0$, means that $\hat{l}_n$ is diagonal. The next relation

$$[\Lambda, l_{n-1}] = nl_n D(\Lambda) - m\Lambda D(l_n) + [\Phi_{m-1}, l_n] ,$$

has a diagonal rhs that implies $[\Lambda, l_{n-1}] = 0$, i.e., l_{n-1} is also diagonal and so on. ∎

It is clear that the gauge transformation (3.2.5) turns the differential operator F_* into a formal series Φ. Nevertheless, the series $\hat{L}$ we shall still call the formal symmetry of the same equation (below the hats will be omitted). It follows from Proposition 3.7 that

$$\Phi = \mathrm{diag}(\Phi^1, \dots, \Phi^M) \quad L = \mathrm{diag}(L^1, \dots, L^M) \tag{3.2.7}$$

and we come back to a scalar "operator"equation

$$L_t^k - [\Phi^k, L^k] = 0 . \tag{3.2.8}$$

Applying the algorithm described in the previous section to (3.2.8), one obtains a set of canonical densities

$$\varrho_k = \mathrm{diag}(\varrho_k^1, \dots, \varrho_k^M) , \quad k = -1, 0, 1, \dots . \tag{3.2.9}$$

A formal symmetry of (3.2.1) is called nondegenerated if the determinant of its leading coefficient does not vanish. A straightforward generalization of Theorem 3.6 is

Theorem 3.8. *Let the matrix (3.2.2) have distinct nonvanishing eigenvalues. Then a nondegenerated formal symmetry of order N of (3.2.1) exists if and only if all canonical densities*

$$\varrho_k^a \ , \quad k = -1, \ldots, N - m - 2 \ , \quad a = 1, \ldots, M$$

are densities of local conservation laws, i.e., the following system of equations

$$D(\sigma_k^a) = \frac{d}{dt} \varrho_k(\sigma_{-1}^a, \sigma_0^a, \ldots, \sigma_{k-m+1}^a) \qquad (3.2.10)$$

is solvable in the class of local functions.

Example 3.9. Let us consider the following coupled equations

$$
\begin{aligned}
u_t &= \lambda u_2 + f(u, v, u_1, v_1) \ , \\
-v_t &= \mu v_2 + g(u, v, u_1, v_1) \ .
\end{aligned}
\qquad (3.2.11)
$$

Operator F_* for (3.38) is

$$
\begin{aligned}
F_* = &\begin{pmatrix} \lambda & 0 \\ 0 & -\mu \end{pmatrix} D^2 + \begin{pmatrix} \partial f/\partial u_1 & \partial f/\partial v_1 \\ -\partial g/\partial u_1 & -\partial g/\partial v_1 \end{pmatrix} D \\
&+ \begin{pmatrix} \partial f/\partial u & \partial f/\partial v \\ -\partial g/\partial u & -\partial g/\partial v \end{pmatrix} \ .
\end{aligned}
$$

We are looking for a gauge transformation (3.2.5) that diagonalizes F_*. Collecting coefficients at $D^k (k = 1, 0)$ in

$$
\begin{pmatrix} F & 0 \\ 0 & -G \end{pmatrix} \circ T = T \circ F_* + T_t \ , \quad \Phi = \operatorname{diag}(F, -G)
$$

where $T = I + T_{-1}D^{-1} + T_{-2}D^{-2} \ldots$, $\operatorname{diag}(T_k) = 0 (k = -1, -2, \ldots)$ $F = D^2 + F_1 D + \ldots$, $G = D^2 + G_1 D + \ldots$, one obtains

$$
F_1 = \frac{\partial f}{\partial u_1} \ , \quad G_1 = \frac{\partial g}{\partial v_1} \ , \quad T_{-1} = \frac{1}{(\lambda + \mu)} \begin{pmatrix} 0 & \partial f/\partial v_1 \\ \partial g/\partial u_1 & 0 \end{pmatrix}
$$

$$
F_0 = \frac{\partial f}{\partial u} - \frac{(\partial f/\partial v_1)(\partial g/\partial u_1)}{(\lambda + \mu)} \ ,
$$

$$
G_0 = \frac{\partial g}{\partial v} - \frac{(\partial f/\partial v_1)(\partial g/\partial u_1)}{(\lambda + \mu)} \ .
$$

Canonical density $\varrho_{-1} = 1$, hence, $\sigma_{-1} = \text{const}$ (without loss of generality we put $\sigma_{-1} = 0$). It follows from (3.1.25, 26) that

$$
\varrho_0^1 = \frac{(\partial f/\partial u_1)}{2\lambda} \ , \qquad (3.2.12)
$$

$$
\begin{aligned}
\varrho_1^1 = &\frac{\lambda^{-1/2}\sigma_0^1}{2} - \frac{\lambda^{-3/2}(\partial f/\partial u_1)^2}{8} + \frac{\lambda^{-1/2}(\partial f/\partial u)}{2} \\
&- \frac{\lambda^{-1/2}(\partial f/\partial v_1)(\partial g/\partial u_1)}{2(\lambda + \mu)} - \frac{\lambda^{-3/2}D(\partial f/\partial u_1)}{4}
\end{aligned}
$$

$$
\varrho_0^2 = \frac{(\partial g/\partial v_1)}{2\mu} \ , \qquad (3.2.13)
$$

$$\varrho_1^2 = \frac{(-\mu)^{-1/2}\sigma_0^2}{2} - \frac{(-\mu)^{-3/2}(\partial g/\partial v_1)^2}{8} - \frac{(-\mu)^{-1/2}(\partial g/\partial v)}{2}$$
$$- \frac{(-\mu)^{-1/2}(\partial g/\partial u_1)(\partial f/\partial v_1)}{2(\lambda + \mu)} + \frac{(-\mu)^{-3/2}D(\partial g/\partial v_1)}{4} . \qquad \blacksquare$$

3.2.2 Canonical Potentials. The fact of the existence of higher conservation laws implies additional restrictions on the rhs of (3.2.1). For instance, if $\lambda \neq \mu$ then (3.2.11) does not possess a conserved density of order higher than one [10]. The regular way to obtain these restrictions is based on an algorithm quite similar to the one described in the previous section.

Definition 3.10. A series

$$R = r_n D^n + r_{n-1} D^{n-1} + \ldots + r_0 + r_{-1} D^{-1} + \ldots , \qquad r_n \neq 0$$

is called *a formal conservation law of order* N if it satisfies the equation

$$R_t + R \circ F_* + F_*^\dagger \circ R = Q \qquad (3.2.14)$$

where the remainder Q is such that

$$\deg(Q) \leq \deg(R) + \deg(F_*) - N . \qquad (3.2.15)$$

The formal conservation law is called nondegenerated if the leading coefficient of R is a nondegenerated matrix, i.e., $\det(r_n) \neq 0$. Below we assume that the formal conservation laws are nondegenerated. In particular, if ϱ is a conserved density of order $N \geq m$, then $R = (\delta\varrho/\delta u)_*$ is a formal conservation law of the same order [see (1.2.9)].

Below we shall study the solvability conditions of the equation

$$R_t + R \circ F_* + F_*^\dagger \circ R = 0 \qquad (3.2.16)$$

and show that they can be cast in a divergent form

$$\omega_k = D(\phi_k)$$

where ω_k, $k = 0, 1, \ldots$ are known differential polynomials of ϕ_j, $j < k$.

Equation (3.2.14) and inequality (3.2.15) are invariant under the gauge transformations (3.2.3),

$$R \to \hat{R} = T^\dagger \circ R \circ T . \qquad (3.2.17)$$

Let the conditions of Proposition 3.7 be valid. Then one can easily verify that in the case of odd m the transformation (3.2.4, 17) diagonalizes the first N coefficients of R ($N = \text{ord}(R)$). If m is even, then the first N coefficients of $\hat{R}$ are block-diagonal 2×2 matrices corresponding to pairs of eigenvalues Λ with the opposite sign $(\lambda_k, -\lambda_k)$ [8]. Hence (3.2.14) splits into $M/2$ equations of the form

$$\hat{R}_t^k + \hat{R}^k \circ \hat{\Phi}^k + (\hat{\Phi}^k)^\dagger \circ \hat{R}^k = Q^k , \tag{3.2.18}$$

where

$$\hat{\Phi}^k = \begin{pmatrix} \lambda_k & 0 \\ 0 & -\lambda_k \end{pmatrix} D^m + \begin{pmatrix} F^k & 0 \\ 0 & -G^k \end{pmatrix} , \quad R^k = \begin{pmatrix} 0 & r^k \\ s^k & 0 \end{pmatrix} . \tag{3.2.19}$$

Consider first the case of even M. To obtain the solvability conditions of (3.2.16) we analyze the following system of two equations

$$R_t + R \circ \hat{\Phi} + \hat{\Phi}^\dagger \circ R = 0 \tag{3.2.20}$$

$$\hat{\Phi} = \begin{pmatrix} F & 0 \\ 0 & -G \end{pmatrix} , \quad R = \begin{pmatrix} 0 & r \\ s & 0 \end{pmatrix}$$

$$F = \lambda D^m + F_{m-1} D^{m-1} + \dots , \quad G = \lambda D^m + G_{m-1} D^{m-1} + \dots .$$

Evidently, these equations are equivalent to one (scalar) equation

$$r_t - r \circ G + F^\dagger \circ r = 0 . \tag{3.2.21}$$

A series r such that

$$\deg(r_t - r \circ G + F^\dagger \circ r) \le \deg(r) + m - N \tag{3.2.22}$$

we shall also call a formal conservation law of order N of (3.1.1).

Theorem 3.11. *Let* (3.2.22) *hold. Then the functions*

$$\omega_n = \begin{cases} \operatorname{res} \log(G^{-1/m} \circ (F^\dagger + r_t r^{-1})^{1/m}) & n = 0 \\ \operatorname{res}(G^{n/m} - (F^\dagger + r_t r^{-1})^{n/m}) & n = 1, 2, \dots, N - 2 \end{cases} \tag{3.2.23}$$

are total derivatives.

Proof. Let us rewrite (3.2.21) in the form

$$(F^\dagger + r_t r^{-1})^{1/m} = r G^{1/m} r^{-1} .$$

Now one can check that the inequality (3.2.15) implies the equations

$$\operatorname{res} \log(G^{-1/m} \circ (F^\dagger + r_t r^{-1})^{1/m}) = \operatorname{res} \log(G^{-1/m} r G^{1/m} r^{-1}) \tag{3.2.24}$$

$$\operatorname{res}(G^{n/m} - (F^\dagger + r_t r^{-1})^{n/m}) = \operatorname{res}([G^{n/m} r^{-1}, r]) , \tag{3.2.25}$$

The right-hand sides of (3.2.24, 25) are total derivatives because of (3.1.10) and Adler's formula (3.1.16).

We define new variables ϕ_k, $k = 0, 1, \dots$ by means of the left-hand sides of (3.2.24, 25):

$$\phi_0 = S(G^{-1/m}, r) , \tag{3.2.26}$$

$$\phi_k = \sigma(G^{k/m} r^{-1}, r) , \quad k = 1, 2, \ldots . \tag{3.2.27}$$

In these new variables equations (3.2.24, 25) can be rewritten in the form

$$\omega_k(\phi_{k-m+1}, \phi_{k-m}, \ldots) = D(\phi_k) , \quad k = 0, 1, \ldots \tag{3.2.28}$$

where

$$\omega_n = \begin{cases} \operatorname{res} \log(G^{-1/m} \circ (F^\dagger + r_t r^{-1})^{1/m}), & n = 0 \tag{3.2.29} \\ \operatorname{res}(G^{n/m} - (F^\dagger + r_t r^{-1})^{n/m}), & n = 1, 2, \ldots . \tag{3.2.30} \end{cases}$$

We want to notice that a change of variables (3.2.26, 27) is invertible. It follows from (3.2.29, 30) that functions ω_k, $0 \le k \le m - 2$ depend on the coefficients of series F, G only

$$\omega_0 = -(m\lambda)^{-1}(F_{m-1} + G_{m-1}) + D(\log(\lambda)) , \quad k = 0 \tag{3.2.31}$$

$$\omega_k = \operatorname{res}(G^{k/m} + (-1)^{k+1} F^{k/m}) , \quad 1 \le k \le m - 2 . \tag{3.2.32}$$

Here we have used the following identities

$$\operatorname{res} \log(B^\dagger) = \operatorname{res} \log(B^{-1}) = -\operatorname{res} \log(B) + n D(\log(b_n))$$

$$\operatorname{res} \log(B^{1/n}) = \operatorname{res} \log(B)/n - [(n-1)/2n] D(\log(b_n))$$

which hold for any formal series B, $\operatorname{ord}(B) = n$.

The following theorem gives a criterion of existence of the N-th order formal conservation law [8].

Theorem 3.12. *A formal series r satisfies (3.2.22) if and only if the finite chain of equations (3.2.28) $0 \le k \le N - 2$ is solvable in the class of local functions.*

Example 3.9. (Continuation) It follows from (3.2.30, 31) that the first two canonical potentials of (3.2.11) are

$$\omega_0 = -(\partial f/\partial u_1 + \partial g/\partial v_1)/2 \tag{3.2.33}$$

$$\omega_1 = \phi_{0t}/2 + (\partial f/\partial u_1)^2/8 - (\partial g/\partial v_1)^2/8$$
$$+ (\partial g/\partial v)/2 - (\partial f/\partial u)/2 + D(\partial f/\partial u_1 - \partial g/\partial v_1)/4 . \tag{3.2.34}$$

Comparison of (3.2.12, 13) with (3.2.33, 34) gives

$$\omega_0 = \varrho_0^1 + \varrho_0^2 , \quad \omega_1 = \varrho_1^1 - \varrho_1^2 . \tag{3.2.35}$$

∎

Consider the case of odd m. One can formulate a criterion of existence of a formal conservation law of a given order [8].

Theorem 3.13. *Equation (3.2.1) of odd degree m possesses a formal conservation law of order $2N$ if and only if*

$$\omega_k(\phi_{k-m+1}, \phi_{k-m}, \dots) = D(\phi_k) , \quad k = 0, 2, \dots 2N - 2 \tag{3.2.36}$$

where the differential polynomials ω_k are defined by

$$\phi_0 = r_1^m F_m^{m-1} ,$$

$$\phi_{2n} = \sigma(F^{2n/m} \circ R^{-1}, R) , \quad k = 1, 2, \dots .$$

$$R = r_1 D + r_0 + r_{-1} D^{-1} + \dots , \quad r_1 \neq 0 \quad R^\dagger = -R , \tag{3.2.37}$$

$$2r_{-2k} + \sum_{i=1}^{2k+1} (-1)^i \binom{i - 2k}{i} D^i(r_{i-2k}) = 0 , \quad k = 0, 1, \dots ,$$

and

$$\omega_{2k} = \begin{cases} 2\mathrm{res}\,\ln(F_*) & \text{if } k = 0 , \\ \mathrm{res}(F_*^{2k/m} - (F_* + R^{-1} \circ R_t)^{2k/m}) & \text{if } k = 1, 2, \dots . \end{cases}$$

3.3 Integrability Conditions

3.3.1 The Third Order Scalar Equations. Now we present the canonical densities for equations of the form

$$u_t = F(x, u, u_1, u_2, u_3) . \tag{3.3.1}$$

First a few canonical densities are

$$\varrho_{-1} = (\partial F / \partial u_3)^{-1/3} , \tag{3.3.2}$$

$$\varrho_0 = (\partial F / \partial u_2)\varrho_{-1}^3 , \tag{3.3.3}$$

$$\varrho_1 = D(2D(\varrho_{-1})\varrho_{-1}^{-2} + \varrho_{-1}^2 \partial F / \partial u_2) + (D(\varrho_{-1}))^2 \varrho_{-1}^{-3} + \varrho_{-1}^5 (\partial F / \partial u_2)^2 / 3$$
$$+ D(\varrho_{-1})\varrho_{-1} \partial F / \partial u_2 - \varrho_{-1}^2 \partial F / \partial u_1 + \varrho_{-1}\sigma_{-1} , \tag{3.3.4}$$

$$\varrho_2 = -D^2(\varrho_{-1})(\partial F / \partial u_2)/3 - D(\varrho_{-1})\,\partial F / \partial u_1 + \varrho_{-1}\partial F / \partial u$$
$$- (D(\varrho_{-1}))^2(\partial F / \partial u_2)/\varrho_{-1} - \varrho_{-1}^4(\partial F / \partial u_1)(\partial F / \partial u_2)/3$$
$$+ D(\varrho_{-1})\varrho_{-1}^3(\partial F / \partial u_2)^2 / 3 + 2\varrho_{-1}^7(\partial F / \partial u_2)^3 / 27 + \varrho_{-1}\sigma_0/3 , \tag{3.3.5}$$

$$\varrho_3 = \varrho_{-1}\sigma_1 - \varrho_1\sigma_{-1} . \tag{3.3.6}$$

It follows from the condition $(\varrho_{-1})_t \in \mathrm{Im}\{D\}$ that any integrable equation (3.3.1) belongs to one of the following classes [30]:

i) quasilinear equations

$$u_t = A_1 u_3 + A_2 \tag{3.3.7}$$

ii) equations of the form

$$u_t = (A_1 u_3 + A_2)^{-2} + A_3 \tag{3.3.8}$$

iii) equations of the form

$$u_t = (2A_1 u_3 + A_2)(A_1 u_3^2 + A_2 u_3 + A_3)^{-1/2} + A_4 \tag{3.3.9}$$

where $A_1 = A_i(x, u, u_1, u_2)$. In the next section we shall give the results of classification of integrable equations of the form of (3.3.7). Classification of essentially nonlinear equations (3.3.8,9) has not been finished yet.

3.3.2 The Fifth Order Scalar Equations. The first six canonical densities for equations of the form

$$u_t = u_5 + f(x, u, u_1, u_2, u_3, u_4) \tag{3.3.10}$$

that can be found by means of the above algorithm (Sect. 3.1.2) are

$$\varrho_0 = \partial f / \partial u_4 , \tag{3.3.11}$$

$$\varrho_1 = 2(\partial f / \partial u_4)^2 - 5 \,\partial f / \partial u_3 , \tag{3.3.12}$$

$$\varrho_2 = 4(\partial f / \partial u_4)^3 - 15(\partial f / \partial u_3) \,\partial f / \partial u_4 + 25 \,\partial f / \partial u_2 , \tag{3.3.13}$$

$$\varrho_3 = (D(\partial f / \partial u_4))^2 - (\partial f / \partial u_3)D(\partial f / \partial u_4) + 7(\partial f / \partial u_4)^4 / 25$$
$$- 7(\partial f / \partial u_3)(\partial f / \partial u_4)^2 / 5 + 2(\partial f / \partial u_4) \,\partial f / \partial u_2 + (\partial f / \partial u_3)^2$$
$$- 5 \,\partial f / \partial u_1 . \tag{3.3.14}$$

$$\varrho_4 = 4(\partial f / \partial u_4)(D(\partial f / \partial u_4))^2 / 5 - D(\partial f / \partial u_3)D(\partial f / \partial u_4)$$
$$+ 3D(\partial f / \partial u_3)(\partial f / \partial u_4)^2 / 10 + 44(\partial f / \partial u_4)^5 / 625$$
$$- 11(\partial f / \partial u_3)(\partial f / \partial u_4)^3 / 25 + 3(\partial f / \partial u_2)(\partial f / \partial u_4)^2 / 5$$
$$+ (\partial f / \partial u_2)D(\partial f / \partial u_4) - (\partial f / \partial u_1) \,\partial f / \partial u_4 - (\partial f / \partial u_2) \,\partial f / \partial u_3$$
$$+ 5 \,\partial f / \partial u_0 + \sigma_0 , \tag{3.3.15}$$

$$\varrho_5 = \sigma_1 . \tag{3.3.16}$$

It follows from conditions (3.3.11–16) that the integrable equation (3.3.10) must be of the following form [31]

$$u_t = u_5 + (A_1 u_2 + A_2)u_4 + A_3 u_3^2 + (A_4 u_2^2 + A_5 u_2 + A_6)u_3$$
$$+ A_7 u_2^4 + A_8 u_2^3 + A_9 u_2^2 + A_{10} u_2 + A_{11} \tag{3.3.17}$$

where $A_i = A_i(x, u, u_1)$. In the last section we present a complete list of integrable equations (3.3.17).

3.3.3 The Case of Coupled Equations. Here we present a list of canonical densities and lists of canonical potentials for the Schrödinger type equations (1.4.23). By applying the foregoing algorithms one can obtain the canonical densities ϱ_k and potentials ω_k:

$$\varrho_0 = (\partial f/\partial u_1 - \partial g/\partial v_1)/2 \tag{3.3.18}$$

$$\varrho_1 = \sigma_0 - [(\partial f/\partial u_1)^2 + (\partial g/\partial v_1)^2]/4 - (\partial f/\partial v_1)\,\partial g/\partial u_1$$
$$+ \partial f/\partial u + \partial g/\partial v \tag{3.3.19}$$

$$\varrho_2 = \sigma_1 \tag{3.3.20}$$

$$\varrho_3 = \sigma_2 + (\varrho_1^2 + \omega_1^2)/2 - \omega_0(\omega_2 - \phi_{1t}) - 4\,\partial f/\partial v\,\partial g/\partial u$$
$$+ (\partial g/\partial u_1)_t\partial f/\partial v_1 - (\partial f/\partial v_1)_t\partial g/\partial u_1 + (\partial f/\partial u - \partial g/\partial v)_t$$
$$- [(\partial f/\partial v_1)(\partial g/\partial u_1)]^2 + 2(\partial f/\partial v_1)(\partial g/\partial u_1)(\partial f/\partial u + \partial g/\partial v)$$
$$- 2D(\partial f/\partial v_1)D(\partial g/\partial u_1) + [(D\omega_0)^2 + (D\varrho_0)^2]/2$$
$$+ \varrho_0[(\partial g/\partial u_1)D(\partial f/\partial v_1) - (\partial f/\partial v_1)D(\partial g/\partial u_1)]$$
$$+ 2(\partial g/\partial u)D(\partial f/\partial v_1) + 2(\partial f/\partial v)D(\partial g/\partial u_1)$$
$$- (\partial g/\partial v)D(\partial g/\partial v_1) - (\partial f/\partial u)D(\partial f/\partial u_1) , \tag{3.3.21}$$

$$\omega_0 = (\partial f/\partial u_1 + \partial g/\partial v_1)/2 , \tag{3.3.22}$$

$$\omega_1 = \phi_{0t} - \phi_0\varrho_0 - (\partial f/\partial v_1)(\partial g/\partial u_1) + \partial f/\partial u - \partial g/\partial v , \tag{3.3.23}$$

$$\omega_2 = \phi_{1t} + 2\omega_0(\partial f/\partial v_1)(\partial g/\partial u_1)$$
$$- 2[(\partial g/\partial u)(\partial f/\partial v_1) + (\partial f/\partial v)(\partial g/\partial u)] , \tag{3.3.24}$$

$$\omega_3 = \phi_{2t} + \varrho_1\omega_1 - \varrho_0(\omega_2 - \phi_{1t}) + (\partial f/\partial u + \partial g/\partial v)_t$$
$$+ \omega_0[(\partial g/\partial u_1)D(\partial f/\partial v_1) - (\partial f/\partial v_1)D(\partial g/\partial u_1)] + D\omega_0 D\varrho_0$$
$$+ (\partial g/\partial v)D(\partial g/\partial v_1) - (\partial f/\partial u)D(\partial f/\partial u_1)$$
$$- 2(\partial g/\partial u)D(\partial f/\partial v_1) + 2(\partial f/\partial v)D(\partial g/\partial u_1) . \tag{3.3.25}$$

It should be noted that in the above formulas [cf. (3.2.35)]

$$\varrho_k = \varrho_k^1 + (-1)^k\varrho_k^2 \quad \omega_k = \varrho_k^1 - (-1)^k\varrho_k^2 .$$

In other words, the existence of a higher conservation law implies that a half of canonical densities are total derivatives.

If (1.4.23) satisfies the integrability conditions

$$d\varrho_k/dt , \quad \omega_k \in \mathrm{Im}\{D\} \quad k = 0,1 \tag{3.3.26}$$

then its rhs can be written in the "canonical" form [8]:

$$F = \varepsilon\exp(\phi)u_1^2 v_1 + (\phi_u + \psi_u)u_1^2/2 + (\phi_v - \psi_v)u_1 v_1$$
$$+ ru_1 + f(u, v, v_1) \tag{3.3.27}$$

$$G = -\varepsilon\exp(\phi)v_1^2 u_1 + (\phi_v + \psi_v)v_1^2/2 + (\phi_u - \psi_u)v_1 u_1$$
$$- rv_1 + g(u, v, u_1) . \tag{3.3.28}$$

One can prove that the functions $f(u, v, v_1)$ and $g(u, v, u_1)$ satisfy the conditions

$$\partial f^3/\partial^3 v_1 = 0 , \quad \partial g^3/\partial^3 u_1 = 0 , \tag{3.3.29}$$

i.e., they are polynomial in the variables v_1, u_1, respectively. The form (3.3.27–29) is conformly invariant. Quite often, to show nonexistence of higher conservation laws it is sufficient to compare the rhs of the equation under investigation with (3.3.27–29). A detailed analysis of the integrability conditions yields a complete list of the integrable Schrödinger type equations that are to be presented in the last section.

4. Integrable Equations

4.1 Scalar Third Order Equations

Below we give the results of classification of quasilinear equations

$$u_t = a_1(x, u, u_1, u_2) \quad u_3 + a_0(x, u, u_1, u_2) \tag{4.1.1}$$

satisfying conditions $(\varrho_k)_t \in \text{Im}\{D\}$, $k = -1, 0, 1, 2, 3$, where canonical densities ϱ_k are defined by (3.3.2–6). These conditions allow one to prove that any such equation must have the form

$$u_t = (u_3 + b_5 u_2^2 + b_4 u_2 + b_3)(b_2 u_2^2 + b_1 u_2 + b_0)^{-3/2} + b_6 , \tag{4.1.2}$$

where $b_i = b_i(x, u, u_1)$.

The contact transformations (1.4.7–9) do not violate the form of (4.1.2). These transformations allow one to reduce (4.1.2) to the one of the following:

$$u_t = u_3 + a_2 u_2^2 + a_1 u_2 + a_0 , \tag{4.1.3}$$

$$u_t = u^{-3} u_3 + a_2 u_2^2 + a_1 u_2 + a_0 , \tag{4.1.4}$$

$$u_t = a_3^{-3} u_3 + a_2 u_2^2 + a_1 u_2 + a_0 , \quad \partial^2 a_3 / \partial u_1^2 \neq 0 \tag{4.1.5}$$

$$u_t = (u_3 + a_4 u_2^2 + a_3 u_2 + a_2)(a_1 u_2 + a_0)^{-3/2} + a_5 , \quad a_1 \neq 0 \tag{4.1.6}$$

where $a_i = a_i(x, u, u_1)$. First, an appropriate contact transformation allows us to make $b_2 = 0$. If $b_1 \neq 0$ we have (4.1.6). In the case of $b_1 = 0$ one obtains

$$u_t = a_3^{-3} u_3 + a_2 u_2^2 + a_1 u_2 + a_0 .$$

If $\partial^2 a_3 / \partial u_1^2 \neq 0$ then we have (4.1.5). In the case $\partial^2 a_3 / \partial u_1^2 = 0$ there exists a point-transformation

$$x' = \phi(x, u) , \quad u' = \psi(x, u) \tag{4.1.7}$$

reducing a_3 into the form $a_3 = a(x, u)$. If $\partial a / \partial u \neq 0$ the change $x' = x$, $u' = a(x, u)$ leads to (4.1.4). Finally, if $a = a(x)$ the obvious change $x' = \int a(x) \, dx$, $u' = u$ gives us (4.1.3).

4.1.1 Equations of the Form (4.1.3). Below we present a complete (up to point-transformations $\tilde{u} = \Psi(x, u)$ and rescaling) list of equations of the form of (4.1.3) satisfying conditions (3.3.2–6) [18].

Integrable Equations of the Form (4.1.3)

$$u_t = u_3 + u u_1 , \tag{4.1.8}$$

---

[18] With the help of computer analytical programs it has been verified that (4.1.9) satisfies the next [after (3.3.6)] integrability condition $(\varrho_4)_t \in \text{Im}\{D\}$ only if $c = 0$.

$$u_t = u_3 + u_1^2 + cx + c_1 , \tag{4.1.9}$$

$$u_t = u_3 + u^2 u_1 + cu_1 , \tag{4.1.10}$$

$$u_t = u_3 + u_1^3 + cu_1 + c_1 , \tag{4.1.11}$$

$$u_t = u_3 - u_1^3/2 + (\lambda_1 \exp(2u) + \lambda_2 \exp(-2u) + c)u_1 , \tag{4.1.12}$$

$$u_t = u_3 - 3u_1^{-1}u_2^2/4 + \lambda_1 u_1^{3/2} + \lambda_2 u_1^2 + cu_1 + c_1 , \tag{4.1.13}$$

$$u_t = u_3 - 3u_1(u_1^2 + 1)^{-1}u_2^2/2 + \lambda_1(u_1^2 + 1)^{3/2} + \lambda_2 u_1^3 + cu_1 + c_1 , \tag{4.1.14}$$

$$u_t = u_3 - 3u_1(u_1^2 + 1)^{-1}u_2^2/2 - 3\mathcal{P}(u)u_1(u_1^2 + 1)/2 + cu_1 , \tag{4.1.15}$$

where $(\mathcal{P}')^2 = 4\mathcal{P}^3 + g_1\mathcal{P} + g_2$,

$$u_t = u_3 - 3u_1^{-1}u_2^2/2 + \lambda_1 u_1^{-1} + \lambda_2 u_1^3 + cu_1 + c_1 , \tag{4.1.16}$$

$$u_t = u_3 - 3u_1^{-1}u_2^2/2 + (4u^3 + g_1u + g_2)u_1^{-1} + cu_1 , \tag{4.1.17}$$

$$u_t = u_3 + \alpha(x)u_1 + \beta(x)u , \tag{4.1.18}$$

$$u_t = u_3 - 3u_1^{-1}u_2^2/4 + \alpha(x)u_1 + cu , \tag{4.1.19}$$

$$u_t = u_3 + 3uu_2 + 3u_1^2 + 3u^2 u_1 + \alpha(x)u_1 + \alpha'(x)u + \beta(x) , \tag{4.1.20}$$

$$u_t = u_3 + 3u^2 u_2 + 9uu_1^2 + 3u^4 u_1 + \alpha(x)u_1 + \alpha'(x)u/2 , \tag{4.1.21}$$

$$\begin{aligned}
u_t = u_3 &- 3u_1^{-1}u_2^2/4 + (3f_u f^{-1}u_1 + 3fu_1^{1/2}/2)u_2 + 3f_u u_1^{5/2} \\
&+ 3f_x u_1^{3/2} + (3f_u^2 f^{-2} + s(u))u_1^3 - (3f_x^2 f^{-2} + s(x))u_1 \\
&+ (\lambda_1 \alpha^2(x) + \lambda_2 \alpha(x) + \lambda_3)(\alpha'(x))^{-1}u_1 \\
&- (\lambda_1 \alpha^2(u) + \lambda_2 \alpha(u) + \lambda_3)(\alpha'(u))^{-1}
\end{aligned} \tag{4.1.22}$$

where

$$f(x, u) = 2(\alpha'(x)\alpha'(u))^{1/2}(\alpha(u) - \alpha(x))^{-1} ,$$

$$s(z) = \alpha'''(z)/\alpha'(z) - 3(\alpha''(z))^2(\alpha'(z))^{-2}/2 .$$

4.1.2 Equations of the Form (4.1.4).
Integrable Equations of the Form (4.1.4)

$$u_t = D(u^{-3}u_2 - 3u^{-4}u_1^2 - 3x) , \tag{4.1.23}$$

$$u_t = D(u^{-3}u_2 - 3u^{-4}u_1^2 + 3x^2/2) , \tag{4.1.24}$$

$$u_t = D(u^{-3}u_2 - 3u^{-4}u_1^2 - 3u^{-1}/2 + cu) , \tag{4.1.25}$$

$$u_t = D(u^{-3}u_2 - 3u^{-4}u_1^2 + u^{-2}/2 + \lambda u^{-1} + cu) , \tag{4.1.26}$$

$$u_t = D(u^{-3}u_2 - 3u^{-4}u_1^2 + u^{-2}/2 + \lambda_1 \exp(2x) + \lambda_2 \exp(-2x)) , \tag{4.1.27}$$

$$\begin{aligned}
u_t = D(u^{-3}u_2 &- 9u^{-4}(\lambda u + 1)^{-1}u_1^2/4 - 3\lambda u^{-3}(\lambda u + 1)^{-1}u_1^2 \\
&- 2u^{-1/2}(\lambda u + 1)^{3/2} + cu) ,
\end{aligned} \tag{4.1.28}$$

$$u_t = D(u^{-3}u_2 - 9u^{-4}(\lambda u + 1)^{-1}u_1^2/4 - 3\lambda u^{-3}(\lambda u + 1)^{-1}u_1^2$$
$$+ 3u^{-1} + \lambda_1 u^{-1/2}(\lambda u + 1)^{3/2} + cu) \tag{4.1.29}$$

$$u_t = D(u^{-3}u_2 - 3u^{-4}u_1^2/2 - 3\lambda u^{-3}(\lambda u + 1)^{-1}u_1^2/2$$
$$+ \lambda_1 u^{-2}(\lambda u + 1)^3 + \lambda_2 u^2(\lambda u + 1)^{-1} + cu) , \tag{4.1.30}$$

$$u_t = D(u^{-3}u_2 - 3u^{-4}u_1^2/2$$
$$- 3u^{-3}(2\lambda_1 u + \lambda_2)(\lambda_1 u^2 + \lambda_2 u + 1)^{-1}u_1^2/4 + \lambda_3(u^{-2} + 3\lambda_2 u^{-1}/2)$$
$$+ \lambda_4 u^{-2}(\lambda_1 u^2 + \lambda_2 u + 1)^{3/2} , \tag{4.1.31}$$

$$u_t = D(u^{-3}u_2 - 3u^{-4}u_1^2/2 - 3u^{-2}(u^2 + 1)^{-1}u_1^2/2$$
$$+ 3\mathcal{P}(x)(u^{-2} + 1)/2 , \tag{4.1.32}$$

where $(\mathcal{P}')^2 = 4\mathcal{P}^3 + g_1\mathcal{P} + g_2$,

$$u_t = D(u^{-3}u_2 - 3u^{-4}u_1^2/2 + u^2(4x^3 + \lambda_1 x + \lambda_2)) , \tag{4.1.33}$$

$$u_t = D(u^{-3}u_2 - 3u^{-4}u_1^2 + cu^{-2}u_1 + c_1 x u + c_2 u) , \tag{4.1.34}$$

$$u_t = D(u^{-3}u_2 - 3u^{-4}u_1^2 + 3xu^{-2}u_1 - 3u^{-1} - 3x^2) , \tag{4.1.35}$$

$$u_t = D(u^{-3}u_2 - 3u^{-4}u_1^2 + 3x^2 u^{-2}u_1 - 9xu^{-1} - 3x^4) , \tag{4.1.36}$$

$$u_t = D(u^{-3}u_2 - 9u^{-4}u_1^2/4 + c_1 x u + c_2 u) , \tag{4.1.37}$$

$$u_t = D(u^{-3}u_2 - 9u^{-4}(\lambda u + 1)^{-1}u_1^2/4 - 3\lambda u^{-3}(\lambda u + 1)^{-1}u_1^2 + cu) , \tag{4.1.38}$$

$$u_t = D(u^{-3}u_2 - 9u^{-4}(u + 1)^{-1}u_1^2/4 - 3u^{-3}(u + 1)^{-1}u_1^2$$
$$+ \lambda u^{-3}(u + 1)u_1 - \lambda^2 u^{-2}/9 - \lambda^2 u^{-1}/3 + cu) , \tag{4.1.39}$$

$$u_t = D(u^{-3}u_2 - 9u^{-4}(u + 1)^{-1}u_1^2/4 - 3u^{-3}(u + 1)^{-1}u_1^2$$
$$+ (\gamma u^{1/2}(u + 1)^{1/2} + 3\gamma'\gamma^{-1}(u + 1))u^{-3}u_1$$
$$- 2\gamma' u^{-3/2}(u + 1)^{3/2} - (2\gamma^2/9 + \gamma'^2\gamma^{-2})u^{-2}$$
$$- (2\gamma^2/3 + 3\gamma'^2\gamma^{-2})u^{-1} - (\gamma^2/3 + 3\gamma'^2\gamma^{-2})) , \tag{4.1.40}$$

where the function $\gamma = \gamma(x)$ satisfies the equation $(\log \gamma)'' = \gamma^2/9$. Without a loss of generality one can choose $\gamma = \pm 3/x$ or $\gamma = 3\lambda/\cos(\lambda x)$.

4.1.3 Equations of the Form (4.1.5).

Let p be the function a_3. In the integrable case of function p satisfies the equation

$$(p''''(p'')^{-5/3}p^2 - 5(p''')^2(p'')^{-7/3}p^2/3 - 2p'''(p'')^{-5/3}p'p$$
$$+ 6(p'')^{1/3}p - 3(p'')^{-2/3}(p')^2)' = 0 ,$$

where $p' = \partial p/\partial u_1$, $p'' = \partial^2 p/\partial u_1^2$ and so on. The function p can be transformed to the one of the following

$$p = u_1/2b + (u_1^2/4b^2 - 1/b)^{1/2} , \quad b = -2\varepsilon(x)u/3 + \beta(x) , \tag{4.1.41}$$

$$p = 1 + 1/2u_1 , \tag{4.1.42}$$

$$p = 1 + u_1/2b + (u_1^2/4b^2 - 1/2b)^{1/2} , \quad b = b(x + u) , \tag{4.1.43}$$

$$p = (u_1 + a(x, u))^{1/2} \tag{4.1.44}$$

by means of a point-transformation (4.1.7).

In the case (4.1.41) with $\varepsilon = 0$ any integrable equation has the form

$$
\begin{aligned}
u_t &= p_2 p^{-5}(\beta p^2 - 1) - p_1^2 p^{-6}(3\beta p_*^2 - 2) - \beta' p_1 p^{-3} - 1/4\lambda^2 p^{-4} \\
&\quad - 1/3\tilde{\lambda} p^{-3} - 1/2(\beta'' - \lambda^2 \beta)p^{-2} + \tilde{\lambda}\beta p^{-1} + 1/2\beta'' \beta \\
&\quad - 1/4\beta'^2 - 1/4\lambda^2 \beta^2 + m(x)u_1 - 2m'(x)u + c_0 \, .
\end{aligned}
$$

where $p_1 = D(p)$, $p_2 = D^2(p)$, etc. There exist integrable equations of this type:

$$
\begin{aligned}
\beta(x) &= \lambda_2 \exp(2\lambda x) + \lambda_1 \exp(\lambda x) + \lambda_0 \\
&\quad + \tilde{\lambda}_1 \exp(-\lambda x) + \tilde{\lambda}_2 \exp(-2\lambda x) \, ,
\end{aligned}
$$

$$\tilde{\lambda} = 0 \, , \quad m(x) = 0 \, ; \tag{4.1.45}$$

$$\beta(x) = \lambda_4 x^4 + \lambda_3 x^3 + \lambda_2 x^2 + \lambda_1 x + \lambda_0 \, ,$$
$$\lambda = 0 \, , \quad \tilde{\lambda} = 0 \, , \quad m(x) = 0 \, ; \tag{4.1.46}$$

$$\beta(x) = \lambda_0 \, , \quad m(x) = 0 \, ; \tag{4.1.47}$$

$$\beta(x) = \lambda_2 x^2 \, , \quad \lambda = 0 \, , \quad \tilde{\lambda} = 0 \, , \quad m(x) = c_1 x \, ; \tag{4.1.48}$$

If $\varepsilon \neq 0$ any integrable equation has the form

$$
\begin{aligned}
u_t &= p_2 p^{-5}(bp^2 - 1) - p_1^2 p^{-6}(3bp^2 - 2) + 1/3\varepsilon p_1 p^{-4} \\
&\quad - b_x p_1 p^{-3} + b\varepsilon p_1 p^{-2} + (2/3\varepsilon' - 1/6Q(x))p^{-3} \\
&\quad - 1/2(b_{xx} + 2/9\varepsilon^2)p^{-2} + (1/3\varepsilon b_x + 2/3\varepsilon' b + 1/2Q(x)b)p^{-1} \\
&\quad - 2/9\varepsilon^2 b + m(x)u_1 - 3/2(m(x)b_x - 2m'(x)b)\varepsilon^{-1} \, .
\end{aligned}
\tag{4.1.49}
$$

There exist integrable equations of the type

$$\beta(x) = \lambda_2 x^2 + \lambda_1 x + \lambda_0 \, ,$$
$$\varepsilon(x) = \lambda \, , \quad Q(x) = 0 \, , \quad m(x) = 0 \, ; \tag{4.1.50}$$

$$\beta(x) = \lambda_1 x + \lambda_0 \, ,$$
$$\varepsilon(x) = \lambda \, , \quad Q(x) = 0 \, , \quad m(x) = c_0 \, ; \tag{4.1.51}$$

$$\beta(x) = \lambda_2 x^2 + \lambda_0 \, ,$$
$$\varepsilon(x) = \lambda \, , \quad Q(x) = 0 \, , \quad m(x) = c_1 x \, ; \tag{4.1.52}$$

$$\beta(x) = -1/27\lambda^2 x^4 + \lambda_2 x^2 + \lambda_1 x + \lambda_0 \, ,$$
$$\varepsilon(x) = \lambda x \, , \quad Q(x) = -2\lambda \, , \quad m(x) = 0 \, ; \tag{4.1.53}$$

$$\beta(x) = 0 \, ,$$
$$\varepsilon(x) = \lambda x^2 \, , \quad Q(x) = -6\lambda x \, , \quad m(x) = 0 \, . \tag{4.1.54}$$

Now let us consider the equations of the type (4.1.42, 43). In this case

$$u_t = p_2 p^{-3}(b - (1/2)(p-1)^{-1}) + (3/4)p_1^2 p^{-4}(p-1)^{-2}$$
$$+ (1/4)p_1^2 p^{-3}(p-1)^{-3} - (3/2)p_1^2 p^{-4}b - (3/2)p_1^2 p^{-3}(p-1)^{-1}b$$
$$- p_1 p^{-2}(p-1)^{-1}b' - (1/4)\lambda^2(p-1)^{-3}$$
$$+ (3\lambda^2 b + l + m - b'')(p-1)^{-1}/2 + l(b - 1/2)p^{-2}$$
$$- (bb'' - b''/2 - b'^2 + 3bl + l/2)p^{-1} + h + mbp, \tag{4.1.55}$$

where

$$b = \lambda_1, \quad l = \lambda_2, \quad m = c_1, \quad h = c_2; \tag{4.1.56}$$

or

$$l = b''/2 - bb'' + (b')^2/2 + 2\lambda^2 b^2 - 2\lambda^2 b + \lambda^2/2,$$
$$m = bb'' - (b')^2/2 - 2\lambda^2 b^2 + c,$$
$$h = (2l - m)b - (b')^2/2 + 2\lambda^2 b^2 - \lambda^2 b + c, \tag{4.1.57}$$
$$b = (1 - (f(x+u))^{-2})/2.$$

In the last case the function f can be defined from

$$(\log(f))'' = g,$$

where g satisfies $(g')^2 = 8g^3/3 + c_1 g + c_2$.

In the case of (4.1.44) any integrable equation has the form

$$u_t = 2p_2 p^{-2} - 4p_1^2 p^{-3} - 2p_1 p^{-2}a_u - 2\lambda p^3 + 2a_{uu}p + (4Q + 2aa_{uu}$$
$$- 2a_u^2 - 2a_{ux})p^{-1} + m(x)u_1 + m'(x)u + h(x). \tag{4.1.58}$$

There exist integrable equations of this type:

$$a = \lambda_4 u^4 + \lambda_3 u^3 + \lambda_2 u^2 + \lambda_1 u + \lambda_0,$$
$$\lambda = 0, \quad Q = -aa_{uu} + 1/2a_u^2, \quad m(x) = c, \quad h(x) = 0; \tag{4.1.59}$$

$$a = 0, \quad \lambda = 0, \quad Q = \tilde{\lambda} \quad m(x) = c_1 x + c_2,$$
$$h(x) = c_3, \quad c_1\tilde{\lambda} = 0; \tag{4.1.60}$$

$$a = \lambda_2 u^2, \quad \lambda = 0, \quad Q = 0,$$
$$m(x) = c_2 x^2 + c_1 x + c_0 \quad h(x) = -c_1; \tag{4.1.61}$$

$$a = \lambda_2 \exp(2u) + \lambda_1 \exp(u) + \lambda_0 + \tilde{\lambda}_1 \exp(-u) + \tilde{\lambda}_2 \exp(-2u),$$
$$\lambda = 1, \quad Q = -aa'' + (a')^2/2 + a^2/2, \quad m = c, \quad h = 0; \tag{4.1.62}$$

$$a = 1, \quad \lambda = 1, \quad Q = \lambda_0, \quad m = c_1, \quad h = c_2; \tag{4.1.63}$$

$$a = s(x)\exp(u) + n(x)\exp(-u), \quad \lambda = 1,$$
$$Q = -6s(x)n(x) + \lambda_0, \quad m = 0, \quad h = 0, \tag{4.1.64}$$

where

$$s'' + 8s^2 n - 2\lambda_0 s = 0 \,,$$
$$n'' + 8n^2 s - 2\lambda_0 n = 0 \,;$$

$$a = s(x)\exp(2u) + n(x)\exp(-2u) \,, \quad \lambda = 1 \,,$$
$$Q = a_u^2/2 - 7a^2/2 + 3a_{ux}/4 - 4s(x)n(x) + \lambda_0 \,, \qquad (4.1.65)$$
$$m = 0 \,, \quad h = 0 \,,$$

where

$$s'' + 32s^2 n - 8\lambda_0 s = 0 \,,$$
$$n'' + 32n^2 s - 8\lambda_0 n = 0 \,.$$

4.1.4 Equations of the Form (4.1.6). Let p be the function $(a_1 u_2 + a_0)^{1/2}$. It is easy to check that the function a_1 can be reduced to the one of the following form: $a_1 = 1$ or $a_1 = a(x, u)u_1^{-1/2}$ via point-transformation (4.1.7).

In the first case any integrable equation has the form

$$u_t = 2p_1 p^{-2} + 2(\partial b/\partial u_1)p^{-1} + H(x, u, u_1) \,, \qquad (4.1.66)$$

where $p = (u_2 + b(x, u, u_1))^{1/2}$, H is a classical symmetry of (4.1.66). The following integrable equations of the form (4.1.66) exist:

$$b = \lambda_4 u_1^4 + \lambda_3 u_1^3 + \lambda_2 u_1^2 + \lambda_1 u_1 + \lambda_0 \,, \quad \lambda_i \in \mathbb{C} \qquad (4.1.67)$$

$$b = u \,, \qquad (4.1.68)$$

$$b = u_1^2 + \exp(-4u) \,, \qquad (4.1.69)$$

$$b = f u_1^2 + \lambda_1 (f^2 - c) + \lambda_2 f \,, \quad \lambda_i \in \mathbb{C} \qquad (4.1.70)$$

where the function $f = f(u)$ satisfies equation $df/du + f^2 + c = 0$. Without loss of generality one can take $c = 0$ or $c = 1$ and $f = u^{-1}$ or $f = 1/\tan(u)$, respectively.

The integrable equations (4.1.6) of the second type have the form

$$u_t = 2\alpha^{-1/2} u_1^{1/2} D(p)p^{-2} - \alpha^{-1}p + 2\alpha^{-1/2} u_1^{1/2}(\partial b/\partial u_1 - b u_1^{-1})$$
$$- \alpha^{-1} u_1 \partial \alpha/\partial u_1 - \alpha^{-1} \partial \alpha/\partial x)p^{-1} + H(x, u, u_1) \,, \qquad (4.1.71)$$

where H is a classical symmetry,

$$p = (\alpha^{1/2} u_1^{-1/2} u_2 + \alpha^{1/2} u_1^{-1/2} b(x, u, u_1))^{1/2} \,,$$

$$b = \lambda_1 \alpha^{3/2} u_1^{5/2} + (\partial \phi/\partial u + \alpha^{-1} \partial \alpha/\partial u)u_1^2 + m \alpha^{1/2} u_1^{3/2}$$
$$- (\partial \phi/\partial x + \alpha^{-1} \partial \alpha/\partial x)u_1 + \lambda_2 \alpha^{3/2} u_1^{1/2} \,.$$

Thus any equation (4.1.71) is defined by the function $\alpha(x, u)$, $\phi(x, u)$, $m(x, u)$ and the constants λ_1, λ_2. The following integrable equations (4.1.71) exist:

$$\alpha = 1 \,, \quad m = c_1 \,, \quad \phi = c_2 x + c_3 u \,, \qquad (4.1.72)$$

λ_1, λ_2, c_1, c_2, c_3 are arbitrary constants;

$$\alpha = 1, \quad m = 0, \quad \phi = \phi(x), \quad \lambda_1 = 0, \quad \lambda_2 = 1$$
$$\phi'' = \phi'^2/2 + \lambda;$$

(4.1.73)

$$\alpha = \alpha(u+x), \quad m = m(u+x), \quad \phi = 0, \quad \lambda_1 = \lambda_2 = 0 \qquad (4.1.74)$$

where

$$m' = \alpha, \quad \alpha'' = \alpha'^2\alpha^{-1} - 3m^2\alpha^2/16 + c\alpha^2, \quad \alpha \neq 0;$$
$$\alpha = \alpha(x,u), \quad m = c, \quad \phi = 0, \quad \lambda_1 = \lambda_2 = 0$$

(4.1.75)

where

$$\partial^2\alpha/\partial x\,\partial u = \alpha^{-1}(\partial\alpha/\partial u)\,\partial\alpha/\partial x + \alpha^2, \quad \alpha \neq 0;$$
$$\alpha = \alpha(u), \quad m = 0, \quad \phi = u, \quad \lambda_1 = \lambda_2 = 0$$

(4.1.76)

where

$$\alpha' = -\alpha/2 + c\alpha^3, \quad \alpha' \neq 0;$$
$$\alpha = \alpha(u+x), \quad m = m(u+x), \quad \phi = \phi(u+x),$$
$$\lambda_1 = \lambda_2 = 0$$

(4.1.77)

where

$$\phi'' = \phi'^2/2 + \alpha^{-1}\alpha'\phi' - c\alpha^2/2,$$
$$m'' = (\phi'/2 + \alpha^{-1}\alpha')m' + (\phi'^2 - c\alpha^2)m,$$
$$\alpha''' = 4\alpha^{-1}\alpha'\alpha'' - 3\alpha^{-2}\alpha'^3 - 3\phi'^2\alpha'/4 - 3\phi'(\alpha'' - \alpha^{-1}\alpha'^2)/2$$
$$\qquad - 3mm'\alpha^2/8 + c\alpha'\alpha^2/4;$$

$$\alpha = 2u(x^2 + u^2)^{-1}, \quad m = 0, \quad \phi = \log(x^2 + u^2), \quad \lambda_1 = \lambda_2 = 0; \quad (4.1.78)$$

$$\alpha = \alpha(u+x), \quad m = m(u+x), \quad \phi = \phi(u+x), \quad \lambda_1 = \lambda_2 = 1 \qquad (4.1.79)$$

where

$$\phi''\alpha^{-1} - \alpha^{-1}\phi'^2/2 + 5\alpha^{-2}\alpha'\phi' + 4\alpha^{-2}\alpha'' - 4\alpha^{-3}\alpha'^2 + 3m^2/4$$
$$\qquad - 14\alpha^2 + 7\alpha m/2 - 5c_1\alpha/2 + c_2 = 0,$$

$$m'' - (\phi'/2 + \alpha^{-1}\alpha')m' + (-2\phi'' + 2\alpha^{-1}\alpha'\phi')m - 4\alpha\phi''$$
$$\qquad - 3\alpha'\phi' - 10\alpha'' + 10\alpha^{-1}\alpha'^2 = 0,$$

$$m = -2\alpha^{-2}\phi'' + \alpha^{-2}\phi'^2 + 2\alpha^{-3}\alpha'\phi' + 10\alpha + c_1;$$
$$\alpha = A(u+x) - B(u-x), \quad m = 10(A(u+x) + B(u-x)) \qquad (4.1.80)$$
$$\phi = 0, \quad \lambda_1 = \lambda_2 = 1$$

where

$$A'^2 + 16A^5 + c_3 A^3 + c_2 A^2 + c_1 A + c_0 = 0, \quad A' \neq 0,$$
$$B'^2 + 16B^5 + c_3 B^3 + c_2 B^2 + c_1 B + c_0 = 0, \quad B' \neq 0;$$

$$\alpha = A(u + x), \quad m = 2A(u + x),$$
$$\phi = -\log(A(u+x)) - \log(B(u-x)), \quad \lambda_1 = \lambda_2 = 1, \tag{4.1.81}$$

where

$$A'^2 + 4A^5 + c_4 A^4 + c_3 A^3 + c_2 A^2 = 0, \quad A' \neq 0,$$
$$B'^2 + c_2 B^2 + c_1 B = 0, \quad B' \neq 0;$$

$$\alpha = A(u + x) - B(u - x), \quad m = 2(A(u + x) + B(u - x))$$
$$\phi = -\log(A(u+x)) - \log(B(u-x)), \quad \lambda_1 = \lambda_2 = 1, \tag{4.1.82}$$

where

$$A'^2 + 4A^5 + c_4 A^4 + c_3 A^3 + c_2 A^2 + c_1 A = 0, \quad A' \neq 0,$$
$$B'^2 + 4B^5 + c_4 B^4 + c_3 B^3 + c_2 B^2 + c_1 B = 0, \quad B' \neq 0.$$

4.1.5 Classification Problems. The above given enormous lists of integrable quasilinear equations are very impressive but hardly convenient in practice. A complete classification requires an extension of class of contact transformations (Sect. 2). The transformations which are connected with potentiation enable us to pick out the so-called *basic* equations. They are indicated in the next two tables.

Table 1. "S-Integrable" Equations

□(4.1.78)	□(4.1.80)	□(4.1.81)	□(4.1.82)
	□(4.1.65)	□(4.1.64)	
	↓	↓	
	□(4.1.32)	□(4.1.33)	
	↓		
	□(4.1.27)		
	↓		
	□(4.1.24)		
	↓		
	□(4.1.23)		

Table 2. "C-Integrable" Equations

□(4.1.22)	□(4.1.53)	□(4.1.54)
↓	↓	↓
□(4.1.20)	□(4.1.35)	□(4.1.36)

Below we describe the way to obtain any equation from the lists given in Sects. 1.4.1–4 starting with the basic equations. Recall [see (2.1.5)] that the equation admits potentiation if it has a conserved density $\varrho(x, u)$.[19] As a rule the basic equations possess such densities. The densities have to be of the following form:

---

[19] In the case $\varrho = \varrho(x, u, u_1)$ the potentiation leads to essentially nonlinear equations (3.3.8,9).

(4.1.23): $\varrho = c_1 u + c_2 xu + c_3 x^2 u + c_4 (u^{-1} - x^3 u)$;

(4.1.24): $\varrho = c_1 u + c_2 xu + c_3 x^2 u + c_4 (u^{-1} + x^4 u/4)$;

(4.1.27): $\varrho = c_1 u + c_2 \exp(x)u + c_3 \exp(-x)u$

$\quad + c_4 [u^{-1} + 2(\lambda_1 \exp(2x) + \lambda_2 \exp(-2x))u/3]$;

(4.1.32): $\varrho = c_1 u + K(x)(u^2 + 1)^{1/2}$,

where $K''' - 3PK' - 3P'K/2 = 0$;

(4.1.33): $\varrho = c_1 u$;

(4.1.64): $\varrho = c_1 (s(x) \exp(u) - n(x) \exp(-u))$;

(4.1.65): $\varrho = \alpha(x) \exp(u) - \beta(x) \exp(-u)$,

where

$$\alpha'' - 4s\beta - 2s'\beta + (8sn - 2\lambda_0)\alpha = 0 ,$$

$$\beta'' - 4n\alpha - 2n'\alpha + (8sn - 2\lambda_0)\beta = 0 ;$$

(4.1.78): $\varrho = c_1 (x^2 + x^2)^{-1/2}$.

Equations (4.1.80–82) do not possess the conserved densities of the form $\varrho = \varrho(x, u)$.

The conserved densities indicated above allow one to construct new equations starting with the basic ones. For example, (4.1.23) possesses the conserved densities

$$\varrho_1 = u , \quad \varrho_2 = xu , \quad \varrho_3 = x^2 u , \quad \varrho_4 = u^{-1} - x^3 u .$$

The potentiations $u_1' = \varrho_k$, $k = 1, 2, 3, 4$ [cf. (2.1.5)] lead to equations that are equivalent to (4.1.8, 25, 29, 46), respectively. It may occur that an equation obtained as a result of the potentiation has a conserved density of the form $\varrho = \varrho(x, u)$ again. In this case the procedure of potentiation must be repeated and so on.

The potentiation establishes the links between equations from different lists. In particular, for any integrable equation of the form of (4.1.4) the function $\varrho = u$ is a conserved density [this density is canonical by virtue of (3.3.2)] and every such equation is reducible to a simpler equation of the form of (4.1.3) via potentiation and point-transformation $x' = u$, $u' = x$. For example, equations (4.1.23, 24, 27, 32, 33) turn into the well-known equations (4.1.8, 10, 12, 15, 17), respectively.

Besides u, equation (4.1.4) may possess other conserved densities of the form $\varrho(x, u)$. If $\varrho = a(x)u + b(x)$, $a' \neq 0$, then the potentiation does not violate the class of (4.1.4). In the case $\partial^2 \varrho / \partial u^2 \neq 0$ the potentiation gives an equation of the form of (4.1.5).

The Miura type transformations (Sect. 2.3.1) enable us to complete the classification of quasilinear equations (4.1.1). In Tables 1, 2 the arrows indicate that the corresponding basic equations are related by a differential substitution of the first order (2.3.11). For example, (4.1.24) rewritten in terms of $y, \tilde{u}$ possesses a nonlocal symmetry of the form of (2.3.14) with

$$\Omega = y\tilde{u} + D(\log(\tilde{u}_1 + y\tilde{u}^2)) .$$

The corresponding substitution

$$x = \tilde{u}^{-1} - y^2/2 , \quad u = \tilde{u}^3/(\tilde{u}_1 + y\tilde{u}^2) ,$$

transforms (4.1.24) into (4.1.23). It seems that all basic equations from Tables 1, 2 can be reduced to (4.1.18, 23 or 33) via the Miura type transformations. This has been verified for all of these equations except (4.1.80, 82).

4.2 Scalar Fifth Order Equations

The classification of evolution equations of the form

$$u_t = u_5 + F(x, u, u_1, u_2, u_3, u_4) , \tag{4.2.1}$$

is based on the integrability conditions $(\varrho_k)_t \in \text{Im}\{D\}$, $k = 0, \ldots, 5$, where canonical densities ϱ_k are given by (3.3.11–16). We present an exhaustive [up to the point-transformations $\tilde{u} = \psi(x, u)$ and scaling] list of integrable equations (4.2.1). The symmetries of the fifth order of the equations from the list in Sect. 4.1.2 and linearizable equations [17] are omitted.

Integrable Equations of the Form (4.2.1)

$$u_t = u_5 + 5uu_3 + 5u_1u_2 + 5u^2u_1 + cu_1 ; \tag{4.2.2}$$

$$u_t = u_5 + 5uu_3 + 25u_1u_2/2 + 5u^2u_1 + cu_1 ; \tag{4.2.3}$$

$$u_t = u_5 + 5u_1u_3 + 5u_1^3/3 + cu_1 + c_1 ; \tag{4.2.4}$$

$$u_t = u_5 + 5u_1u_3 + 15u_2^2/4 + 5u_1^3/3 + cu_1 + c_1 ; \tag{4.2.5}$$

$$u_t = u_5 + 5(u_1 - u^2)u_3 + 5u_2^2 - 20uu_1u_2 - 5u_1^3 + 5u^4u_1 + cu_1 ; \tag{4.2.6}$$

$$u_t = u_5 + 5(u_2 - u_1^2)u_3 - 5u_1u_2^2 + u_1^5 + cu_1 + c_1 ; \tag{4.2.7}$$

$$u_t = u_5 + 5(u_2 - u_1^2 + \lambda_1 \exp(2u) - \lambda_2^2 \exp(-4u))u_3$$
$$\quad - 5u_1u_2^2 + 15(\lambda_1 \exp(2u) + 4\lambda_2^2 \exp(-4u))u_1u_2 + u_1^5$$
$$\quad - 90\lambda_2^2 \exp(-4u)u_1^3 + 5(\lambda_1 \exp(2u) - \lambda_2^2 \exp(-4u))^2 u_1 + cu_1 ; \tag{4.2.8}$$

$$u_t = u_5 + 5(u_2 - u_1^2 - \lambda_1^2 \exp(2u) + \lambda_2 \exp(-u))u_3$$
$$\quad - 5u_1u_2^2 - 15\lambda_1^2 \exp(2u)u_1u_2 + u_1^5$$
$$\quad + 5(-\lambda_1^2 \exp(2u) + \lambda_2 \exp(-u))^2 u_1 + cu_1 ; \tag{4.2.9}$$

$$u_t = u_5 - 5u_1^{-1}u_2u_4/2 - 5u_1^{-1}u_3^2/4 + (5u_1^{-2}u_2^2$$
$$\quad + 5u_1^{-1/2}u_2/2 - 5u_1 + 10\lambda u_1^{1/2} + 5\lambda^2)u_3 - 35u_1^{-3}u_2^4/16$$
$$\quad - 5u_1^{-3/2}u_2^3/3 + (5\lambda u_1^{-1/2} - 15\lambda^2 u_1^{-1}/4 + 5/4)u_2^2 + 5u_1^3/3$$
$$\quad - 8\lambda u_1^{5/2} - 15\lambda^2 u_1^2 - 10\lambda^3 u_1^{3/2}/3 + cu_1 + c_1 ; \tag{4.2.10}$$

$$u_t = u_5 - 5u_1^{-1}u_2u_4 + 5(u_1^{-2}u_2^2 + \lambda_1 u_1^{-1} + \lambda_2 u_1^2)u_3$$
$$- 5(\lambda_1 u_1^{-2} + \lambda_2 u_1)u_2^2 - 5\lambda_1^2 u_1^{-1}$$
$$+ 5\lambda_1\lambda_2 u_1^2 + \lambda_2^2 u_1^5 + cu_1 + c_1 \ ; \tag{4.2.11}$$

$$u_t = u_5 - 5u_1^{-1}u_2u_4 - 15u_1^{-1}u_3^2/4 + 5(13u_1^{-2}u_2^2/4 + \lambda_1 u_1^{-1} + \lambda_2 u_1^2)u_3$$
$$- 135u_1^{-3}u_2^4/16 - 5(7\lambda_1 u_1^{-2}/4 - \lambda_2 u_1/2)u_2^2$$
$$- 5\lambda_1^2 u_1^{-1} + 5\lambda_1\lambda_2 u_1^2 + \lambda_2^2 u_1^5 + cu_1 + c_1 \ ; \tag{4.2.12}$$

$$u_t = u_5 - 15(\exp(5f) + 2\exp(2f))(\exp(3f) - 1)^{-2}u_2u_4/2$$
$$- 45\exp(2f)(\exp(3f) - 1)^{-2}u_3^2/4 + 45(\exp(10f) + 22\exp(7f)$$
$$+ 13\exp(4f))(\exp(3f) - 1)^{-4}u_2^2u_3/4 - 3645(2\exp(12f)$$
$$+ 4\exp(9f) + \exp(6f))(\exp(3f) - 1)^{-6}u_2^4/16 + 5\lambda\exp(2f)u_3$$
$$- 15\lambda(2\exp(7f) + 7\exp(4f))(\exp(3f) - 1)^{-2}u_2^2/4$$
$$+ 2\lambda^2\exp(5f)/3 - 5\lambda^2\exp(2f)/3 + cu_1 + c_1 \ ; \tag{4.2.13}$$

where the function $f = f(u_1)$ is defined by algebraic equation

$$2\exp(3f) - 3u_1\exp(2f) + 1 = 0 \ ; \tag{4.2.14}$$

$$u_t = u_5 - 15(\exp(5f) + 2\exp(2f))(\exp(3f) - 1)^{-2}u_2u_4/2$$
$$- 45\exp(2f)(\exp(3f) - 1)^{-2}u_3^2/4 + 45(\exp(10f) + 22\exp(7f)$$
$$+ 13\exp(4f))(\exp(3f) - 1)^{-4}u_2^2u_3/4 - 3645(2\exp(12f)$$
$$+ 4\exp(9f) + \exp(6f))(\exp(3f) - 1)^{-6}u_2^4/16$$
$$- 5\beta(5\exp(2f) + 2\exp(-f) + 2\exp(-4f))u_3/9$$
$$+ 5\beta(10\exp(7f) + 39\exp(4f) + 36\exp(f) - 4\exp(-2f))$$
$$\times (\exp(3f) - 1)^{-2}u_2^2/12 - 5\beta'(10\exp(3f) - 3 + 12\exp(-3f)$$
$$+ 8\exp(-6f))u_2/54 - 5\beta^2(14\exp(-f) + 39\exp(-4f)$$
$$+ 24\exp(-7f) + 4\exp(-10f))(\exp(3f) - 1)^2/243 + cu_1 \ , \tag{4.2.15}$$

where the function $f(u_1)$ satisfies (4.2.14) and $\beta(u)$ is defined by

$$\beta'^2 = 4\beta^3 + \lambda \ , \quad \beta' \neq 0 \ ; \tag{4.2.16}$$

$$u_t = u_5 - 15(\exp(5f) + 2\exp(2f))(\exp(3f) - 1)^{-2}$$
$$\times u_2u_4/2 - 45\exp(2f)(\exp(3f) - 1)^{-2}u_3^2/4$$
$$+ 45(\exp(10f) + 22\exp(7f) + 13\exp(4f))(\exp(3f) - 1)^{-4}u_2^2u_3/4$$
$$- 3645(2\exp(12f) + 4\exp(9f) + \exp(6f))(\exp(3f) - 1)^{-6}$$
$$\times u_2^4/16 - 10\beta(16\exp(2f) + \exp(-f) + \exp(-4f))u_3/9$$
$$+ 5\beta(16\exp(7f) + 57\exp(4f) + 9\exp(f) - \exp(-2f))$$
$$\times (\exp(3f) - 1)^{-2}u_2^2/3 - 10\beta'(16\exp(3f) + 6 + 3\exp(-3f)$$
$$+ 2\exp(-6f))u_2/27 + 20\beta^2(64\exp(-f) + 24\exp(-4f) - 6\exp(-7f)$$
$$- \exp(-10f))(\exp(3f) - 1)^2/243 + cu_1 \ ; \tag{4.2.17}$$

$$u_t = u_5 - 15(\exp(5f) + 2\exp(2f))(\exp(3f) - 1)^{-2}$$
$$\times u_2 u_4/2 - 45\exp(2f)(\exp(3f) - 1)^{-2}u_3^2/4 + 45(\exp(10f)$$
$$+ 22\exp(7f) + 13\exp(4f))(\exp(3f) - 1)^{-4}u_2^2 u_3/4$$
$$- 3645(2\exp(12f) + 4\exp(9f) + \exp(6f))(\exp(3f) - 1)^{-6}$$
$$\times u_2^4/16 - 10\beta(16\exp(2f) + \exp(-f) + \exp(-4f))u_3/9$$
$$+ 5\beta(16\exp(7f) + 57\exp(4f) + 9\exp(f) - \exp(-2f))$$
$$\times (\exp(3f) - 1)^{-2}u_2^2/3 - 10\beta'(16\exp(3f) + 6 + 3\exp(-3f)$$
$$+ 2\exp(-6f))u_2/27 + 20\beta^2(64\exp(-f) + 24\exp(-4f)$$
$$- 6\exp(-7f) - \exp(-10f))(\exp(3f) - 1)^2/243$$
$$- 5\mu\beta^{-2}(5\exp(2f) + 2\exp(-f) + 2\exp(-4f))u_3/9$$
$$+ 5\mu\beta^{-2}(10\exp(7f) + 39\exp(4f) + 36\exp(f) - 4\exp(-2f))$$
$$\times (\exp(3f) - 1)^{-2}u_2^2/12 - 5\mu\beta'\beta^{-2}(10\exp(3f) - 3$$
$$+ 12\exp(-3f) + 8\exp(-6f))u_2/27 + 40\mu\beta^{-1}(10\exp(-f)$$
$$- 3\exp(-4f) - 6\exp(-7f) - \exp(-10f))(\exp(3f) - 1)^2/243$$
$$- 5\mu^2\beta^{-4}(14\exp(-f) + 39\exp(-4f)$$
$$+ 24\exp(-7f) + 4\exp(-10f))(\exp(3f) - 1)^2/243 + cu_1 . \qquad (4.2.18)$$

The functions $f(u)$, $\beta(u)$ are defined by (4.2.14, 16).

The basic equations of the form (4.2.1) are indicated in Table 3.

Table 3.

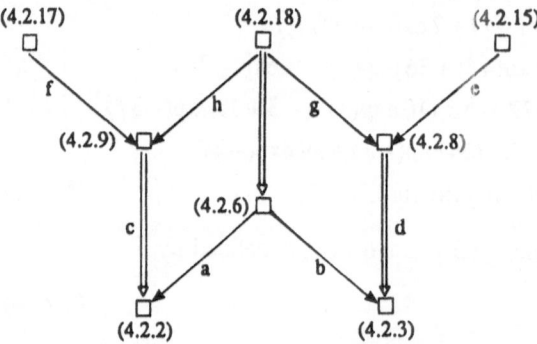

The differential substitutions that connect vertices of the graph in Table 3 have the form

a) (4.2.6) $\to$ (4.2.2): $\tilde{u} = -u_1 - u^2$;

b) (4.2.6) $\to$ (4.2.3): $\tilde{u} = 2u_1 - u^2$;

c) (4.2.8) $\to$ (4.2.3): $\tilde{u} = 2u_2 - u_1^2 + 6\lambda_2\exp(-2u)u_1 + \lambda_1\exp(2u) - \lambda_2^2\exp(-4u)$;

d) (4.2.9) $\to$ (4.2.2): $\tilde{u} = -u_2 - u_1^2 + 3\lambda_1\exp(u)u_1 - \lambda_1^2\exp(2u) + \lambda_2\exp(-u)$.

The substitutions e), f), g) h) have the form

$$\tilde{u} = f(u_1) + \alpha(u) ,$$

where the function $f(u_1)$ satisfies (4.2.14) and $\alpha(u)$ is connected with $\beta(u)$, λ, λ_1, λ_2 [see (4.2.16, 8, 9)] in the following way:

e) (4.2.15) $\rightarrow$ (4.2.8): $\beta = -\lambda_1 \exp(2\alpha) + 4\lambda_2^2 \exp(-4\alpha)$, $\quad \lambda = -108\lambda_1^2\lambda_2^2$;

f) (4.2.17) $\rightarrow$ (4.2.9): $\beta = \lambda_1^2 \exp(2\alpha)/4 + \lambda_2 \exp(-\alpha)/2$, $\quad \lambda = 27\lambda_1^2\lambda_2^2/16$;

g) (4.2.18) $\rightarrow$ (4.2.8): $\beta = -\lambda_1 \exp(2\alpha)/4$, $\quad \lambda = \lambda_1^2\lambda_2^2/4$;

h) (4.2.18) $\rightarrow$ (4.2.9): $\beta = \lambda_2 \exp(-\alpha)/2$, $\quad \lambda = \lambda_1^2\lambda_2^2/4$;

We note that an attempt to classify equations of the form $u_t = u_5 + F(u, u_1, u_2, u_3)$ possessing a symmetry of order seven has been undertaken [32]. In that paper (4.2.9) has been discovered, but (4.2.8) missed. The fact that (4.2.8) possesses a symmetry of order seven has been verified [33].

4.3 Schrödinger Type Equations

Equations which are related by invertible transformations should be considered to be equivalent. The above classification of scalar equations has shown that complete lists of equations up to invertible transformations have proved to be enormous, and to obtain a visible description one has to involve a wider class of transformations. In many applications there occurs a situation when an equation possesses a continuous point group, and it is sufficient to restrict ourselves to a reduced subset of dynamical variables consisting of the group invariants, without loss of essential information. Therefore it is natural to consider two equations to be equivalent if their reduced subsets of dynamical variables are related by an invertible transformation. From this point the linearizable (Burgers type) equations considered in Sect. 2 are equivalent to linear ones.

We have already mentioned (Sect. 1.4.2) that a system of two equations of the form

$$u_t = A(u)u_2 + F(u, u_1), \quad \det(A(u)) \neq 0, \quad u = (u^1, u^2), \tag{4.3.1}$$

possessing higher conservation laws can be reduced to

$$
\begin{aligned}
u_t &= u_2 + F(u, v, u_1, v_1), \\
-v_t &= v_2 + G(u, v, u_1, v_1),
\end{aligned}
\tag{4.3.2}
$$

where the functions F, G are of the form (3.3.27–29). Invertible transformations which do not violate the form of (4.3.2) are a composition of the elementary transformations

$$x \rightarrow ax + bt + c, \quad t \rightarrow a^2 t + d, \quad a, b, c, d \in \mathbb{C}, \tag{4.3.3}$$

the conformal transformations

$$u \rightarrow \tilde{u}(u), \quad v \rightarrow \tilde{v}(v), \tag{4.3.4}$$

and the involution

$$u \to v, \quad v \to u, \quad t \to -t, \quad x \to -x . \tag{4.3.5}$$

Equations (4.3.2) which are invariant under a continuous subground of transformations (4.3.4) can be put in the form

$$\begin{aligned} u_t &= u_2 + F(\varepsilon u + v, u_1, v_1) , \\ -v_t &= v_2 + G(\varepsilon u + v, u_1, v_1) , \end{aligned} \quad (\varepsilon = 0 \quad \text{or} \quad 1), \tag{4.3.6}$$

via a conformal transformation and possibly (4.3.5). For instance, the famous nonlinear Schrödinger equation

$$i\psi_\tau = \psi_2 + |\psi|^2 \psi \tag{4.3.7}$$

is reduced to the form of (4.3.6) with $\varepsilon = 1$:

$$\begin{aligned} u_t &= u_2 + u_1^2 + \exp(u + v) \\ -v_t &= v_2 + v_1^2 + \exp(u + v) \end{aligned} \tag{4.3.8}$$

by the obvious change

$$x = x, \quad t = -i\tau, \quad \psi = \exp(u), \quad \psi^* = \exp(v) .$$

If the continuous conformal group of (4.3.2) is nonabelian, the resulting equation (4.3.6) may depend on a particular choice of a one parameter subgroup. The Heisenberg model

$$S_t = S \times S_2 , \quad S = (S^1, S^2, S^3), \quad (S)^2 = 1 \tag{4.3.9}$$

may serve as an example of an equation that has a different representation in the form of (4.3.6):

$$\begin{aligned} u_t &= u_2 - 2u_1^2 \tanh(u + v) \\ -v_t &= v_2 - 2v_1^2 \tanh(u + v) \end{aligned} \tag{4.3.10}$$

and

$$\begin{aligned} u_t &= u_2 - 2u_1^2/(u + v) \\ -v_t &= v_2 - 2v_1^2/(u + v) . \end{aligned} \tag{4.3.11}$$

4.3.1 Symmetrical Equations and Transformations. Equation (4.3.6) is called symmetrical if it is invariant under the involution (4.3.5). Many applicable equations (for instance (4.3.8, 10, 11)) are symmetrical. Symmetrical equations have the form:

$$\begin{aligned} u_t &= u_2 + F(u + v, u_1, v_1) , \\ -v_t &= v_2 + F(u + v, -v_1, -u_1) . \end{aligned} \tag{4.3.12}$$

If (4.3.12) possesses an invariant[20] conserved density ϱ of the form

$$\varrho = p'(u+v)(u_1 - v_1) + 2q(u+v) \quad p' \neq 0 \tag{4.3.13}$$

we can define a transformation

$$\overline{u} + \overline{v} = p(u+v) , \tag{4.3.14}$$
$$\overline{u}_1 = p'(u+v)u_1 + q(u+v)$$

that does not violate the symmetrical structure. Relations (4.3.14) define the one-to-one correspondence of the sets

$$\{\overline{u} + \overline{v}, \overline{u}_1, \overline{v}_1, \overline{u}_2, \overline{v}_2, \ldots\} \quad \text{and} \quad \{u+v, u_1, v_1, u_2, v_2, \ldots\} .$$

Expressing the lhs of

$$\overline{u}_t + \overline{v}_t = p_t \quad \overline{u}_t - \overline{v}_t = \sigma$$

in terms of the new variables, we obtain the transformed equation. Transformation (4.3.14) defines the equivalence relation on the set of symmetrical equations [34]. Therefore equations linked by (4.3.14) we shall call symmetrically equivalent. To construct the equivalence class of a symmetrical equation one should find all conserved densities of the form (4.3.13). If there are no nontrivial ones, then the equivalence class is determined by the substitutions

$$\overline{u} + \overline{v} = \alpha u + \alpha v , \quad \overline{u}_1 = \alpha u_1 + \beta , \quad \overline{v}_1 = \alpha v_1 - \beta , \quad \alpha, \beta \in \mathbb{C} .$$

Example 4.1. An equation of the form (4.3.2) with

$$F = 2auvu_1 + bu^2 v_1 + b(a-b)u^3 v^2/2 + cu^2 v , \quad g = F^* \tag{4.3.15}$$

[were F^* denotes the result of the involution (4.3.5)] by the conformal change $\overline{u} = \log(u)$, $\overline{v} = \log(v)$ is reduced to the form of (4.3.12) where

$$F = \overline{u}_1^2 + (2a\overline{u}_1 + b\overline{v}_1 + c)\exp(\overline{u} + \overline{v}) + b(a-b)\exp[2(\overline{u} + \overline{v})]/2 .$$

Using the conserved densities $\alpha \overline{u}_1 + \beta \exp(\overline{u} + \overline{v})$, $\alpha, \beta \in \mathbb{C}$, it is easy to check that

i) if $b = 2a$, $c = 0$, this equation is symmetrically equivalent to a linear one;
ii) if $b = 2a$, $c \neq 0$, this equation is symmetrically equivalent to the nonlinear Schrödinger equation (4.3.7);
iii) if $b \neq 2a$, this equation is symmetrically equivalent to the well-known derivative nonlinear Schrödinger equation [i.e., to (4.3.2, 15) with $a = b = 1$, $c = 0$.

∎

---

[20] A function $h(u, v, u_1, v_1, \ldots)$ is called invariant if it is not changed under the involution (4.2.5) and $\partial h/\partial u = \partial h/\partial v$. If a conserved density ϱ is invariant, then the function σ (which determines by $\varrho_t = D(\sigma)$) is also invariant [34].

Example 4.2. The well-known equations [4]

$$\bar{u}_t = \bar{u}_2 - \bar{u}_1^2 + 2\bar{u}_1\bar{v}_1 \ ,$$
$$-\bar{v}_t = \bar{v}_2 - \bar{v}_1^2 + 2\bar{v}_1\bar{u}_1 \tag{4.3.16}$$

and the Heisenberg model (4.3.11) are symmetrically equivalent. The link is given by

$$\bar{u} + \bar{v} = \log(u + v) \ ,$$
$$\bar{u}_1 = u_1/(u + v) \ .$$

These examples show that equations not related to each other at first glance might prove to be symmetrically equivalent (for a simple criterion of such equivalence see [34]).

4.3.2 Classification of Integrable Equations. Below we present a complete list[21] of equations of the form of (4.3.2), which satisfies the explicit integrability conditions (3.3.18–26). This list has been obtained in [8, 34, 35]. Moreover, all equations of the list prove to be integrable.

Integrable Equations of the Form of (4.3.2)

$$u_t = u_2 + D(u^2 + v) \ ,$$
$$-v_t = v_2 - 2D(uv) \ ; \tag{A}$$

$$u_t = u_2 + u_1^2 + v_1 \ ,$$
$$-v_t = v_2 - 2u_1v_1 \ ; \tag{a}$$

$$u_t = u_2 + u^2 v \ ,$$
$$-v_t = v_2 + v^2 u \ ; \tag{b}$$

$$u_t = u_2 + (u + v)u_1 \ ,$$
$$-v_t = v_2 - (u + v)v_1 \ ; \tag{c}$$

$$u_t = u_2 + D(u^2 v - 4v) \ ,$$
$$-v_t = v_2 - D(v^2 u - 4u) \ ; \tag{D}$$

$$u_t = u_2 + u_1^2 v_1 - 4v_1 \ ,$$
$$-v_t = v_2 - v_1^2 u_1 + 4u_1 \ ; \tag{d}$$

$$u_t = u_2 - (u + v)^{-2}u_1^2 v_1 - 2(u + v)^{-1}u_1^2 \ ,$$
$$-v_t = v_2 + (u + v)^{-2}v_1^2 u_1 - 2(u + v)^{-1}v_1^2 \ ; \tag{d'}$$

---

[21] Of course, it is complete up to symmetrical and conformal equivalence. We do not include in this list the triangular and decoupled equations.

$$u_t = u_2 + \text{sech}^2(u+v)u_1^2 v_1 - 2\tanh(u+v)u_1^2 \,,$$
$$-v_t = v_2 - \text{sech}^2(u+v)v_1^2 u_1 - 2\tanh(u+v)v_1^2 \,;$$
(d'')

$$u_t = u_2 - 2\tanh(u+v)(u_1^2 - 4) \,,$$
$$-v_t = v_2 - 2\tanh(u+v)(v_1^2 - 4) \,;$$
(e)

$$u_t = u_2 - 2(u+v)^{-1}u_1^2$$
$$\qquad - 4(u+v)^{-2}(2(1+uv)u_1 + (1-u^2)v_1) \,,$$
$$-v_t = v_2 - 2(u+v)^{-1}v_1^2$$
$$\qquad + 4(u+v)^{-2}(2(1+uv)v_1 + (1-v^2)u_1) \,;$$
(f)

$$u_t = u_2 + D(u^2 v) \,,$$
$$-v_t = v_2 - D(v^2 u + u) \,;$$
(G)

$$u_t = u_2 + u_1^2 v_1 \,,$$
$$-v_t = v_2 - v_1^2 u_1 - u_1 \,;$$
(g)

$$u_t = u_2 + D(u^2 - 2uv) \,,$$
$$-v_t = v_2 + D(v^2 - 2uv) \,;$$
(H)

$$u_t = u_2 + u_1^2 - 2u_1 v_1 \,,$$
$$-v_t = v_2 + v_1^2 - 2u_1 v_1 \,;$$
(h)

$$u_t = u_2 - 2(u+v)^{-1}u_1^2 \,,$$
$$-v_t = v_2 - 2(u+v)^{-1}v_1^2 \,;$$
(h')

$$u_t = u_2 - 2\tanh(u+v)u_1^2 \,,$$
$$-v_t = v_2 - 2\tanh(u+v)v_1^2 \,;$$
(h'')

$$u_t = u_2 + D(u^2 v) \,,$$
$$-v_t = v_2 - D(v^2 u) \,;$$
(I)

$$u_t = u_2 + u_1^2 v_1 \,,$$
$$-v_t = v_2 - v_1^2 u_1 \,;$$
(i)

$$u_t = u_2 + \exp(u+v)u_1^2 v_1 + u_1^2 \,,$$
$$-v_t = v_2 - \exp(u+v)v_1^2 u_1 + v_1^2 \,;$$
(i')

$$u_t = u_2 - 2(u+v)^{-1}(u_1^2 + 1) \,,$$
$$-v_t = v_2 - 2(u+v)^{-1}(v_1^2 + 1) \,;$$
(j)

$$u_t = u_2 - 2(u+v)^{-1}u_1^2 - 4(u+v)^{-2}((u-v)u_1 + uv_1) \,,$$
$$-v_t = v_2 - 2(u+v)^{-1}v_1^2 - 4(u+v)^{-2}((u-v)v_1 - vu_1) \,;$$
(k)

$$u_t = u_2 + r(y)u_1^2v_1 + r'(y)u_1^2 - 2(r''(y) - 2c)u_1/3 + r'''(y)/3 ,$$
$$-v_t = v_2 - r(y)v_1^2u_1 + r'(y)v_1^2 + 2(r''(y) - 2c)v_1/3 + r'''(y)/3 , \qquad (1)$$

where

$$y = y(u + v) , \quad y' = r(y) \neq 0 ,$$
$$r(y) = c_4y^4 + c_3y^3 + c_2y^2 + c_1y + c_0 ;$$

$$u_t = u_2 - 2(u + v)^{-1}u_1^2 - 4(u + v)^{-2}(p(u, v)u_1 + r(u)v_1) ,$$
$$-v_t = v_2 - 2(u + v)^{-1}v_1^2 + 4(u + v)^{-2}(p(u, v)v_1 + r(-v)u_1) ; \qquad (m)$$

$$u_t = u_2 - 2(u + v)^{-1}(u_1^2 + r(u)) + r'(u)/2 ,$$
$$-v_t = v_2 - 2(u + v)^{-1}(v_1^2 + r(-v)) + r'(-v)/2 . \qquad (n)$$

[In (n), (m) we denote $r(y) = c_4y^4 + c_3y^3 + c_2y^2 + c_1y + c_0$ and $p(u, v) = 2c_4u^2v^2 + c_3(uv^2 - vu^2) - 2c_2uv + c_1(u - v) + 2c_0$.]

$$u_t = u_2 + \exp(\phi)(u_1^2 + 1)v_1 + (\partial\phi/\partial u)u_1^2 + 2ru_1 ,$$
$$-v_t = v_2 - \exp(\phi)(v_1^2 + 1)u_1 + (\partial\phi/\partial v)v_1^2 - 2rv_1 ; \qquad (o)$$

$$u_t = u_2 + \exp(\phi)(u_1^2 + 1)v_1 + (\partial\phi/\partial u)(u_1^2 + 1) ,$$
$$-v_t = v_2 - \exp(\phi)(v_1^2 + 1)u_1 + (\partial\phi/\partial v)(v_1^2 + 1) . \qquad (p)$$

[In (o), (p): $\exp(\phi) = y(u + v) - y(u - v)$, $r = y(u + v) + y(u - v)$, where $y' \neq 0$, and $(y')^2 = -4y^2 + c_3y^3 + c_2y^2 + c_1y + c_0$ or $(y')^2 = -y^4 + c_3y^3 + c_2y^2 + c_1y + c_0$ respectively.]

$$u_t = u_2 + v_1 ,$$
$$-v_t = v_2 - D(u^2) ; \qquad (Q)$$

$$u_t = u_2 + v_1 ,$$
$$-v_t = v_2 - u_1^2 ; \qquad (q)$$

$$u_t = u_2 + (u + v)^2 ,$$
$$-v_t = v_2 - (u + v)^2 ; \qquad (r)$$

$$u_t = u_2 + (u + v)v_1 - (u + v)^3/6 ,$$
$$-v_t = v_2 - (u + v)u_1 - (u + v)^3/6 ; \qquad (s)$$

$$u_t = u_2 + v_1 ,$$
$$-v_t = v_2 - u_1^2 - (v + u^2/2)u_1 ; \qquad (t)$$

$$u_t = u_2 + 2vv_1 ,$$
$$-v_t = v_2 + 2uu_1 ; \qquad (U1)$$

$$u_t = u_2 + v_1^2 ,$$
$$-v_t = v_2 + u_1^2 , \qquad (u1)$$

$$u_t = u_2 + v_1^2 + b\exp(u+v) - 2c\exp(-2u-2v) \,,$$
$$-v_t = v_2 + u_1^2 + b\exp(u+v) - 2c\exp(-2u-2v) \,;$$
(u2)

$$u_t = u_2 + v_1^2 - (a\exp(-u-v)$$
$$\quad + \omega a_1 \exp(-\omega u - \omega^* v) + \omega^* a_2 \exp(-\omega^* u - \omega v))v_1 \,,$$
$$-v_t = v_2 + u_1^2 + (a\exp(-u-v)$$
$$\quad + \omega^* a_1 \exp(-\omega u - \omega^* v) + \omega a_2 \exp(-\omega^* u - \omega v))u_1 \,;$$
(u3)

$$u_t = u_2 + v_1^2 - 2c\exp(-2u-2v)$$
$$\quad - 2\omega^* c_1 \exp(-2\omega u - 2\omega^* v) - 2\omega c_2 \exp(-2\omega^* u - 2\omega v) \,,$$
$$-v_t = v_2 + u_1^2 - 2c\exp(-2u-2v)$$
$$\quad - 2\omega c_1 \exp(-2\omega u - 2\omega^* v) - 2\omega^* c_2 \exp(-2\omega^* u - 2\omega v) \,;$$
(u4)

$$u_t = u_2 + v_1^2 + b\exp(u+v)$$
$$\quad + \omega^* b_1 \exp(\omega u + \omega^* v) + \omega b_2 \exp(\omega^* u + \omega v) \,,$$
$$-v_t = v_2 + u_1^2 + b\exp(u+v)$$
$$\quad + \omega b_1 \exp(\omega u + \omega^* v) + \omega^* b_2 \exp(\omega^* u + \omega v) \,;$$
(u5)

$$u_t = u_2 + v_1^2 - (a\exp(-u-v)$$
$$\quad + \omega a_1 \exp(-\omega u - \omega^* v) + \omega^* a_2 \exp(-\omega^* u - \omega v))v_1$$
$$\quad - a_1 a_2 \exp(u+v)/6 - \omega^* a a_2 \exp(\omega u + \omega^* v)/6$$
$$\quad - \omega a a_1 \exp(\omega^* u + \omega v)/6 - a^2 \exp(-2u-2v)/6$$
$$\quad - \omega^* a_1^2 \exp(-2\omega u - 2\omega^* v)/6 - \omega a_2^2 \exp(-2\omega^* u - 2\omega v)/6 \,,$$
$$-v_t = v_2 + u_1^2 + (a\exp(-u-v)$$
$$\quad + \omega^* a_1 \exp(-\omega u - \omega^* v) + \omega a_2 \exp(-\omega^* u - \omega v))u_1$$
$$\quad - a_1 a_2 \exp(u+v)/6 - \omega a a_2 \exp(\omega u + \omega^* v)/6$$
$$\quad - \omega^* a a_1 \exp(\omega^* u + \omega v)/6 - a^2 \exp(-2u-2v)/6$$
$$\quad - \omega a_1^2 \exp(-2\omega u - 2\omega^* v)/6 - \omega^* a_2^2 \exp(-2\omega^* u - 2\omega v)/6 \,;$$
(u6)

$$u_t = u_2 - (u+v)^{-1}(u_1^2 + 2u_1 v_1)/2 + a(u+v) \,,$$
$$-v_t = v_2 - (u+v)^{-1}(v_1^2 + 2v_1 u_1)/2 + b(u+v) \,,$$
(v)

$$u_t = u_2 + D(u^2 + v^{-1}) \,,$$
$$-v_t = v_2 - 2D(uv) - 1 \,.$$
(w)

Equations denoted by the same letters with differing numbers of primes are conformal or symmetrically equivalent. For instance, (h) and (h') are symmetrically equivalent but are not related to each other via conformal transformation, (h') and (h'') are conformal equivalent but belong to different symmetrical classes.

4.3.3 Differential Substitutions.
We can relate some equations of the above list via the differential substitutions. As we have already mentioned in Sect. 2,

these substitutions are connected with existence of point group symmetries or conservation laws of zero or first order.

For equations (4.3.6) we define an important class of invertible transformations of the form

$$\bar{u} = Z(\varepsilon u + v, u_1) , \quad \bar{v} = W(\varepsilon u + v) , \tag{4.3.17}$$

where Z, W, are any functions $((\partial Z/\partial u_1) \cdot \partial W/\partial v \neq 0)$. In general, substitution (4.3.17) violates the form of (4.3.6) or the original equation. This form is to be preserved if

$$\partial^2 f/\partial v_1^2 = 0 , \quad 2\partial Z/\partial v = (\partial f/\partial v_1) \cdot \partial Z/\partial u_1 . \tag{4.3.18}$$

The substitutions of the form (4.3.17, 18) allow us to establish some additional relationships between the equations of the list. We shall use the following notations

$$(x) \rightarrow (y) \quad u \rightarrow s(u, u_1) ,$$

which means that $u' = s(u, u_1)$ and u satisfies (x), but u' satisfies (y).

Differential Substitutions

(a) $\rightarrow$ (b)	$u \rightarrow \exp(u)$,	$v \rightarrow \exp(-u)v_1$;
(b) $\rightarrow$ (A)	$u \rightarrow u_1/u$,	$v \rightarrow uv$;
(c) $\rightarrow$ (A)	$u \rightarrow (u+v)/2$,	$v \rightarrow -v_1$;
(a) $\rightarrow$ (c)	$u \rightarrow 2u_1 + v$,	$v \rightarrow -v$;
(d) $\rightarrow$ (e)	$u \rightarrow \mathrm{atanh}(u_1/2) - v$,	$v \rightarrow v$;
(e) $\rightarrow$ (D)	$u \rightarrow 2\tanh(u+v)$,	$v \rightarrow v_1$;
(d'') $\rightarrow$ (f)	$u \rightarrow \tanh(u+v)$,	$v \rightarrow -\tanh(u+v) - 2v_1^{-1}$;
(g) $\rightarrow$ (j)	$u \rightarrow 2u_1^{-1}$,	$v \rightarrow v$;
(j) $\rightarrow$ (G)	$u \rightarrow 2(u+v)^{-1}$,	$v \rightarrow v_1$;
(g) $\rightarrow$ (h'')	$u \rightarrow -iu/2$,	$v \rightarrow iu/2 + \mathrm{atanh}(-iv_1)$;
(h'') $\rightarrow$ (G)	$u \rightarrow 2iu_1$,	$v \rightarrow i\tanh(u+v)$;
(h') $\rightarrow$ (I)	$u \rightarrow 2(u+v)^{-1}$,	$v \rightarrow v_1$;
(I) $\rightarrow$ (H)	$u \rightarrow -uv/2$,	$v \rightarrow -uv/2 - v_1/v$;
(i) $\rightarrow$ (h')	$u \rightarrow 2u_1^{-1} - v_1$,	$v \rightarrow v$;
(i') $\rightarrow$ (k)	$u \rightarrow \exp(u+v)$,	$v \rightarrow -2v_1^{-1} - \exp(u+v)$;
(l) $\rightarrow$ (m)	$u \rightarrow y(u+v)$,	$v \rightarrow -2v_1^{-1} - y(u+v)$;
(q) $\rightarrow$ (r)	$u \rightarrow (2u_1+v)/4$,	$v \rightarrow -v/4$;
(r) $\rightarrow$ (Q)	$u \rightarrow 2(u+v)$,	$v \rightarrow -4v_1$;
(s) $\rightarrow$ (t)	$u \rightarrow -(u+v)$,	$v \rightarrow 2v_1 - (u+v)^2/2$.

The following relations link (v) and (w) with linear and split equations

$$u_t = u_2 + v_1 + (a-b)u/2 ,$$
$$-v_t = v_2 - 2bu_1 + (a-b)v/2 ; \quad \text{(linear)}$$

$$u_t = u_2 + v^{-1}\,,$$
$$\quad\text{(split)}$$
$$-v_t = v_2\,;$$

respectively:

(v) →   (linear) $u \to 2(u+v)^{1/2}\,,\quad v \to -2v_1(u+v)^{-1/2}\,;$

(split) → (w)$u \to u_1/u\,,\quad v \to uv\,.$

The following substitution relates any pair of equations denoted by the same lower and upper case letter, of which the lower case letter equation does not contain v, u explicitly:

(a) ⇒ (A) $\bar{u} = u_1\,,\quad \bar{v} = v_1\,.$ (4.3.19)

A convenient graphical representation of the above substitutions is as follows: equations we shall denote by open blocks □ (black blocks ■ correspond to symmetrical equations) and the substitutions of variables of the form (4.3.17, 18) by arrows —→ [double arrows ⇒ correspond to substitutions (4.3.19)]. As a result we get:

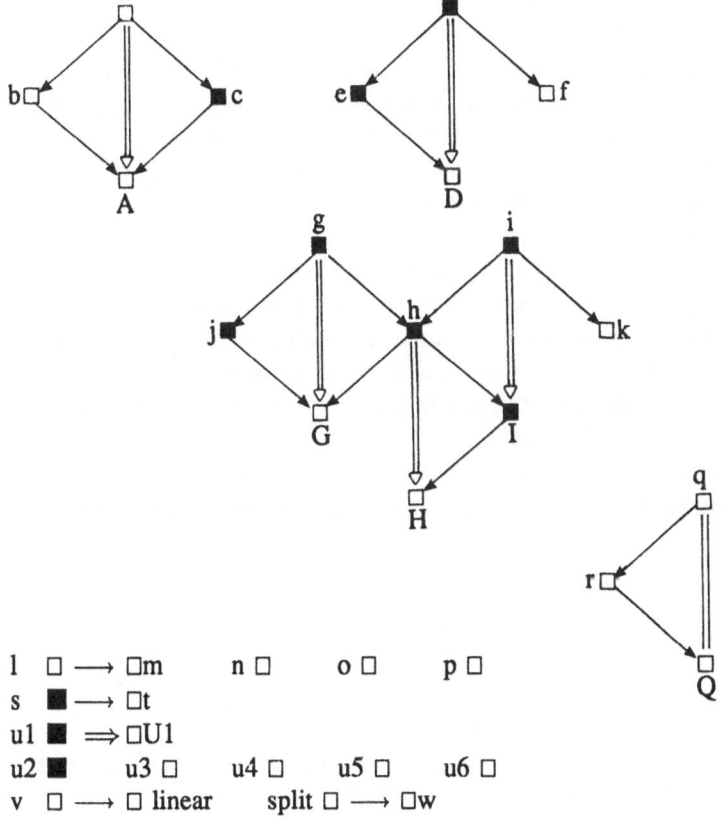

l □ —→ □m   n □   o □   p □
s ■ —→ □t
u1 ■ ⇒ □U1
u2 ■   u3 □   u4 □   u5 □   u6 □
v □ —→ □ linear   split □ —→ □w

We notice in conclusion that is is just enough to study one of the equations from each connected graph: for the other equations of this graph all results can be reproduced via the above-given differential substitutions.

Historical Remarks

The concept of the symmetry approach based on explicit integrability conditions was formulated at the first Kiev conference (September 1979) by one of the authors (A. B. S.)[22] Somewhat earlier, in the same year, these ideas had been applied to the problem of classification of the nonlinear Klein-Gordon equations (1.1.9) [1]. The symmetry approach was also discussed by *Fokas* [38] (1980), where, in particular, all equations of the form $u_t = u_3 + f(u, u_1)$ possessing one nonclassical symmetry of the fixed order were found.

The first publication on the formal symmetries and explicit integrability conditions for evolution type equations is [30] (1980). In Ref. 25 it has been shown that the existence of two higher local conservation laws implies that of a formal symmetry and integrable equations of the form

$$u_t = u_3 + f(u, u_1, u_2) .$$

The symmetry approach was extended by *Yamilov* [39] (1982) to involve nonlinear lattice equations in the area. A complete list of integrable equations of the form $u_{nt} = f(u_{n-1}, u_n, u_{n+1})$ has been obtained in [1] (1983). The classical theory of contact transformations has been attached to the symmetry approach in the paper of *Svinolupov* [17] (1985) devoted to the problem of the classification of the Burgers type equations. The generalization of the scheme on the vector case has been given in [10] (1985). After that the exhaustive list of the nonlinear integrable Schrödinger type equations was obtained [35] (1986), [34] (1988). A sharpening of the concept of formal symmetry, an algorithmic way to compute explicit integrability conditions, development of the classical theory of transformations and a number of other original results were included in review articles [27] (1984), [8] (1987), [21] (1988).

Many interesting results concerning the symmetry properties of integrable equations and other points of view one can find in [2, 6, 11, 19, 26, 33, 39–60].

---

[22] A very similar approach had been proposed [36]. However, we have not found references where the proper development of this promising paper is obtained.

References

1   A. V. Zhiber, A. B. Shabat: Dokl. Acad. Nauk. SSSR **247**, 1104–1107 (1979) [in Russian]
2   L. A. Takhtajan, L. D. Faddeev: *Hamiltonian Approach to Soliton Theory* (Nauka, Moscow (1986) [in Russian]
3   P. D. Lax: Comm. Pure Appl. Math. **28**, 141–188 (1975)
4   D. Levi: J. Phys. A **14**, 1083–1098 (1981)
5   A. B. Shabat, R. I. Yamilov: Phys. Lett. A **130**, 271–275 (1988)
6   A. Newell: *Solutions in Mathematics and Physics* (SIAM, Philadelphia 1985)
7   V. V. Sokolov: Dokl. Acad. Nauk. SSSR **294**, 1065–1068 (1987)
8   A. V. Mikhailov, A. B. Shabat, R. I. Yamilov: Usp. Mat. Nauk **42**, 3–53 (1987) [in Russian]
9   G. Rosen: Phys. Rev. B **19**, 2392–2399 (1979)
10 A. V. Mikhailov, A. B. Shabat: Teor. Mat. Fiz. **62**, 163–158 (1985) [in Russian]
11 F. Kh. Muckminov: Teor. Mat. Fiz. (1988) [in Russian]
12 J. D. Cole: Quart. Appl. Math. **2**, 225–236 (1951)
13 E. Hopf: Comm. Pure Appl. Math. **3**, 201–230 (1965)
14 A. V. Samokhin: Dokl. Acad. Nauk. SSSR **263**, 274–280 (1982)
15 A. E. Borovik, V. Yu. Popkov, V. N. Robuk: Dokl. Acad. Nauk. SSSR (in press 1988)
16 S. I. Svinolupov: Usp. Mat. Nauk **40**, 263–264 (1985)
17 S. I. Svinolupov: Teor. Mat. Fiz. **65**, 303–307 (1985)
18 S. I. Svinolupov: Phys. Lett. A (in press) (Preprint, BFAN SSSR, Ufa 1987)
19 F. Kh. Muckminov, V. V. Sokolov: Mat. Sbor. **133**, 392–414 (1987) [in Russian]
20 S. I. Svinolupov: "Evolution equations possessing higher symmetries"; Ph. D. Thesis, BFAN SSSR Ufa (1985) [in Russian]
21 V. V. Sokolov: Usp. Mat. Nauk (in press 1988) [in Russian]
22 G. Bluman, S. Kumai: J. Math. Phys. **21**, 1019–1023 (1980)
23 V. G. Drinfeld, V. V. Sokolov: Dokl. Acad. Nauk. SSSR **284**, 29–33 (1985) [in Russian]
24 V. V. Sokolov: Funk. analiz **21**, 45–54 (1988) [in Russian]
25 S. I. Svinolupov, V. V. Sokolov: Funk. Analiz **16**, 86–87 (1982) [in Russian]
26 V. G. Drinfeld, V. V. Sokolov: "Lie algebras and the Korteveg de-Vries type equations", (in *Itogi Nauki i Tekhniki (Ser. Sovremennye problemy matematiki. Noveiishye dostizheniya* VINITI, Moskva) **24**, 81–180 (1984) [in Russian]
27 V. V. Sokolov, A. B. Shabat: "Classification of integrable evolution equations" in Soviet Scientific Reviews, Section C, **4**, 221–280 (1984)
28 D. Mumford: *Tata Lectures on Theta I, II* (Birkhauser, Boston 1983, 1984)
29 M. Adler: Inventiones Math. **50**, 219–248 (1979)
30 N. Kh. Ibragimov, A. B. Shabat: Funk. analiz **14**, 79–80 (1980) [in Russian]
31 V. G. Drinfeld, S. I. Svinolupov, V. V. Sokolov: Dokl. Acad. Nauk. UkSSR A **10**, 7–10 (1985) [in Russian]
32 A. Fujimoto, Y. Watanabe: Math. Jap. **28**, 43–65 (1983)
33 H. D. Wahlquist, F. B. Estabrook: J. Math. Phys. **17**, 1293–1297 (1976)
34 A. V. Mikhailov, A. B. Shabat, R. I. Yamilov: Comm. Math. Phys. **115**, 1–19 (1988)
35 A. V. Mikhailov, A. B. Shabat: Teor. Mat. Fiz. **66**, 47–65 (1986) [in Russian]
36 H. H. Chen, Y. C. Lee, C. S. Liu: Physica Scripta **20**, 490–492 (1979)
37 A. S. Fokas: J. Math. Phys. **21**, 1318–1325 (1980)
38 R. I. Yamilov: Usp. Mat. Nauk **38**, 155–156 (1983)
39 A. B. Shabat (ed.): *Integrable Systems* (BFAN, Ufa 1982)
40 A. B. Shabat, R. I. Yamilov: "Exponential Systems of the Type I and Cartan Matrices". (Preprint BFAN Ufa 1981)
41 L. Abellanas, A. Galindo: J. Math. Phys. **24**, 504–509 (1983)
42 F. Calogero, A. Degasperis: *Spectral Transform and Solitons* (North-Holland, Amsterdam 1982) pp. 58–59
43 V. N. Chetverikov: Dokl. Acad. Nauk. SSSR **286**, 54–57 (1986)
44 R. K. Dodd, A. P. Fordy: J. Phys. A **17**, 3249–3266 (1984)
45 R. K. Dodd, R. K. Bullough: Proc. R. Soc. Lond. A **352**, 481–503 (1977)
46 A. S. Fokas: Stad. Appl. Math. **77**, 253 (1987)

47   J. Hietarinta: J. Math. Phys., in press (Preprint, University of Turku, TURKU-FTL-R102, R115, 1986)
48   N. Kh. Ibragimov, A. B. Shabat: Funk. analiz 14, 25–36 (1980) [in Russian]
49   P. P. Kulish: Teor. Mat. Phys. 26, 198–205 (1976)
50   A. N. Leznov: Teor. Mat. Fiz. 42, 343–349 (1980)
51   A. N. Leznov, V. G. Smirnov, A. B. Shabat: Teor. Mat. Fiz. 51, 10–22 (1982)
52   V. V. Lychagin: Usp. Mat. Nauk 34 137–165 (1979)
53   A. G. Meshkov, B. B. Mikhalyaev: Teor. Mat. Fiz. 72, 163–171 (1987) [in Russian]
54   A. V. Mikhailov: Physica 3D, 73–117 (1981)
55   A. V. Mikhailov, A. B. Shabat: Phys. Lett. A116, 191–194 (1986)
56   S. I. Svinolupov, V. V. Sokolov, R. I. Yamilov: Dokl. Acad. Nauk. SSSR 271, 802–805 (1983) [in Russian]
57   A. M. Vinogradov, I. S. Krasilschik: Dokl. Acad. Nauk. SSSR 253, 1289–1293 (1980)
58   H. D. Wahlquist, F. B. Estabrook: J. Math. Phys. 16, 1–7 (1975)
59   R. I. Yamilov: "On the discrete equations of the form $du_n/dt = F(u_{n-1}, u_n, u_{n+1})$ possessing infinitely many local conservation laws"; Ph. D. Thesis, BFAN SSSR Ufa (1985) [in Russian]
60   A. V. Zhiber, A. B. Shabat: Dokl. Acad. Nauk. SSSR 277, 29–33 (1984) [in Russian]

Integrability of Nonlinear Systems and Perturbation Theory

V.E. Zakharov and E.I. Schulman

1. Introduction

The theory of so-called integrable Hamiltonian wave systems arose as a result of the inverse scattering method discovery by *Gardner, Green, Kruskal* and *Miura* [1] for the Korteveg–de Vries equation. This discovery was initiated by the pioneering numerical experiment by *Kruskal* and *Zabusky* [2]. After a pragmatic phase, which was devoted to finding new soliton equations, the theory became rather complicated. One of its branches may be called the "qualitative theory of infinite-dimensional Hamiltonian systems", to which the results reviewed in this paper belong. We consider only Hamiltonian systems possessing Hamiltonians with a quadratic part which may be transformed in normal variables to the form

$$H_0 = \sum_{\alpha=1}^{N} \int \omega_k^{(\alpha)} a_k^{(\alpha)} a_k^{*(\alpha)} dk \ . \tag{1.1.1}$$

Here, $a_k^{(\alpha)}$ are normal coordinates of the α-th linear mode (usually simply expressed through Fourier components of physical fields): $k = (k_1, \ldots, k_d)$ is the wave vector; d is the dimension of space; and $\omega_k^{(\alpha)}$ is the dispersion law of the α-th mode. Corresponding Hamiltonian systems, i.e., those having Hamiltonians of the form

$$H = H_0 + H_{\text{int}} \ , \tag{1.1.2}$$

are called "Hamiltonian wave systems". The majority of nonlinear wave theory problems may be mapped into this class. The crucial property of systems (1.1.2) is that they make a weak nonlinear approximation possible. Our approach is based on treating H_{int} as a perturbation; besides, we assume that H_{int} is an analytic functional of the fields $a_k^{(\alpha)}$. This is not very limiting since it is usually true at the weak nonlinear approximation.

The qualitative theory of infinite-dimensional Hamiltonian systems now being developed stems from the qualitative theory of ordinary differential equations; all existing methods can somehow be linked with this theory. The approach used in the papers by *Newell, Tabor* and by *Siggia*, and *Ercolani*, presented in this volume, actually originates from the analytic theory of ordinary differential equations, while Makhailov, Sokolov and Shabat's method can be traced to the Sophus Lie symmetry theory. Our own work stems from Poincaré's proof of the nonexistence of the invariants of motion, analytic in a small parameter, and from Birkhoff's results on the canonical transformations of Hamiltonians, to the normal form near the equilibrium. The main theorem in Sect. 2.2 is an infinite-dimensional generalization of the well-known *Poincaré* theorem [3] which determines the sufficient conditions for the nonexistence of an additional motion invariant; the theorem in Sect. 2.6 should be considered as a theorem which in analogy with Birkhoff's result determines the conditions for a Hamiltonian wave system to be reducible to the form of the Birkhoff's infinite-dimensional integrable chain. In the infinite-dimensional case a new notion arises, which is absent in the finite-dimensional case: the degenerative dispersion laws.

The Painlevé test method is based on the study of solution singularities and works effectively both in one-dimensional and in multidimensional cases. It may be used to determine whether a given equation is solvable exactly. If the equation satisfies the test, a Lax representation may be found for it. The "Lie-Bäcklund symmetry approach" is used for one-dimensional systems with functional freedom: it permits conclusions about the existence or absence of additional *local* motion invariants and symmetries, thus making possible a choice of "good" equations among those of a given functional form. This method is, however, inappropriate for finding L-A pairs.

Our approach does not permit functional arbitrariness in an equation but effectively proves the nonexistence of additional motion invariants analytic in $a_k^{(\alpha)}$ independent of its locality or nonlocality and the dimensionality. For reasons which will be explained below, this method is often simpler in multidimensional spaces.

An approach based on perturbation theory has another important advantage. It concerns the definition of the content of the concept "integrable equations". It leads to a natural subdivision into two classes of all systems of the form (1.1.2) with additional integrals: i) exactly solvable but not integrable in Liouville's sense and ii) exactly solvable and completely integrable. For example, the Kadomtsev-Petviashvili (KP) equation

$$(u_t + 6uu_x + u_{xxx})_x = 3\alpha^2 u_{yy} \tag{1.1.3}$$

with $\alpha^2 = 1$, belongs to the first class, while this equation with $\alpha^2 = -1$ and the well-known Davey-Stewartson equation (DS),

$$i\Psi_t + \Psi_{yy} - \Psi_{xx} + \Phi\Psi = 0$$
$$\Phi_{xx} + \Phi_{yy} = \left(\frac{\partial^2}{\partial x^2} - \frac{\partial^2}{\partial y^2} \right) |\Psi|^2 , \tag{1.1.4}$$

belong to the second class [4–7].

This method of classification, properties of the equations from the first and the second classes, interrelations between solvability (existence of commutation representation and infinite number of conservation laws) and complete integrability (introduction of virtual action-angle variables which do not disappear at periodic boundary conditions) are considered in Sect. 2.6 for the general case of periodic boundary conditions.

Besides the above-mentioned direct methods, other approaches may be effectively applied in some cases: the Walquist-Estabrooq method of finding L-A pairs, the method of searching for alternative commutation representation (when a linear operator defining time dynamics arises as the Gateux derivative of the original equation; see *Chen, Lie, Lin* [8]), etc.

Our paper is organized as follows: Chap. 2 is self-contained; it is devoted to the description of the general theory in the case of zero boundary conditions at infinity with the exception of Sect. 2.6, in which periodic boundary conditions are explored. Chapter 3 contains some information about the physics giving rise to various universal, exactly solvable equations (Sect. 3.1) and their properties from the viewpoint of the general theory (Sects. 3.2, 3); it also offers examples of verification of the integrability of some particular systems (Sects. 3.4, 5). The appendices contain proofs of the most important theorems.

2. General Theory

2.1 The Formal Classical Scattering Matrix in the Solitonless Sector of Rapidly Decreasing Initial Conditions [6]

Consider a homogeneous medium of d dimensions, where the waves of N types can propagate, and their dispersion laws are $\omega_k^{(\alpha)}$, $\alpha = 1, \ldots, N$. The Hamiltonian of such a medium can be represented in the form (1.2) (see Sect. 3.1), with H_0 of the form (1.1.1) and H_{int} practically always being the functional series in the complex normal coordinates $a_k^{(\alpha)}$, $a_k^{*(\alpha)}$, $\alpha = 1, \ldots, N$. The $a_k^{(\alpha)}$ indicate the wave amplitudes for corresponding linear modes with wave vector k. Amplitudes $a_k^{(\alpha)}$ obey the equations

$$i\dot{a}_k^{(\alpha)} = \omega_k^{(\alpha)} a_k^{(\alpha)} + \frac{\delta H_{\text{int}}}{\delta a_k^{*(\alpha)}} . \tag{2.1.1}$$

In analogy with the quantum scattering theory, let us consider the system with interaction, adiabatically decreasing as $t \to \pm\infty$:

$$H = H_0 + H_{\text{int}} e^{-\epsilon|t|} . \tag{2.1.2}$$

For the system (2.1.1), the global solvability theorem may not be fulfilled, and asymptotic states may not exist as $t \to \pm\infty$. However, for the system with the Hamiltonian (2.1.2) at finite and sufficiently small a_k, they do exist, i.e., the solution of (2.1.1) turns asymptotically into the solution of the linear equation:

$$a_k^{(\alpha)}(t) \longrightarrow \left[a_k^{(\alpha)}(t)\right]^{\pm} = \left[C_k^{(\alpha)}\right]^{\pm} e^{-i\omega_k^{(\alpha)}t} . \tag{2.1.3}$$

Furthermore, asymptotic states may contain solitons, which certainly cannot exist at finite ε. So our consideration should be restricted to the class of initial states without solitons and with smooth C_k^-. We shall call this class the solitonless sector. Although our consideration is restricted to a special class of initial states, the result will be very useful because the structure obtained for the formal series for the S-matrix provides us with the structure of motion invariants (Sect. 2.5).

Functions $C_k^{(\alpha)\pm}$ are not independent; there is a nonlinear operator $S_\varepsilon^{(\alpha)}[C^-]$, transforming them into each other. To study this operator we go as usual to the interaction representation:

$$a_k^{(\alpha),s}(t) = b_k^{(\alpha),s}(t)e^{-is\omega_k^{(\alpha)}t} . \tag{2.1.4}$$

Here, $s = \pm 1$, $a_k^1(t) = a_k(t)$, $a_k^{-1}(t) = a_k^*(t)$. The motion equations now take the form

$$sib_k^{j(\alpha),s} = \frac{\delta H_{\text{int}}}{\delta b_k^{(\alpha),-s}} . \tag{2.1.5}$$

In (2.1.5), H_{int} is the interaction Hamiltonian expressed in the variables b_k^s. Equation (2.1.5) is equivalent to an integral equation:

$$b_k^{(\alpha),s} = \left[C_k^{(\alpha),s}\right]^- - \frac{is}{2}\int_{-\infty}^t dt_1 \frac{\delta H_{\text{int}}}{\delta b_k^{(\alpha),-s}(t_1)}e^{-\varepsilon|t_1|} . \tag{2.1.6}$$

Equation (2.1.6) gives a map $C_k^{(\alpha),-} \to b_k^{(\alpha),s}(t)$ which may be written in the form

$$b_k^{(\alpha),s} = S_\varepsilon^{(\alpha),s}(-\infty,t)\left[\{C_k^-\}\right] . \tag{2.1.7}$$

As $t \to +\infty$ in (2.1.7), one finds

$$C_k^{(\alpha),+} = S_\varepsilon\left[\{C_k^-\}\right] , \tag{2.1.8}$$

where $S_\varepsilon^{(\alpha)} = S_\varepsilon^{(\alpha)}(-\infty,\infty)$.

At finite ε and sufficiently small $a_k^{(\alpha)}$, operators $S_\varepsilon^{(\alpha)}(-\infty,\infty)$ and $S_\varepsilon^{(\alpha)}$ may be obtained in the form of a convergent series by iterations of (2.1.6). Let $\varepsilon \to 0$ now in each term of the series. As we shall see, the expression obtained is finite in the sense of generalized functions. The series obtained for the operator $S_\varepsilon^{(\alpha)}(-\infty,t)$ as $\varepsilon \to 0$ will be called the classical transition matrix. We shall refer to the corresponding series for S as the formal classical scattering matrix. Let us designate

$$S^{(\alpha)}(-\infty,t) = \lim_{\varepsilon \to 0} S_\varepsilon^{(\alpha)}(-\infty,t) \tag{2.1.9}$$

$$S^{(\alpha)} = \lim_{\varepsilon \to 0} S_\varepsilon^{(\alpha)}(-\infty,\infty) ,$$

where the limits are to be understood in the above-mentioned sense.

Before proceeding, let us introduce a more convenient notation. For the function $\Pi_{k_1 \ldots k_n}^{\pm s_1 \ldots \pm s_n}$ we will write simply $\Pi_{\pm 1, \ldots, \pm n}$. Moreover, we will deisgnate

$$\Pi_{\pm 1, \ldots, \pm n} \delta(\pm s_1 k_1 \pm \ldots \pm s_n K_n) = \hat{\Pi}_{\pm 1, \ldots, \pm n}$$

and

$$\hat{\Pi}_{\pm 1, \ldots, \pm n} \delta(\pm s_1 \omega_{k_1} \pm \ldots \pm s_n \omega_{k_n}) = \hat{\hat{\Pi}}_{\pm 1, \ldots, \pm n} .$$

This notation reduces the length of the formulae and makes their structure visible. In addition we will use the special notation

$$
\begin{aligned}
E_{\pm 1, \ldots, \pm n} &= \pm s_1 \omega_{k_1} \pm \ldots \pm s_n \omega_{k_n} \\
P_{\pm 1, \ldots, \pm n} &= \pm s_1 k_1 \pm \ldots \pm s_n k_m .
\end{aligned}
\tag{2.1.10}
$$

As $\varepsilon \to 0$, the series for $S_\varepsilon(-\infty, t)$ and for S_ε are generally speaking divergent and formal. Consider the structure of the classical scattering matrix in the simplest case of a cubic interaction Hamiltonian H_{int} and only one mode:

$$
\begin{aligned}
H_{\text{int}} = \frac{1}{3!} \sum_{s s_1 s_2} \int V_{k k_1 k_2}^{s s_1 s_2} a_k^s a_{k_1}^{s_1} a_{k_2}^{s_2} \\
\times \delta(sk + s_1 k_1 + s_2 k_2) dk\, dk_1 dk_2 .
\end{aligned}
\tag{2.1.11}
$$

From the fact that the Hamiltonian is real, it follows that

$$V_{*k k_1 k_2}^{-s-s_1-s_2} = V_{k k_1 k_2}^{s s_1 s_2} .
\tag{2.1.12}$$

Besides, coefficient functions V possess an evident symmetry,

$$V_{k k_1 k_2}^{s s_1 s_2} = V_{k k_2 k_1}^{s s_2 s_1} = {}_{k_1 k k_2}^{s_1 s s_2} .
\tag{2.1.13}$$

In the interaction representation, we have the integral equation

$$
\begin{aligned}
is\left(b_k^s(t) - c_k^s\right) = \frac{1}{2} \sum_{s_1 s_2} \int_{-\infty}^t dt_1 \int dk_1 dk_2 V_{k k_1 k_2}^{-s s_1 s_2}(t_1) \\
\times b_{k_1}^{\delta_1}(t_1) b_{k_2}^{\delta_2}(t_2) \delta(-sk - s_1 k_1 + s_2 k_2)
\end{aligned}
\tag{2.1.14}
$$

$$V_{k k_1 k_2}^{s s_1 s_2}(t) = V_{k k_1 k_2}^{s s_1 s_2} \exp\left(i E_{k k_1 k_2}^{s s_1 s_2} t - \varepsilon|t|\right)
\tag{2.1.15}$$

$$E_{k k_1 k_2}^{s s_1 s_2} = s\omega_k + s_1 \omega_{k_1} + s_2 \omega_{k_2} .
\tag{2.1.16}$$

Equation (2.1.14) may be symbolically represented in graphical form:

$$s =\!=\!=\!=\!= \, s -\cdot-\cdot- \; -\frac{i}{2} \; ----0 \overset{=\,=\,=\,=}{\underset{=\,=\,=\,=}{}} \; ,
\tag{2.1.17}$$

where $=\!=\!=\!=\!=$ indicates the two-component over the index s unknown value b_k^s, $s = \pm 1$; $-\cdot-\cdot-$ designates c_k^{-s}; $----$ corresponds to the factor $\exp\{-i E_{k k_1 k_2}^{-s s_1 s_2}\}$; 0 indicates $V_{k k_1 k_2}^{-s s_1 s_2} \delta(-sk + s_1 k_1 + s_2 k_2)$, and summation is assumed over s_1 and s_2. Using (2.1.17), certain graphical expressions (diagrams) may be attributed to each term of the series arising when iterating (2.1.14). These graphical expressions are connected graphs, having no loops; they are, in other words, "trees".

Each graph consists of two types of elements: lines and vertices; the former are subdivided into inner and external lines. One of the external lines is different from the others (we shall call it a "root"); the other ones may be called "leaves". Each tree, corresponding to the n-th iteration, contains exactly n vertices and $n+2$ leaves. Inner lines are usually called "branches". They correspond to both the external and internal lines, a certain value of wave vector k_i and the index s_i. The "external" value of k and s corresponds to the root. Integration goes over all k_i except $k_i = k$; the summation goes over all s_i except $s_i = s$. To each leaf with the wave vector k_q and index s_q corresponds a factor $c_{k_q}^{-s_q}$.

The graph corresponding to the N-th iteration contains N integrations over time variables $t_1, \ldots, t_N$. Each time variable t_i in the diagram for the transition matrix corresponds to its own branch. The external time t corresponds to the root. The presence of the root leads to partial ordering of the graph elements. From each vertex in which three lines meet there is a unique path to the root. The line leading to the root we shall designate as the exiting line. Let the corresponding wave vector and index be k_α and s_α. The other two lines are entering. Let them correspond to the wave vector k_β, k_γ and indices s_β, s_γ. It is important that both entering lines correspond with one and the same time variable t_q. Corresponding to this vertex factor is

$$V_{k_\alpha k_\beta k_\gamma}^{-s_\alpha s_\beta s_\gamma} \exp\left[iE_{k_\alpha k_\beta k_\gamma}^{-s_\alpha s_\beta s_\gamma} - \varepsilon|t_q|\right] \delta\left(-s_\alpha k_\alpha + s_\beta k_\beta + s_\gamma k_\gamma\right) . \qquad (2.1.18)$$

Let us cut the graph across the line exiting from the vertex. Now that part of the graph which is cut off from the root is to be integrated over the variable t_q in the limits $-\infty < t_q \leq t_p$. In fact this method of ordering is equivalent to the chronological ordering used in quantum field theory.

To conclude our description of the diagram technique let us note that the set of digarams which correspond to the n-th iteration consists of all possible trees containing n-vertices and fixed roots. In front of each diagram there is a numerical factor i/p. The number p is equal to the number of the symmetry group elements for the diagram under consideration, i.e. the number of rotations at different vertices which leaves the diagram unchanged, identity transformation included.

At finite $\varepsilon > 0$, the actual calculation of diagrams is a rather difficult task. However, it becomes much simpler as $\varepsilon \to 0$. We shall refer to integration over the time variable t_1 closest to the root as outer integration; all the other integrations will be called inner integrations. It is important that when integrating over any inner variable t_q, one may make the replacement

$$e^{-\varepsilon|t_q|} \to e^{\varepsilon t_q} . \qquad (2.1.19)$$

We shall not prove this statement here. The analogous statement has been proved in the quantum field theory (see [10], for example). It is important to notice that using (2.1.19), all the integrations over inner times may be carried out explicitly, greatly simplifying the diagram technique.

Consider an inner branch with the wave vector k_p and the index s_p such that when cutting it, we may separate a tree having m leaves ($m \geq 2$) from the root. Let these leaves have wave vectors k_i and indices s_i, $i = 1, \ldots, m$. Let the vertex, from which this tree grows, be entered from the other sides by lines (branches or leaves) with the wave vectors and indices k_q, k_r and s_q, s_r. Then the expression corresponding to this vertex is as follows (the line with k_q, s_q is the exiting line):

$$V_{k_q k_r k_p}^{-s_q s_r s_p} \delta \left(-s_q k_q + s_r k_r + s_p k_p \right) , \tag{2.1.20}$$

while the expression corresponding to the branch with the wave vector k_p and the index s_p is

$$G_m = \lim_{\varepsilon \to 0} \frac{\exp(iE_m t + m\varepsilon t)}{i(E_m - im\varepsilon)} = \frac{\exp(iE_m t)}{i(E_m - i0)} . \tag{2.1.21}$$

Consider now the last (outer) integration over t_1. We have

$$S_{N\varepsilon}(-\infty, t) = W_N \int_{-\infty}^{t} \exp \left[-\varepsilon|t_1| + iE_N t_1 \right] dt_1 . \tag{2.1.22}$$

Here,

$$W_N = W_{k, k_1 \ldots k_N}^{-s, s_1 \ldots s_N} \delta \left(-sk + s_1 k_1 + \ldots + s_N k_N \right) \tag{2.1.23}$$

is some expression which tends to the constant in the limit $\varepsilon \to 0$. At finite t we have, from (2.1.22),

$$S_N(-\infty, t) = \lim_{\varepsilon \to 0} S_{N\varepsilon}(-\infty, t) = \frac{W_N e^{iE_N t}}{i(E_N - i0)} ; \tag{2.1.24}$$

as $t \to +\infty$, we have

$$S_N = \lim_{\varepsilon \to 0} S_{N\varepsilon}(-\infty, \infty) = 2\pi \delta(E_N) W_N . \tag{2.1.25}$$

So the expressions for the $S_N(-\infty, t)$ and S_N have the singularity on a manifold defined by the equations

$$P_N = -sk + s_1 k_1 + \ldots + s_N k_N = 0$$
$$E_N = -s\omega_k + s_1 \omega_{k_1} + \ldots + s_k \omega_{k_N} = 0 . \tag{2.1.26}$$

Equation (2.1.26), depending on the choice of the $s, s_1, \ldots, s_N$, splits into a set of relations:

$$k + k_1 + \ldots + k_n = k_{n+1} + \ldots + k_{n+m} \tag{2.1.27}$$

$$\omega_k + \omega_{k_1} + \ldots + \omega_{k_n} = \omega_{k_{n+1}} + \ldots + \omega_{k_{n+m}} .$$

Equation (2.1.27) determines a manifold which we shall call the resonant manifold $\Gamma^{n+1,m}$. We designate the corresponding entity W_N via

$$W^{n+1,m}_{k,k_1,\dots,k_{n+1},\dots,k_{n+m}} = W^{n+1,m} .$$

It is important to notice that W_N is regular on the manifold (2.1.27) in the points of a general position. However it has singularities on the submanifolds of lower dimension on which at least one of the entities E_m becomes zero, which corresponds to one of the inner lines of any diagram constituting the $W^{n+1,m}$. As can be seen from (2.1.21), these singularities may be of two types, in agreement with the two terms in (2.1.21). The first item in (2.1.21) is distributed over all of $\Gamma^{n+1,m}$, while the second one is localized on a manifold (to be more precise, on a set of manifolds):

$$-s_p\omega_{k_p} + s_1\omega_{k_1} + \dots + s_m\omega_{k_m} = 0 \tag{2.1.28}$$

$$-s_p k_p + s_1 k_1 + \dots + s_m k_m = 0 .$$

Manifolds (2.1.28) may be considered the youngest resonant manifolds in comparison with (2.1.27). Equations (2.1.28) together with (2.1.27) determine a set of submanifolds of $\Gamma^{n+1,m}$ having the codimension unity. The division of two items in (2.1.21) has a certain physical meaning. One may say that the first item describes processes which go via virtual waves while the second item describes processes going via real intermediate particles. The elements of a classical S-matrix with interactions going via real waves may be called singular. They decompose on the singularity powers, depending on the number of inner lines in which the Green function G_m denominator becomes zero and on the correspondent codimension of the younger resonant manifold. For any concrete dispersion law there is an element of the scattering matrix possessing maximal singularity.

Let us now set some additional symmetry property of the amplitudes of the classical scattering matrix, i.e., let us consider the equation

$$is\dot{a}^s_k = \omega_k a^s_k + \frac{\delta H_{int^*}}{\delta a^{-s}_k} , \tag{2.1.29}$$

where H_{int^*} may be obtained from H_{int} in (1.1.2) by the substitution of complex conjugated Hamiltonian coefficients, for example, into (2.1.6): $V^{ss_1s_2}_{kk_1k_2} \rightarrow V^{-s-s_1-s_2}_{kk_1k_2}$. As before, we shall assume the interaction to be the adiabatically switched on and off. Then as $t \rightarrow \pm\infty$, the solutions of (2.1.29) and of (1.1.2) as well will degenerate into those of the linear equation.

Let us consider the solution of (2.1.29), which becomes $C^{*+}_k \exp(-i\omega_k t)$ as $t \rightarrow -\infty$:

$$a_k \rightarrow C^-_{*k} e^{-i\omega_k t} = C^{*+}_k e^{-i\omega_k t} .$$

As in (1.1.2), (2.1.29) possesses a classical scattering matrix, $C^+_{*k} = S_*[C^-_{*k}]$. One should note here that (2.1.29) is derived from (1.1.2) by complex conjugation and change of the time sign. So, on account of the unique solution of the Cauchy problem for (1.1.2) and also for (2.1.29), $S_*[C_k^{*+}] = C_k^{*-}$.

Substituting the definition of the classical scattering matrix (2.1.8), we get

$$S_* \left[S^* \left[C_k^- \right] \right] = C_k^{*-} . \tag{2.1.30}$$

Identity (2.1.30) is analogous to the unitarity condition for the scattering matrix in quantum mechanics.

Nonlinear operator S_* can be easily calculated. It coincides with the operator S, where the Hamiltonian coefficient function V is substituted for the complex conjugated in each vertex of a diagram. It is convenient for us to introduce operator R by the following formula:

$$S = 1 + R . \tag{2.1.31}$$

Then from (2.1.30) we obtain the following condition for R:

$$R_* \left[C_k^{*-} \right] + R^* \left[C_k^- \right] + R_* \left[R^+ \left[C_k^- \right] \right] = 0 . \tag{2.1.32}$$

One may also simply verify that

$$W_{m,n+1}^* = -\frac{m}{n+1} W_{n+1,m} . \tag{2.1.33}$$

It follows from (2.1.33) in particular that the amplitude $W_{m,n}$ is asymmetric relative to the permutation of the m-indices, so that the diagram root does not really occur as a marked line. From physical considerations it is clear that the classical scattering matrix we have constructed coincides with the quantum one, were radiation corrections are not taken into account, and only diagrams of the "tree type" are retained.

Formulae for the case of many modes can be obtained from those above by ascribing mode numbers α, $\alpha = 1, \dots, N$, to the field variables, coefficient functions V of the Hamiltonian and other objects. We will do so in what follows without further explanation.

2.2 Infinite-Dimensional Generalization of Poincaré's Theorem. Definition of Degenerative Dispersion Laws [4, 5, 6]

The classical scattering matrix introduced in Sect. 2.1 may be used to understand what restrictions should be imposed on the Hamiltonian system in order for additional motion invariants to exist. Indeed, let the system (2.1.1) have a Hamiltonian

$$H_i = \frac{1}{3!} \sum_{s_1 s_2 s_3} \int \hat{V}_{k_1 k_2 k_3}^{(\alpha_1), s_1(\alpha_2), s_2(\alpha_3), s_3} a_1 a_2 a_3 dk_1 dk_2 dk_3 . \tag{2.2.1}$$

The cubic term in

$$I[a] = I_0 + \ldots = \sum_{\alpha_j} f_{\boldsymbol{k}}^{\alpha_j} \left| a_{\boldsymbol{k}}^{\alpha_j} \right|^2 dk + \ldots \tag{2.2.2}$$

is

$$I_1 = \frac{1}{3!} \int \hat{I}_{123} a_1 a_2 a_3 dk_1 dk_2 dk_3 .$$

Using the condition $dI/dt = 0$ and motion equations (2.1.1), we find, after collecting terms cubic in a_k:

$$E_{123} I_{123} = \hat{V}_{123} F_{123} , \tag{2.2.3}$$

where

$$F_{123} = s_1 f_{\boldsymbol{k}_1}^{\alpha_1} + s_2 f_{\boldsymbol{k}_2}^{\alpha_2} + s_3 f_{\boldsymbol{k}_3}^{\alpha_3} . \tag{2.2.4}$$

The existence of the integral $I[a]$ depends on the presence of the limit of the right-hand side of (2.2.3) as $E_{123} = s_1 \omega_{\boldsymbol{k}_1}^{\alpha_1} + \ldots + s_3 \omega_{\boldsymbol{k}_3}^{\alpha_3} \to 0$. We remember that $\hat{V} = V \delta(P_{123})$.

Now two cases are possible. Consider a system of equations,

$$P_{123} = s_1 \boldsymbol{k}_1 + s_2 \boldsymbol{k}_2 + s_3 \boldsymbol{k}_3 = 0 \tag{2.2.5}$$

$$E_{123} = s_1 \omega_{\boldsymbol{k}_1}^{\alpha_1} + s_2 \omega_{\boldsymbol{k}_2}^{\alpha_2} + s_3 \omega_{\boldsymbol{k}_3}^{\alpha_3} = 0 . \tag{2.2.6}$$

If this system has no solution, the formula (2.2.4) gives the nonsingular expression for I_{123} and there is no nontrivial information available in this order. If the system (2.2.5, 6) has nontrivial solutions, it determines the simplest possible resonant surface on which the coefficient functions of a new motion invariant may have singularities. One of the following alternatives should take place in the absence of this singularity on the resonant surface (2.2.5, 6): either

$$V_{123} = V_{\boldsymbol{k}_1 \boldsymbol{k}_2 \boldsymbol{k}_3}^{(\alpha_1), s_1(\alpha_2), s_2(\alpha_3), s_3} = 0 , \tag{2.2.7}$$

or

$$F_{123} \equiv s_1 f_{\boldsymbol{k}_1}^{\alpha_1} + s_2 f_{\boldsymbol{k}_2}^{\alpha_2} + s_3 f_{\boldsymbol{k}_3}^{\alpha_3} = 0 . \tag{2.2.8}$$

In the latter case, if a nontrivial solution of (2.2.8) exists, we call the set of dispersion laws $\{\omega_{\boldsymbol{k}}^{\alpha_1}, \omega_{\boldsymbol{k}}^{\alpha_2}, \omega_{\boldsymbol{k}}^{\alpha_3}\}$ degenerate with respect to the process (2.2.5, 6).

If there is only one type of waves in the system with the dispersion law ω_k satisfying $\omega_k > 0$ (the absence of waves with negative energy), the system (2.2.5, 6) is reduced to the equation

$$\omega(\boldsymbol{k}_1 + \boldsymbol{k}_2) = \omega(\boldsymbol{k}_1) + \omega(\boldsymbol{k}_2) . \tag{2.2.9}$$

If this equation is solvable, the dispersion law is called decaying.

The alternative (2.2.7, 8) allows a generalization to higher orders of perturbation theory. To do this it is necessary to use the classical scattering matrix introduced in Sect. 2.1. The result stated below is really the infinite-dimensional generalization of the well-known Poincaré theorem [3].

Theorem 2.2.1. For the existence of an additional motion invariant of (2.1.1) $I[a]$ of the form [6]

$$I[a] = I_0[a] + I_1[a] + \dots \ , \quad I_0 = \sum_{\alpha_j} f_k^{\alpha_j} \left| a_k^{\alpha_j} \right|^2 dk$$

it is necessary that on each resonant surface,

$$E_{1\dots q} = 0 \ , \quad P_{1\dots q} = 0 \ , \tag{2.2.10}$$

in the points of general position, the following alternative occurs: either the amplitude W of the classical scattering matrix, corresponding to (2.2.10), equals zero,

$$W_{1\dots q} = 0 \ , \tag{2.2.11a}$$

or the following condition holds:

$$F_{1\dots q} \equiv \sum_{1}^{q} s_j f_{k_j}^{\alpha_i} = 0 \ . \tag{2.2.11b}$$

Proof. The conservation of the integral $I[a]$ results in the equality of its limit values as $t \to \pm\infty$:

$$\lim_{t\to-\infty} I\left[b_k e^{-i\omega_k t}\right] = \lim_{t\to+\infty} I\left[b_k e^{-i\omega_k t}\right] \ , \tag{2.2.12}$$

where the limits in (2.2.12) should be understood in terms of distributions.

By definition of the classical scattering and transition matrices (2.1.7,8) we have: $b_k(t) = S_k(-\infty,t)[C^-] \ C_k^+ = S_k[C^-]$. Now let us insert this formula into (2.2.12), taking into account (2.1.32) and the explicit form of the integral quadratic part I_0. By doing so we reduce two limit points, $t = \pm\infty$, to only one point, $t = -\infty$, and obtain

$$\lim_{t\to-\infty} \sum^{n} f_k^{(\alpha)} \left[C_k^{-(\alpha)^*} R_k^{(\alpha)}[C^-](t) \right.$$
$$\left. + C_k^{*-(\alpha)} R_k^{(\alpha)}[C^-](t) \right] dk = \mathcal{D}_k \ . \tag{2.2.12a}$$

Here we have already used the fact that $\lim_{t\to-\infty} b_k = C_k^-$. In (2.2.12a) we keep an explicit dependence of $R_k(t) = S(-\infty, t) - 1$ on t, because this dependence leads to the important fact that each term in (2.2.12a) is localized on the corresponding resonant surface.

The $\mathcal{D}_k$ contains the term $\sum \int f_k^{(\alpha)} R_{\alpha k}^{(\alpha)^*} R_k^{(\alpha)} dk$ resulting from the I_0 and all terms resulting from the higher orders in $I[a]$. As we have already seen in

Sect. 2.1, R_k is a series, and each of its terms, $S^{n,m}_{k_1 \ldots k_{n+m}}$ corresponding to some nonlinear processes "$n \to m$", has the structure (2.1.24).

Recalling now (2.1.3) and using the well-known identity from distribution theory,

$$\lim \frac{e^{iEt}}{E - i0} = \pi \delta(E) \, ,$$

we see that the integrand in $S^{n,m}_k$ takes the form

$$\hat{W}^{n,m}_{k_1 \ldots k_{n+m}} C_1 \ldots C_{n+m} \delta(s_1 \omega_1 + \ldots + s_{n+m} \omega_{n+m}) \, ,$$

and each term resulting from the left-hand side of (2.2.12a) is localized on the resonant surface. As to $\mathcal{D}_k$, each contributing it term contains at least one additional δ-function of frequencies and is therefore localized on the submanifold of codimensionality 1 or more.

To see this, consider an arbitrary term in $\mathcal{D}_k$, for example one resulting from the cubic part of $I[a]$:

$$\lim_{\to -\infty} \frac{1}{3!} \sum_{s_i, \alpha_i} \int f_k \hat{V}^{(\alpha), s(\alpha_1), s_1(\alpha_2), s_2}_{kk_1 k_2} R^{(s, \alpha)}_k C^{-(s_1, \alpha_1)}_{k_1} C^{-(s_2, \alpha_2)}_{k_2}$$

$$\times \left[\exp \left[i \left(E^{s s_1 s_2}_{kk_1 k_2} \right) t \right] \left(E^{s s_1 s_2}_{kk_1 k_2} \right)^{-1} \right] dk dk_1 dk_2 \, .$$

This term has two δ-functions of frequencies: one resulting from $R^{(s, \alpha)}_k$ and the other from the expression in squared brackets. Certainly, the integrand is localized on a submanifold of a codimensionality 1 of the whole resonant manifold, and in points of a general position this term should not be taken into account. Analogously, each term constituting $\mathcal{D}_k$ possesses the property.

Now consider points of a general position ($\mathcal{D}_k = 0$) of a resonant surface for terms (on the left-hand side) which contain a combination of fields (C for C^-):

$$C^{(\alpha_1)}_{k_1} \ldots C^{(\alpha_n)}_{k_n} C^{*(\tilde{\alpha}_1)}_{\tilde{k}_1} \ldots C^{*(\tilde{\alpha}_m)}_{\tilde{k}_m} \, .$$

By symmetrizing these terms we obtain

$$\int \left[f^{(\alpha_1)}_{k_1} + \ldots + f^{(\alpha_n)}_{k_n} - f^{(\tilde{\alpha}_1)}_{\tilde{k}_1} - \ldots - f^{(\tilde{\alpha}_m)}_{\tilde{k}_m} \right] W^{\alpha_1 \ldots \alpha_n \tilde{\alpha}_1 \ldots \tilde{\alpha}_m}_{k_1 \ldots k_n \tilde{k}_1 \ldots \tilde{k}_m}$$

$$\times c^{(\alpha_1)}_{k_1} \ldots c^{*(\tilde{\alpha}_m)}_{\tilde{k}_m} \delta \left(k_1 + \ldots + k_n - \tilde{k}_1 - \ldots - \tilde{k}_m \right)$$

$$\times \delta \left(\omega^{(\alpha_1)}_{k_1} + \ldots + \omega^{(\alpha_n)}_{k_n} - \omega^{(\tilde{\alpha}_1)}_{\tilde{k}_1} - \ldots - \omega^{(\tilde{\alpha}_m)}_{\tilde{k}_m} \right)$$

$$\times dk_1 \ldots dk_n d\tilde{k}_1 \ldots d\tilde{k}_m = 0 \, .$$

Hence, due to the arbitrariness of C^-_k, we obtain that in points of a general position of the resonant surface

$$k_1 + \ldots - \tilde{k}_m = 0 \, , \quad \omega^{(\alpha_1)}_{k_1} + \ldots - \omega^{(\tilde{\alpha}_m)}_{\tilde{k}_m} = 0 \, , \qquad (2.2.13)$$

the following equality should hold true:

$$\left[f_{k_1}^{(\alpha_1)} + \ldots + f_{k_n}^{(\alpha_n)} - \ldots - f_{k_1}^{(\tilde{\alpha}_1)} \right] W_{k_1 \ldots k_n \tilde{k}_1 \ldots \tilde{k}_m}^{\alpha_1 \ldots \alpha_n \tilde{\alpha}_1 \ldots \tilde{\alpha}_m} = 0 \, ,$$

from which the alternative (2.11a–b) follows. (End of proof.)

We now present a more general definition of degenerative dispersion laws.

Definition. The set of dispersion laws

$$\left\{ \omega_k^{\alpha_1} , , \omega_k^{\alpha_q} \right\} , \quad \alpha_j = 1, \ldots , N \tag{2.2.14}$$

is called degenerative with respect to the process (2.2.13) in the point Q of a manifold (2.2.12), if (2.2.11b) in the neighbourhood of the point Q on (2.2.13) has a nontrivial solution, i.e., $f_k^\alpha \neq (v, k) + A\omega_k^\alpha + \text{const}$. The set (2.2.14) is called degenerative in the domain Ω in (2.2.13) if it is degenerative in each point of Ω. And the set (2.2.14) is called completely degenerative (or simply "degenerative") on (2.2.13) if it is degenerative in each point of the manifold. If the domain Ω does not exist, the set (2.2.14) is called nondegenerative with respect to (2.2.13).

If Ω exists but does not coincide with (2.2.13), the set (2.2.14) is called particularly degenerative and if an additional integral exists, the scattering amplitude outside Ω should become zero according to (2.2.11a). If all functions $\omega_k^{\alpha_j}$ from the degenerative set of dispersion laws coincide, the correspondent dispersion law is called degenerative.

Degenerative and particularly degenerative dispersion laws and degenerative sets represent in themselves exclusive phenomena. The properties of such exclusive ω_k^α will be described in the next paragraph.

2.3 Properties of Degenerative Dispersion Laws [6]

Properties of degenerative dispersion laws differ strongly in spaces of dimensionality $d = 1$, $d = 2$ and $d \geq 3$. For this reason we shall describe them separately.

2.3.1 Dimension $d = 1$ In this case any three functions $\omega_k^{\alpha_i}$, $i = 1, 2, 3$, $\alpha = 1, \ldots , N$ form a degenerative set with respect to the process

$$\begin{aligned} k &= k_1 + k_2 \\ \omega_{k_1}^{\alpha_1} &= \omega_{k_2}^{\alpha_2} + \omega_{k_3}^{\alpha_3} \, , \end{aligned} \tag{2.3.1}$$

if such a process is possible.

Actually, (2.3.1) defines the one-dimensional manifold in a three-dimensional space (k_1, k_2, k_3) so that locally $k_i = k_i(\xi)$, $i = 1, 2, 3$. Consider any two functions $f_k^{(2)}$ and $f_k^{(3)}$. On the surface (2.3.1), we define $f_k^{(1)}(\xi)$ by the equality $f_k^{(1)}(\xi) = f_{k_2}^{(2)}(\xi) + f_{k_3}^{(3)}(\xi)$. Then we have to invert the equality $k_1 = k_1(\xi)$ to obtain the function $f_{k_1}^{(1)} = f^{(1)}(\xi(k_1))$, which, together with $f_k^{(2)}$ and $f_k^{(3)}$, forms a nontrivial solution of (2.2.8).

Any dispersion law ω_k in a one-dimensional case is degenerative with respect to the scattering process "2 waves into 2 waves" ("2 → 2"):

$$k_1 + k_2 = k_3 + k_4$$
$$\omega_1 + \omega_2 = \omega_3 + \omega_4 .$$

(2.3.2)

In fact, (2.3.2) defines the two-dimensional manifold in the four-dimensional space $(k_i, i = 1, \ldots, 4)$. On the other hand it is obvious that (2.3.2) is satisfied by the substitution

$$\left.\begin{array}{c} k_1 = k_3 \\ k_2 = k_4 \end{array}\right\} \quad \text{or} \quad \left.\begin{array}{c} k_1 = k_4 \\ k_2 = k_3 \end{array}\right\} ,$$

(2.3.3)

corresponding to the trivial scattering. The manifolds of trivial scattering prove to be very important when constructing action-angle variables.

Manifolds (2.3.2) and (2.3.3) obviously coincide. But on (2.3.3), any function f_k obeys the corresponding equation (2.2.11b), namely,

$$f_1 + f_2 = f_3 + f_4 ;$$

(2.3.4)

this is proof of nondegeneracy. For the process "2 → 2" with several modes, this is in general not so. For example, a set $\omega_k^{(1)} = k^2$, $\omega_k^{(2)} = c|k|$ is only degenerative to the process

$$k_1 + q_1 = k_2 + q_2$$
$$\omega_{k_1}^{(1)} + \omega_{q_1}^{(2)} = \omega_{k_2}^{(1)} + \omega_{q_2}^{(2)} .$$

(2.3.5)

The manifold (2.3.5) is split onto two parts, $\Gamma_1^{\pm}$ and $\Gamma_2^{\pm}$. The first corresponds to the forward scattering of a sound wave and the second, to backward scattering. Corresponding parametrization has the form [11]:
for $\Gamma_1^{\pm}$,

$$\begin{array}{ll} k_1 = \tfrac{1}{2}(\pm c + \xi) & q_1 = \tfrac{1}{2}(\eta - \xi) \\ k_2 = \tfrac{1}{2}(\pm c - \xi) & q_2 = \tfrac{1}{2}(\eta + \xi) , \end{array}$$

(2.3.6a)

and for $\Gamma_2^{\pm}$,

$$\begin{array}{ll} k_1 = \tfrac{1}{2}(\eta \pm 2c\xi) & q_1 = \xi(\eta \mp c) \\ k_2 = \tfrac{1}{2}(\eta \mp 2c\xi) & q_2 = \xi(\eta \pm c) . \end{array}$$

(2.3.6b)

It happens that the set $\omega^{(1)}$, $\omega^{(2)}$ is degenerative on (2.3.6a) and nondegenerative on (2.3.6b). The solution of a corresponding equation (2.2.11b) on $\Gamma_1^{\pm}$,

$$f_{k_1}^{(1)} + f_{q_1}^{(2)} = f_{k_2}^{(1)} + f_{q_2}^{(2)} ,$$

(2.3.7)

has the form

$$f^{(1)}(\xi) = \mu\left(\xi - \frac{c}{2}\right) + A\xi^2 + (B - Ac)\xi$$

$$f^{(2)}(\xi) = B\xi , \quad \forall \mu(\xi) = \mu(-\xi) .$$

Consider two dispersion laws: $\omega_k^{(1)} = c_1 k^2$, $\omega_k^{(2)} = c_2 k^2$. When $c_1 \neq \pm c_2$, the manifold (2.3.5) is nondegenerative. Indeed, let $\varrho = c_2/c_1 \neq \pm 1$. Manifold (2.3.5) then allows the following rational parametrization [12]:

$$k_1 = \frac{\varrho - 1}{2} q_1 + \frac{\varrho + 1}{2} q_2$$

$$k_2 = \frac{\varrho + 1}{2} q_1 + \frac{\varrho - 1}{2} q_2 . \tag{2.3.8}$$

Substituting (2.3.8) into (2.3.7), differentiating two times in q_1 and one in q_2 and setting $q_1 = q_2 = \xi/\varrho$, we obtain

$$(\varrho^2 - 1)(\varrho - 1) f^{(1)'''}(\xi) = (\varrho^2 - 1)(\varrho + 1) f^{(1)'''}(\xi) .$$

Hence, at $\varrho \neq \pm 1$, $f^{(1)'''} = 0$:

$$f^{(1)} = A^{(1)} \xi^2 + B^{(1)} \xi + C^{(1)} ; \tag{2.3.9}$$

i.e., the set $\{c_1 k^2, c_2 k^2, c_1/c_2 \neq \pm 1\}$ is nondegenerative to (2.3.5). At $\varrho = \pm 1$, (2.3.5) is degenerative.

Processes with more than four waves have not been very well studied, in spite of some special results. It is certainly clear that degeneracy in such processes is an exclusive phenomenon. For example, the same set $(c_1 k^2, c_2 k^2)$ is nondegenerative with respect to a "3 → 3" process:

$$k_1 + k_2 + k_3 = k_4 + k_5 + k_6$$

$$\omega_1 + \omega_2 + \omega_3 = \omega_4 + \omega_5 + \omega_6 \tag{2.3.10}$$

at any ϱ. The proof can be performed by using a rational parametrization of (2.3.10) of the form [12]:

$$k_1 = \frac{3P\varrho}{1 + 2\varrho} + R \left[u + \frac{1}{u} - \frac{1}{v} + (1 + 2\varrho)v \right]$$

$$k_2 = \frac{3P\varrho}{1 + 2\varrho} + R \left[u + \frac{1}{u} + \frac{1}{v} - (1 + 2\varrho)v \right]$$

$$k_3 = \frac{3P}{1 + 2\varrho} - \frac{2R}{u} - 2Ru$$

$$k_4 = \frac{3P}{1 + 2\varrho} + \frac{2R}{u} - 2Ru \tag{2.3.11}$$

$$k_5 = \frac{3P\varrho}{1 + 2\varrho} + R \left[u - \frac{1}{u} + \frac{1}{v} + (1 + 2\varrho)v \right]$$

$$k_6 = \frac{3P\varrho}{1 + 2\varrho} + R \left[u - \frac{1}{u} - \frac{1}{v} - (1 + 2\varrho)v \right] .$$

Parametrization (2.3.11) should be substituted into the condition corresponding to (2.2.11b):

$$f^{(1)}_{k_1} + f^{(1)}_{k_2} + f^{(2)}_{k_3} = f^{(2)}_{k_4} + f^{(1)}_{k_5} + f^{(1)}_{k_6} .\qquad(2.3.12)$$

After this, the proof of nondegeneracy can be obtained by three differentiations and by subsequently taking a corresponding limit to obtain a differential equation from the functional one.

2.3.2 Dimensionality $d = 2$. Consider the simplest nonlinear process: decay of the wave into two waves of the same type. If the correspondent dispersion law is decaying, corresponding manifold (2.2.9) defines a three-dimensional manifold $\Gamma^{1,2}$ in a four-dimensional space (k_1, k_2). As an example of a decaying dispersion law, one can consider an isotropic function,

$$\omega_k = \omega(|k|) , \quad \omega(0) = 0 , \quad \omega' > 0 .\qquad(2.3.13)$$

The equation (2.2.8) then takes a simple form,

$$f_{k_1+k_2} = f_{k_1} + f_{k_2} .\qquad(2.3.14)$$

Let us now show that the degenerative decaying dispersion laws exist at $d = 2$. We designate components of a vector k via (p, q) and let $\omega(p, q)$ be defined parametrically by formulae

$$p = \xi - \xi ; \quad q = a(\xi) - a(\xi) ; \quad \omega_k = b(\xi) - b(\xi) ,\qquad(2.3.15)$$

where $a(\xi)$ and $b(\xi)$ are arbitrary functions of one variable. (The natural appearance of a parametrization of this type in exactly solvable systems from an underlying linear problem was shown by *Manakov* in [38].) We consider the three-dimensional manifold $\tilde{\Gamma}^{1,2}$ defined by

$$\begin{aligned} p_1 &= \xi_1 - \xi_3 & p_2 &= \xi_3 - \xi_2 \\ q_1 &= a(\xi_1) - a(\xi_3) & q_2 &= a(\xi_3) - a(\xi_2) . \end{aligned}\qquad(2.3.16)$$

Now

$$p = p_1 + p_2 = \xi_1 - \xi_2$$

$$q = q_1 + q_2 = a(\xi_1) - a(\xi_2) ,$$

and in accordance with (2.3.15),

$$\begin{aligned} \omega_{k_1+k_2} &= b(\xi_1) - b(\xi_2) = b(\xi_1) - b(\xi_3) \\ &\quad + b(\xi_3) - b(\xi_2) = \omega_{k_1} - \omega_{k_2} . \end{aligned}\qquad(2.3.17)$$

Thus, the manifold $\tilde{\Gamma}^{1,2}$ is a domain in $\tilde{\Gamma}^{1,2}$.

Consider now a function $f(p, q)$ parametrized by

$$p = \xi_1 - \xi_2 , \quad q = a(\xi_1) - a(\xi_2) , \quad f = c(\xi_1) - c(\xi_2) ,\qquad(2.3.18)$$

where $c(\xi)$ is an arbitrary function. Obviously $f(p, q)$ obey (2.3.14) on $\tilde{\Gamma}^{1,2}$, and the law (2.3.15) is at least particulary degenerative. Its complege degeneracy should be considered separately.

Let $a(\xi) = \xi^2$, $b(\xi) = 4\xi^3$ in (2.3.15). Then

$$\omega(p, q) = p^3 + \frac{3q^2}{p} \, . \tag{2.3.19}$$

This is a dispersion law of the Kadomtsev-Petviashvily equation (1.1.3) (referred to in the following as KP-1) with $\alpha^2 = 1$.
Equation (2.3.13) now takes the form:

$$(p_1 + p_2)^2 = \left(\frac{q_1}{p_1} - \frac{q_2}{p_2} \right)^2 \, , \tag{2.3.20}$$

and it is clear that it consists of two parts. Simple analysis shows that $\tilde{\Gamma}^{1,2}$ coincides with the $\Gamma_+^{1,2}$ part given by the formulae

$$p_1 + p_2 = \frac{q_1}{p_1} - \frac{q_2}{p_2} \, . \tag{2.3.21}$$

Dispersion law (2.3.19) can also be obtained from a parametrization $a(\xi) = -\xi^2$, $b(\xi) = 4\xi^3$. Now $\tilde{\Gamma}^{1,1}$ coincides with $\Gamma_-^{1,2}$ when

$$p_1 + p_2 = -\frac{q_1}{p_1} + \frac{q_2}{p_2} \, . \tag{2.3.22}$$

Thus the dispersion law (2.3.19) is proved to be completely degenerative.
Now let $\xi_1 - \xi_2 = \delta \ll 1$ in (2.3.15–18). Then in the first order in δ relations,

$$\frac{q}{p} = a'(\xi_2) \, , \quad \frac{\omega}{p} = b'(\xi_2) \tag{2.3.23}$$

also define the degenerative dispersion law, and it is the homogeneous function of degree one,

$$\omega = p\Phi \left(\frac{q}{p} \right) \, . \tag{2.3.24}$$

We should note that (2.3.24) together with the function (2.3.15) are not analytic at $p \to 0$. Thus, the homogeneous function of degree one dispersion law is degenerative. The manifold $\Gamma^{1,2}$ for dispersion law (2.3.24) is

$$\frac{q_1}{p_1} = \frac{q_2}{p_2} = \frac{q}{p} \, ,$$

which means that k_1 and k_2 are parallel and unidirected.
When many modes exist, there are three sets of dispersion laws $\{\omega^{(\alpha_1)}, \omega^{(\alpha_2)}, \omega^{(\alpha_3)}\}$ degenerative with respect to decay processes, too:

$$\omega_{k_1}^{(\alpha_1)} = \omega_{k_2}^{(\alpha_2)} + \omega_{k_3}^{(\alpha_3)}$$
$$k_1 = k_2 + k_3 \, . \tag{2.3.25}$$

They are defined parametrically by the formulae

$$p_1 = \xi_1 - \xi_2 \ , \quad p_2 = \xi_1 - \xi_3 \ , \quad p_3 = \xi_3 - \xi_2$$

$$q_1 = a_1(\xi_1) - a_2(\xi_2) \ , \quad q_2 = a_1(\xi_1) - a_3(\xi_3) \ , \quad q_3 = a_3(\xi_3) - a_2(\xi_2)$$

$$\omega^{(\alpha_1)} = b_1(\xi_1) - b_2(\xi_2) \ , \quad \omega^{(\alpha_2)} = b_1(\xi_1) - b_3(\xi_3) \ , \quad \omega^{(\alpha_3)} = b_3(\xi_3) - b_2(\xi_2) \ .$$

$$(2.3.26)$$

Now the solutions corresponding to (2.2.11b),

$$f_{k_1}^{(\alpha_1)} = f_{k_2}^{(\alpha_2)} + f_{k_3}^{(\alpha_3)} \ , \tag{2.3.27}$$

have a parametric form in (2.3.26):

$$\begin{aligned}
f_{k_1}^{(\alpha_1)} &= c_1(\xi_1) - c_2(\xi_2) \\
f_{k_2}^{(\alpha_2)} &= c_1(\xi_1) - c_3(\xi_3) \\
f_{k_3}^{(\alpha_3)} &= c_3(\xi_3) - c_2(\xi_2) \ .
\end{aligned} \tag{2.3.28}$$

We should recall the fact mentioned above with respect to the specific case of the KP-1 equation. Namely, if in (2.3.15) we replace $p \rightarrow p$, $a(\xi) \rightarrow -a(-\xi)$; $b(\xi) \rightarrow -b(-\xi)$, such a dispersion law will be also degenerative with respect to the process (2.3.13). In the case of KP-1 these two parametrizations together cover the entire manifold $\Gamma^{1,2}$. It is still an open question as to whether these two parametrizations cover the whole degenerative piece of resonant manifold in all cases.

In addition, all homogeneous dispersion laws of the weight 1 (2.3.24) are degenerative to any decay processes $1 \rightarrow n$,

$$\begin{aligned}
\omega &= \omega_1 + \ldots + \omega_n \\
k &= k_1 + \ldots + k_n \ .
\end{aligned} \tag{2.3.29}$$

The question naturally arises as to whether degenerative dispersion laws exist which differ from (2.3.15). The following theorems are true.

Theorem 2.3.1 (Local Uniqueness Theorem) [6]. All dispersion laws of the form

$$p = \xi_1 - \xi_2 \ , \quad q = a(\xi_1) - a(\xi_2) \tag{2.3.30a}$$

$$\omega = b(\xi_1) - b(\xi_2) + \sum_{1}^{\infty} \varepsilon^n \omega_n(\xi_1, \xi_2) \ ,$$

satisfying the degeneracy condition (2.3.14) with

$$f = c(\xi_1) - c(\xi_2) + \sum_{1}^{\infty} \varepsilon^n f_n(\xi_1, \xi_2) \ , \tag{2.3.30b}$$

$\varepsilon \ll 1$, will belong to the class (2.3.15).

This statement means that there are variables $\eta_1(\xi_1, \xi_2)$ and $\eta_2(\xi_1, \xi_2)$ in which terms the degenerative dispersion law (2.3.30a), with (2.3.30b) holds true, taking the form (2.3.15).

Theorem 2.3.2. If $w(p, q)$, degenerative with respect to the dispersion law (2.2.9), is analytic in the neighborhood of $p = q = 0$, then the corresponding function $f(p, q)$ cannot be analytic in the same domain.

Theorem 2.3.3. Let the dispersion law $w(k)$ near the point k_0 admit the expansion

$$w(k_0 + \kappa) = w(k_0) + (\nu, \kappa) + \sum A_{ij}\kappa_i\kappa_j . \qquad (2.3.31)$$

Then in some domain near $k_1 = k_2 = k_3 = k_4 = k_0$, the dispersion law (2.3.31) is nondegenerative to the process

$$k_1 + k_2 = k_3 + k_4 \qquad (2.3.32)$$

$$w_{k_1} + w_{k_2} = w_{k_3} + w_{k_4} ;$$

i.e., the equation

$$f_{k_1} + f_{k_2} = f_{k_3} + f_{k_4} \qquad (2.3.33)$$

does not have nontrivial solutions.

Theorem 2.3.4 (Global Theorem). If $w^{(\alpha i)}(p, q)$ is a system of dispersion laws, degenerative with respect to the process (2.3.25), and if the equation (2.3.27) has at least *three* independent nontrivial solutions, the system $w^{(\alpha i)}(p, q)$ either belongs to the class (2.3.26) or could be obtained from it by some limiting process.

Now let $w(0) = 0$. From Theorem 2.3.3 it follows that the dispersion laws admitting expansions (2.3.31) are nondegenerative to the process

$$\sum_{j}^{n} k_j = \sum_{i}^{m} k_i , \quad \sum_{i}^{n} w_i = \sum_{l}^{m} w_l , \qquad (2.3.34)$$

$$n \geq 2, \quad m \geq 2 .$$

To see this, one can consider the neighbourhood of the manifold (2.3.32), putting the "extra" wave vectors equal to zero. Thus, only homogeneous functions of degree one dispersion laws can be degenerative and only to the processes (2.3.29).

From the above it follows that there is no unique dispersion law completely degenerative with (2.3.32). It is very doubtful that w_k exist which are degenerative to (2.3.32) in particular.

Let a dispersion law w_k be decaying. Then the manifold $\Gamma^{2,2}$ contains a submanifold $\Gamma_M^{2,2}$ of codimension one given by the equations

$$k_1 + k_2 = k_3 + k_4 = q \qquad (2.3.35)$$

$$w_{k_1} + w_{k_2} = w_{k_3} + w_{k_4} = w_q .$$

If the dispersion law is degenerative to a "$1 \rightarrow 2$" process, then the correspondent degenerative function $f(\mathbf{k})$ obeys on $\Gamma_M^{2,2}$ the equation

$$f_{\mathbf{k}_1} + f_{\mathbf{k}_2} = f_{\mathbf{k}_3} + f_{\mathbf{k}_4} = f_q . \tag{2.3.36}$$

This certainly does not mean that dispersion law ω_k is even particulary degenerative. For degeneracy to take place it is necessary that (2.3.33) be true on (2.3.32) in the neighbourhood of at least one point of the manifold (2.3.35). Degeneration, as we know, is possible only at $d = 2$, and the corresponding dispersion laws are given by (2.3.15).

Consider now the neighbourhood of any point on $\Gamma_M^{2,2}$. It can be defined as

$$p_1 = \xi_1 - \xi_2 , \quad p_2 = \xi_2 - \xi_3 , \quad p_3 = \xi_1 - \xi_4 , \quad p_4 = \xi_4 - \xi_3 ,$$

$$q_1 = a(\xi_1) - a(\xi_2) , \quad q_2 = a(\xi_2 + \nu_1) - a(\xi_3 + \nu_1)$$

$$q_3 = a(\xi_1 + \nu_2) - a(\xi_4 + \nu_2) , \quad q_4 = a(\xi_4 + \nu_3) - a(\xi_3 + \nu_3)$$

$$\omega_1 = b(\xi_1) - b(\xi_2) , \quad \omega_2 = b(\xi_2 + \nu_1) - b(\xi_3 + \nu_1)$$

$$\omega_3 = b(\xi_1 + \nu_2) - b(\xi_4 + \nu_2) , \quad \omega_4 = b(\xi_4 + \nu_3) - b(\xi_3 + \nu_3) .$$

The resonant conditions are

$$\left[a'(\xi_2) - a'(\xi_3)\right] \nu_1 = \left[a'(\xi_1) - a'(\xi_3)\right] \nu_2 + \left[a'(\xi_4) - a'(\xi_3)\right] \nu_3$$

$$\left[b'(\xi_2) - b'(\xi_3)\right] \nu_1 = \left[b'(\xi_1) - b'(\xi_3)\right] \nu_2 + \left[b'(\xi_4) - b'(\xi_3)\right] \nu_3 .$$

The condition of degenerativeness gives another relation:

$$\left[c'(\xi_2) - c'(\xi_3)\right] \nu_1 = \left[c'(\xi_1) - c'(\xi_3)\right] \nu_2 + \left[c'(\xi_4) - c'(\xi_3)\right] \nu_3 .$$

If functions a, b, c are linear independent, this equation has only zero solutions. It follows from this that the manifold $\Gamma_M^{2,2}$ cannot be locally extended with preservation of degeneracy.

Consider now any process "n into m" given by the resonant conditions (2.3.34), and let ω_k be decaying and degenerative to "1 into 2" [see (2.3.15)]. In the corresponding manifold $\Gamma^{n,m}$ one can point out a set of manifolds $\Gamma_M^{n,m}$ which we can call minimal. To describe these manifolds, we recall that the scattering amplitude $W^{n,m}$ is given by a diagram of the tree type with a finite number of vertices and inner lines.

Let us designate via p_i, s_i the outer wave vectors and their directions. Let some vertex contain vectors p_i, s_i, p_j, s_j, p_l, s_l. Then we have

$$s_i \cdot p_i + s_j \cdot p_j + s_l \cdot p_l = 0 . \tag{2.3.37}$$

We require another condition to be fulfilled:

$$s_i \omega_{p_i} + s_j \omega_{p_j} + s_l \omega_{p_l} = 0 . \tag{2.3.38}$$

From (2.3.37, 38), it follows naturally that (2.3.34) is true but (2.3.37, 38) determine the manifold of smaller dimensionality – one of the minimal ones, $\Gamma_M^{n,m}$.

If ω_k is degenerative, in each vertex the equation

$$s_i f_{p_i} + s_j f_{p_j} + s_l f_{p_l} = 0$$

will be true and, as a consequence, so will (2.2.11b), but it is impossible to enlarge the dimensionality of $\Gamma_M^{n,m}$ under condition (2.2.11b).

2.3.3 Dimensionality $d \geq 3$.

In higher dimensions the possibility of degeneracy is strongly limited in comparison with $d = 1, 2$. Only the homogeneous functional of degree one dispersion law

$$\omega(c\boldsymbol{k}) = c\omega(\boldsymbol{k}) \tag{2.3.39}$$

is degenerative to (2.3.29) only. Its degeneracy does not depend on the space dimensionality: (2.3.29) is solvable only if all $\boldsymbol{k}_i$, $i = 1, 2, \ldots$ are collinear to $\boldsymbol{k}$. So the corresponding manifold has smaller dimensionality than for decaying dispersion laws of general form; e.g., at $d = 3$, the dimensionality of (2.2.9) equals five while that of (2.3.29) equals four. On the basis of the following local theorem, it can be stipulated that no other degenerative laws exist.

Theorem 2.3.5. Let the dispersion law ω_k, $\boldsymbol{k} = (p, q, r)$ be parametrized in the neighbourhood of $r = 0$ by

$$p = \xi_1 - \xi_2 , \quad q = a(\xi_1) - a(\xi_2) \tag{2.3.40}$$

$$\omega(p, q, r) = b(\xi_1) - b(\xi_2) + r \sum_0^\infty r^n \omega_n(\xi_1, \xi_2)$$

and the dimensionality of $\Gamma^{1,2}$ be equal to five. Then $\omega_0 = const$, $\omega_n = 0$, $n > 0$. For the proof, see Appendix I.

2.4 Properties of Singular Elements of a Classical Scattering Matrix. Properties of Asymptotic States [6]

Let us examine the "$n \rightarrow m$" process. We shall choose another notation for wave vectors, and designate the nonlinear part of the amplitude via $S^{n,m}_{k_1 \ldots k_n ; \tilde{k}_1 \ldots \tilde{k}_m}$.

We consider a diagram describing the process (2.3.34) in which the inner line (Green function) with the wave vector q is replaced by a δ-function. Let vector q be directed from the "root" of the diagram and "to the right of it"; i.e., further away from the root, there are external lines with vectors $k_1, \ldots, k_{n_1}$, $\tilde{k}_1, \ldots, \tilde{k}_{m_1}$, $n_1 < n$, $m_1 < m$. Now the following equations are added to (2.3.34):

$$k_1 + \ldots + k_{n_1} = \tilde{k}_1 + \ldots + \tilde{k}_{m_1} + q \tag{2.4.1}$$

$$\omega_{k_1} + \ldots + \omega_{k_{n_1}} = \omega_{\tilde{k}_1} + \ldots + \omega_{\tilde{k}_{m_1}} - \omega_q .$$

Moreover, it holds that

$$k_{n_1+1} + \ldots + k_n = \tilde{k}_{m_1+1} + \ldots + \tilde{k}_m - q \qquad (2.4.2)$$

$$\omega_{k_{n_1+1}} + \ldots + \omega_{k_n} = \omega_{\tilde{k}_{m_1+1}} + \ldots + \omega_{\tilde{k}_m} - \omega_q .$$

Let us designate via $\tilde{S}^{n,m}_{k_1 \ldots k_n ; \tilde{k}_1 \ldots \tilde{k}_m}$ the singular part of the amplitude of the "n into m" process corresponding to equations (2.4.1), (2.4.2).

We obtain the expression for $\tilde{S}^{n,m}$ as a result of summing of all diagrams of the form

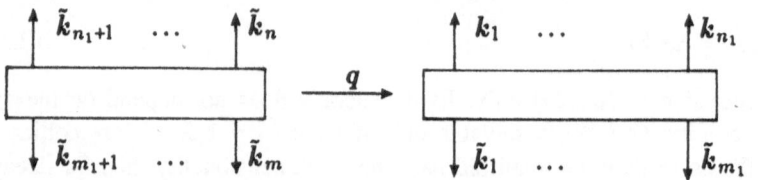

Then we have the relation:

$$\tilde{S}^{n,m}_{k_1 \ldots k_n , \tilde{k}_1 \ldots \tilde{k}_m}$$

$$= \pi i \int S^{n_1,m_1+1}_{k_1 \ldots k_{n_1} , \tilde{k}_1 \ldots \tilde{k}_{m_1} q} S^{n-n_1+1,m-m_1}_{k_{n_1+1} \ldots k_n , q ; \tilde{k}_{m_1+1} \ldots k_m} dq .$$

The formula shows that the singular amplitude $\tilde{S}^{n,m}$ is factorized through the composition of the two nonsingular amplitudes of lower order. It is clear that the analogous statement holds for the amplitude of any degree of singularity when there are several additional equations of the structure (2.4.1). All of them are factorized in the form of the composition of the finite number of the nonsingular amplitudes of lower orders. In particular, the maximum singular elements of the scattering matrix defined by the diagrams where all "Green functions" of the internal lines are substituted for δ-functions, are factorized in the form of the composition of the simplest scattering amplitudes "one into two".

These facts have a simple physical meaning. The substitution of one of the internal "Green functions" of a δ-function means that the corresponding wave is the eigenoscillation of the system (a "real particle"), and the process with such a wave occurs stage by stage, combined out of the process of the lowest order.

Now let the dynamical system under consideration possess the additional motion integral, and let the dispersion law be nondegenerative relative to all nonlinear processes. Then all nonsingular elements of the scattering matrix on the resonance surfaces are vanishing. As mentioned above, all singular amplitudes are vanishing too. Thus, in this case, the classical scattering matrix is trivial and the asymptotic states coincide, i.e.,

$$c_k^+ = c_k^- . \qquad (2.4.3)$$

In particular, this holds for the Kadomtsev-Petviashvili equation with $\alpha^2 = -1$, or KP-2, as was first noted in [9].

We have seen in Sect. 2.3 that in the two-dimensional case the situation in which the dispersion law is degenerative relative to the lowest-order process "one into two" and nondegenerative relative to all higher-order processes is typical. All degenerative dispersion laws constructed in Sect. 2.3 possess this property. In such a situation the classical scattering matrix S is nontrivial, but only its most singular part is nonvanishing, factorizing into the composition of the three-wave processes. This applies to the KP-1 equation, too.

It is very important in this case to find the scattering matrix in the explicit form in some sense. Let us note that for the most singular part of the S-matrix, one can cancel all inner Green functions and make a replacement in every vorticity:

$$V_{k_p k_q k_r}^{-s_p s_q s_r} \delta\left(-s_p k_p + s_q k_q + s_r k_r\right)$$
$$\longrightarrow \pi i V_{k_p k_q k_r}^{-s_p s_q s_r} \delta\left(-s_p k_p + s_q k_q + s_r k_r\right) \delta\left(-s_p \omega_{k_p} + s_q \omega_{k_q} + s_r \omega_{k_r}\right) .$$
$$(2.4.4)$$

This modified vorticity will be denoted symbolically as $\hat{\hat{V}}$.

Now we must rememeber that the entire set of diagrams has the factor $2\pi i$. So we can write symbolically

$$c^+ = c^- + 2\left\{\hat{\hat{V}}\left[c^-, c^-\right] + \ldots\right\} . \tag{2.4.5}$$

The expression in curly brackets is the whole set of diagrams. Formulae (2.4.5) can be rewritten in the form:

$$\frac{c^+ + c^-}{2} = c^- \hat{\hat{V}}\left[c^-, c^-\right] + \ldots , \tag{2.4.6}$$

or

$$c^- = \frac{c^+ + c^-}{2} - \hat{\hat{V}}\left[\frac{c^+ + c^-}{2}, \frac{c^+ + c^-}{2}\right] . \tag{2.4.7}$$

Finally, we have

$$\frac{c^+ - c^-}{2} = \hat{\hat{V}}\left[\frac{c^+ + c^-}{2}, \frac{c^+ + c^-}{2}\right] ;$$

or, more detailed:

$$c_k^{+s} - c_k^{-s} = \frac{\pi i}{2} \int \sum_{s_1 s_2} V_{kk_1 k_2}^{-ss_1 s_2} \delta(sk - s_1 k_1 - s_2 k_2)$$
$$\times \delta\left(s\omega_k - s_1 \omega_{k_1} - s_2 \omega_{k_2}\right)\left(c_{k_1}^{+s_1} + c_{k_1}^{-s_1}\right)$$
$$\times \left(c_{k_2}^{+s_2} + c_{k_2}^{-s_2}\right) dk_1 dk_2 . \tag{2.4.8}$$

Formula (2.4.8) gives a direct connection between asymptotic states in the case of the degenerative dispersion law.

The equation similar to (2.4.8) applies in the one-dimensional case if one type of wave is involved. In the one-dimensional case, any dispersion law is degenerative to the process of two-particle scattering. For simplicity we consider the Hamiltonian (2.1.11); we have

$$\frac{c_k^+ - c_k^-}{2} = \pi \int T_{kk_1k_2k_3} \delta(k + k_1 - k_2 - k_3)$$

$$\times \delta(\omega_k + \omega_{k_1} - \omega_{k_2} - \omega_{k_3}) \left(\frac{c_{k_1}^{*+} + c_{k_1}^{*-}}{2}\right) \left(\frac{c_{k_2}^+ + c_{k_2}^-}{2}\right)$$

$$\times \left(\frac{c_{k_3}^+ + c_{k_3}^-}{2}\right) dk_1 dk_2 dk_3 . \tag{2.4.9}$$

It follows from (2.4.9) that the squared module of the classical S-matrix is equal to unity:

$$|c_k^+|^2 = |c_k^-|^2 ;$$

but in general, $\arg c_k^+ \neq \arg c_k^-$. Actually, it is well known that in such one-dimensional systems, the interaction is reduced to a phase shift only.

Now let us return to the two-dimensional case with the decaying degenerative dispersion law, and consider the amplitude of the "two into two" process with the resonant conditions

$$k_1 + k_2 = k_3 + k_4 \tag{2.4.10}$$

$$\omega_{k_1} + \omega_{k_2} = \omega_{k_3} + \omega_{k_4} .$$

This amplitude is described by three diagrams:

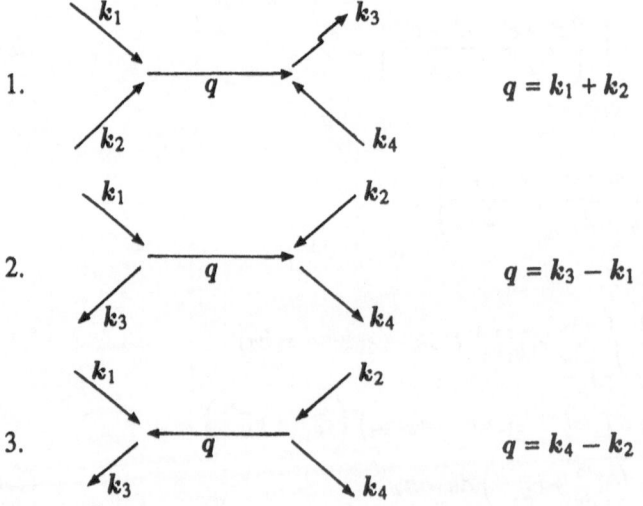

1. $q = k_1 + k_2$

2. $q = k_3 - k_1$

3. $q = k_4 - k_2$

As we have stated above, the nonsingular part of the amplitude localized on the whole manifold (2.4.7) must be identically zero. On the other hand, this amplitude becomes infinity near resonant manifolds corresponding to an interaction via real waves. (The singular part of the amplitude is localized on these very manifolds). These manifolds are different for the three diagrams above. They are defined by the formulae:

$$\omega_{k_1+k_2} = \omega_{k_1} + \omega_{k_2} = \omega_{k_3} + \omega_{k_4} \tag{2.4.11a}$$

for diagram 1;

$$\omega_{k_1-k_3} = \omega_{k_1} - \omega_{k_3} = \omega_{k_4} - \omega_{k_2} \tag{2.4.11b}$$

for diagram 2; and

$$\omega_{k_3-k_1} = \omega_{k_3} - \omega_{k_1} = \omega_{k_2} - \omega_{k_4} \tag{2.4.11c}$$

for the diagram 3. Since the amplitude of the process (2.4.7) becomes zero, the singularities localized near manifolds (2.4.11) must cancel each other. This cancellation can only occur if the manifolds coincide, at least partially.

The resonant surface "one into two" for KP-1 consists of two connected parts [see (2.3.21–22)]. A simple analysis shows that each of the two parts described by one of the equations (2.4.11a–c) coincides with some parts described by another of these three equations. This results in the number of connected manifolds, defined by (2.4.11a–c), being equal to three, but not six. The statement about pair compatibility of (2.4.11a–c) is a general one for the degenerative dispersion laws and could be used for enumeration of such laws. It is worth noticing that the coincidence of manifolds (2.4.11a–c) (in the above-mentioned sense) is only a necessary but not sufficient condition for the singularities in (2.4.7) to cancel each other. Rather rigid conditions imposed on the coefficient functions $V^{ss_1 s_2}_{kk_1 k_2}$ of the three-wave Hamiltonian (2.1.11) should be satisfied. We have checked these conditions for KP-1 equations. We should also note that checking for the cancellation of singularities is a useful and simple way to analyse the existence of the additional motion invariants for the particular systems.

2.5 The Integrals of Motion [5]

One of the important statements in the present paper is that the existence of an infinite set of additional integrals follows from the existence of one such integral of system (2.1.1). Let us prove this fact and find the integral of motion in the form of a formal integropower series:

$$G = \int g_k \, |a_k|^2 \, dk$$

$$+ \sum_q \sum_{s \ldots s_q} \int G^{s \ldots s_q}_{k \ldots k_q} a_k \, s \ldots a_{k_q}^{s_q} \, \delta \left(P_q \right) dk \ldots dk_q \, . \tag{2.5.1}$$

Here,

$$P_q = sk + s_1 k_1 + \ldots + s_q k_q \, ,$$

and g_k is some function of wave numbers. Substituting (2.5.1) into (2.1.1), it is clear that the functions $G_{k \ldots k_q}^{s \ldots s_q}$ are expressed from the recurrent formulae

$$G_{kk_1 k_2}^{ss_1 s_2} = \frac{s g_k + s_1 g_1 + s_2 g_2}{E_{kk_1 k_2}^{ss_1 s_2}} V_{kk_1 k_2}^{ss_1 s_2} \tag{2.5.2}$$

$$G_{k \ldots k_q}^{s \ldots s_q} = \frac{P_{k \ldots k_q}^{s \ldots s_q}}{E_{k \ldots k_q}^{s \ldots s_q}} \, . \tag{2.5.3}$$

In these formulae,

$$E_{k \ldots k_q}^{s \ldots s_q} = s\omega_k + s_1 \omega_{k_1} + \ldots + s_q \omega_{k_2}$$

and the function $P_{k \ldots k_q}^{s \ldots s_q}$ is linearly expressed via $G_{k \ldots k_{q-1}}^{s \ldots s_{q-1}}$. It is not necessary for us to write out this dependence.

It follows from (2.5.2, 3) that the coefficient functions in the integrals possess singularities on all possible resonance manifolds of the form

$$E_n = E_{k \ldots k_n}^{s \ldots s_n} = 0 \, , \quad P_n = 0 \, . \tag{2.5.4}$$

We may continue as follows: Let the wave field $a(r)$ in a physical space be a rapidly decreasing function. Then its Fourier transformation – the field a_k – is a smooth function. This makes it possible to peform the regularization in the expression (2.5.1), but not in a unique way. For example, in all denominators one can perform the substitution:

$$E_{k \ldots k_q}^{s \ldots s_q} \rightarrow E_{k \ldots k_q}^{+ s \ldots s_q} = E_{k \ldots k_q}^{s \ldots s_q} + i0 \, , \tag{2.5.5}$$

or the substitution

$$E_{k \ldots k_q}^{s \ldots s_q} \rightarrow E_{k \ldots k_q}^{- s \ldots s_q} = E_{k \ldots k_q}^{s \ldots s_q} - i0 \, . \tag{2.5.6}$$

Generally speaking, in this case we obtain different integrals of motion; let us designate them as $G^{\pm}$. Any linear combination of these integrals may be the integral of motion; particularly, the difference $(1/2\pi i)G^0 = G^+ - G^-$. The integral G^0 does not have a quadratic part; its expansion in powers of a_k^s starts from the term:

$$G^0 = \sum_{s, s_i} \int \left(s g_k + s_1 g_{k_1} + s_2 g_{k_2} \right) \delta \left(s\omega_k + s_1 \omega_{k_1} + s_2 \omega_{k_2} \right)$$

$$\times \delta \left(sk + s_1 k_1 + s_2 k_2 \right) V_{kk_1 k_2}^{ss_1 s_2} a_k^s a_{k_1}^{s_1} a_{k_2}^{s_2} dk \, dk_1 dk_2 \, . \tag{2.5.7}$$

The integral G^0 can be called an essentially nonlinear one. It is one of a large number of such integrals. The linear equation

$$\dot{a}_k^s + i s \omega_k a_k^s = 0$$

allows an essentially nonlinear integral of the form:

$$I = \int \Phi_{k \ldots k_q}^{s \ldots s_q} \delta \left(E_{k \ldots k_q}^{s \ldots s_q} \right) \delta \left(P_q \right) a_k^s \ldots a_{k_q}^{s_q} dk \ldots dk_q . \tag{2.5.8}$$

Here, q is an arbitrary integer; $\Phi_{k \ldots k_q}^{s \ldots s_q}$ is an arbitrary function. In the nonlinear system (2.1.1), one can search for the integral in the form of the integropower series in a_k^s, the first term of which is the expression (2.5.8). In this case the regularization problem of the denominators of the form $E_{k \ldots k_r}^{s \ldots s_r}$, $r > q$ again exists, and cannot be solved uniquely. The different integrals obtained will differ by the essentially nonlinear integrals of higher orders.

One can attach a simple physical sense to the integrals $G^\pm$ occurring as a result of the regularizations (2.5.5) and (2.5.6). It is easy to see that

$$G^\pm = \int g_k \left| a_k^\pm \right|^2 dk . \tag{2.5.9}$$

Here, $a_k^\pm$ are asymptotic states of the wave field at $t \to \pm\infty$. Formula (2.5.9) shows that an arbitrary system (2.1.1) in the rapidly decreasing case is completely integrable. Actually, the change $a_k^s(t)$ in time is a canonical transformation, so the variables $a_k^{s\pm}(t) = c_k^\pm \exp(-i s \omega_k t)$ are canonical. It is now evident that the variables

$$I_k^\pm = \left| a_k^\pm \right| \quad \text{and} \quad \varphi_k^\pm = \arg a_k^\pm$$

are the action-angle variables for the system (2.1.1), irrespective of the form of its Hamiltonian. This rather impressive statement is based on a rapid decrease of the function $a(r)$ and, respectively, on the smoothness of the function $a(k)$. In the periodic case, when the function $a(k)$ represents a set of δ-functions,

$$a(k) = \sum_n a_n \delta(k - n k_0) \tag{2.5.10}$$

(k_0 being the vector of the reverse lattice and n a multiindex); integrals (2.5.1) in a general position make no sense (become infinite) and as a rule, integrability vanishes. In the periodic case only those integrals still make sense, the coefficient functions of which remain finite on all resonance manifolds, i.e., where reduction of singularities occurs. For further discussion of the periodic case, see Sect. 2.6.

To observe the singularities, let us introduce the operators $R^\pm$, inverse with respect to the operator of the transition (2.1.6), taken for simplicity at $t < 0$:

$$a_k^\pm = R_\varepsilon^\pm [a_k] \tag{2.5.11}$$

$$a_k^{\pm s} = a_k^s + \sum_q \sum_{s \ldots s_q} \int R_{\varepsilon\, k k_1 \ldots k_q}^{\pm -s s_1 \ldots s_q} a_{k_1}^{s_1} \ldots a_{k_q}^{s_q} \delta \left(P_q \right) dk_1 \ldots dk_q .$$

The coefficients $R_{\varepsilon\, k\, \ldots\, k_q}^{\pm\, s\, \ldots\, s_q}$ at $\varepsilon = 0$ do not depend on time. They have singularities on all possible resonance surfaces $E_q = 0$. Let us put

$$\lim_{\varepsilon \to 0} R_{\varepsilon\, k k_1 \ldots\, k_q}^{\pm\, s s_1 \ldots\, s_q} = \frac{\tilde{R}_{k k_1 \ldots\, k_q}^{\pm\, s s_1 \ldots\, s_q}}{E_q \pm i0} \; . \tag{2.5.12}$$

Expression $\tilde{R}_{k k_1 \ldots\, k_q}^{\pm\, s s_1 \ldots\, s_q}$ is regular on the resonance manifold $E_q = 0$, $P_q = 0$, but it can possess singularities on various "junior" resonance manifolds.

Let us consider the operator $R^{\pm}$ and let $t \to -\infty$ in (2.5.11). In this case, $a_k \to a_k^{-}$, and operator R^{+} is to be transformed into a classical scattering matrix. That means that on the resonance surface $E_q = 0$, $P_q = 0$, the numerator in (2.5.12) coincides with the corresponding element in the scattering matrix,

$$\tilde{R}_{k \ldots\, k_q}^{+\, s \ldots\, s_q} = S_{k \ldots\, k_q}^{s \ldots\, s_q} \; . \tag{2.5.13}$$

Now let us represent the integral of the motion G^{+} in the form

$$G^{+} = \int g_k a_k a_k^{*} dk + \int g_k a_k^{*} \left(a_k^{+} - a_k \right) dk$$

$$+ \int g_k a_k \left(a_k^{*+} - a_k^{*} \right) dk + \int g_k \left(a_k^{+} - a_k \right) \left(a_k^{*+} - a_k^{*} \right) dk \; , \tag{2.5.14}$$

and substitute (2.5.11) into (2.5.14). We collect the terms in (2.5.14) having the singularity on the whole resonant manifold (2.5.14) and having a complete power q; such terms are only contained in the second and third terms in (2.5.14). After symmetrization they are reduced to the form

$$\frac{1}{N} \int \frac{L_{k \ldots\, k_q}^{s \ldots\, s_q}}{E_{k \ldots\, k_q}^{s \ldots\, s_q}} \tilde{R}_{k \ldots\, k_q}^{+\, s \ldots\, s_q} a_k^{s} \ldots a_{k_q}^{s_q} \delta\left(P_q \right) dk \ldots dk_q \; . \tag{2.5.15}$$

N is some integer,

$$L_{k \ldots\, k_q}^{s \ldots\, s_q} = s g_k + \ldots + s_q g_{k_q} \; . \tag{2.5.16}$$

Comparing (2.5.14) with (2.5.3), it is clear that $\mathcal{P}_{k \ldots\, k_q}^{s \ldots\, s_q}$ can be represented as follows:

$$\mathcal{P}_{k \ldots\, k_q}^{s \ldots\, s_q} = \frac{1}{N} L_{k \ldots\, k_q}^{s \ldots\, s_q} \tilde{R}_{k \ldots\, k_q}^{s \ldots\, s_q} + A_{k \ldots\, k_q}^{s \ldots\, s_q} E_{k \ldots\, k_q}^{s \ldots\, s_q} \; , \tag{2.5.17}$$

where $A_{k \ldots\, k_q}^{s \ldots\, s_q}$ is regular on $E_q = 0$, although it probably has singularities on the "junior" resonant surfaces.

Let the dispersion law $\omega(k)$ be nondegenerative and the system (2.1.1) have an additional integral of motion with continuous coefficients. As we have already seen, this leads to the triviality of the scattering matrix and the coincidence of asymptotic states $a_k^{\pm}$. Now on the resonance manifold $E_q = 0$, $P_q = 0$ the matrix element $\tilde{R}_{k \ldots\, k_q}^{s \ldots\, s_q} = 0$. This means that on the resonance surface $E_q = 0$, $P_q = 0$, the singularity in the motion integral is cancelled. It can be seen directly from

(2.5.2) that the singularity is cancelled in the junior term of expansion (2.5.1) as well.

Now applying induction, we observe that generally all the singularities are cancelled. Thus, in the case under consideration, one can use an arbitrary function g_k in order to construct the motion invariants of the system (2.1.1). Roughly speaking, in this case there are as many integrals with continuous coefficients having a quadratic part as there are in the linear problem. All these integrals are conserved in the periodic case as well; i.e., the periodic system (2.1.1) is quite integrable. In particular, the periodic equation KP-2 is integrable. *Krichever* has recently come to this conclusion on the basis of his algebrogeometric approach [13]. We should keep in mind that our results have been obtained on the level of a formal series, the convergency of which we still do not know.

Now let the dispersion law be degenerative. We restrict ourselves to a case in the form of (2.3.15) at $d = 2$. Now the scattering matrix is different from unity, $S_{k \ldots k_q}^{s \ldots s_q} \neq 0$. However, the nonvanishing scattering matrix is concentrated on the minimal manifold $\Gamma_M^{n,m}$, when all the scattering occurs with the participation of real intermediate waves only.

Now in the expression (2.5.17), $\tilde{R}_{k \ldots k_q}^{s \ldots s_q} \neq 0$ and, generally speaking, the integral of the form (2.5.1) is singular. The only way out of this situation is to require the vanishing of the expression $L_{k \ldots k_q}^{s \ldots s_q}$. It is possible to do this on the manifold $\Gamma_M^{n,m}$ by requiring $g(k) = f(k)$; i.e., the function itself should represent the degenerative dispersion law, permitting parametrization:

$$p = \xi_1 - \xi_2, \qquad q = a(\xi_1) - a(\xi_2)$$
$$\omega = b(\xi_1) - b(\xi_2), \quad g = c(\xi_1) - c(\xi_2) .$$

Here, the function $c(\xi)$ is arbitrary.

Thus, in the given case, system (2.1.1) also has an infinite set of integrals of motion with continuous coefficients, but this set is sufficiently narrower than in the previous case. Instead of an arbitrary function of two variables at our disposal, we have only an arbitrary function of one variable. This is not quite enough for the integrability in the periodic case. So the systems with a degenerative dispersion law under periodic boundary conditions are nonintegrable [14], although they might possess an infinite set of integrals of motion. In the following, we study the periodic boundary conditions and search for the action-angle variables.

2.6 The Integrability Problem in the Periodic Case.
Action-Angle Variables [5, 7]

2.6.1 Canonical Transformations
The formulation of the problem of integrability in the periodic case differs from its analogue discussed above, because of the discreteness of all wave vectors and the absence of asymptotic states and a scattering matrix. Therefore we have to find the appropriate language with which to study it. This language does exist and is the infinite-dimensional analogue of the Birkhoff method of finding canonical transformations to the normal form.

We shall see that the Hamiltonian wave system with an additional integral and nondegenerate dispersion law can be reduced by such a transformation to the form of the infinite-dimensional Birkhoff chain.

In spite of all these differences we can get some useful information from the case of smooth a_k. Consider system (2.1.1) with one type of wave and then equations (2.1.14–17). We have used them to represent the "current" fields b_k via asymptotic fields c_k^- and to find the transition matrix operator.

Now we go from interaction representation to the usual fields $a_k = b_k e^{-i\omega_k t}$ and $a_k^- = c_k^- e^{-i\omega_k t}$. We see that as $\varepsilon \to 0$, the explicit dependence on time variable t in (2.1.14, 17) disappears and the transformation between $a_k(t)$ and $a_k^-(t)$ becomes a time-independent formal canonical transformation. (The formal scattering marix defines the formal transformation from a_k^- to a_k^+.)

Certainly, these transformations are generally divergent, due to the classical problem of resonances. In each order the corresponding terms in this transformation have the structure (2.1.24). If our system has an additional integral of motion (and we shall take this for granted in what follows), Theorem 2.2.1 holds. So, if the dispersion law is degenerative with respect to decays, our canonical transformation has unequivocally the resonance in the first order and does not exist. Naturally, it does not exist in the periodic case either.

If the dispersion law is nondegenerative, all resonances vanish and the canonical transformations $a_k \to a_k^-$ and $a_k^- \to a_k$ exist; the first of them map equation (2.1.1) to its linear part.

2.6.2 Small Denominators.

Let us try to find a periodic analogue to the abovementioned canonical transformations in the nondegenerative case. All wave vectors belong to the lattice

$$k = k_n = (2\pi\nu_1/l_1, \dots, 2\pi\nu_d/l_d) , \qquad (2.6.1)$$

where ν_i, $i = 1, \dots, d$ are integrals, l_i are space periods, and $n = (\nu_1, \dots, \nu_d)$ is an integer-valued vector. Sometimes we shall write n for k_n to simplify the formulae. In our notation (Sect. 2.2) the Hamiltonian of (2.2.1) takes the form

$$H^{(3)} = \frac{1}{3!} \sum_{1,2,3} \hat{U}_{1,2,3} a_1 a_2 a_3 . \qquad (2.6.2)$$

Consider the canonical transformation $a_k \to a_k^-$:

$$a_k^-(t) = a_k(t) + \sum_{p \geq 2} \sum_{0,1,\dots,p} \hat{\Psi}_{-0,1,\dots,p} a_1 \dots a_p \qquad (2.6.3)$$

with all fields defined on the lattice (2.6.1). Coefficients $\Psi_{-0,1,\dots,p}$ are coefficients of the inverse of (2.1.8), rewritten via $a_k(t)$, $a_k^-(t)$. This transformation, as we have seen, generates a motion invariant $I = \sum_n f_n |a_n|^2$, which is the same integral as in Sect. 2.5, restricted on the lattice (2.6.1).

If we now consider lattice values of $k_1, \dots, k_p$ in (2.6.3) obeying

$$s_1 k_1 + \ldots + s_p k_p = 0 , \tag{2.6.4a}$$

they generally do not belong to the resonant surface:

$$s_1 \omega_{k_1} + \ldots + s_p \omega_{k_p} = 0 . \tag{2.6.4b}$$

However there are special values which always belong to (2.6.4) and correspond to trivial billiard scattering with $p = 2q$, $s_1, \ldots, s_q = 1$, $s_{q+1}, \ldots, s_{2q} = -1$ where the set $(k_1, \ldots, k_q)$ is a transposition of the set $(k_{q+1}, \ldots, k_{2q})$. We shall see that these billiard scattering processes play an important role in the construction of the normal form.

Regardless of the special values of periods l_i, $i = 1, \ldots, d$, at large $|k_i|$, $i = 1, \ldots, p$, the corresponding set $(k_1, \ldots, k_p)$ can satisfy (2.6.4) with great accuracy, and we come to the problem of small denominators. However, in our case, Theorem 2.2.1 guarantees that coefficients of (2.6.3) are finite at these points; thus we only have to deal with trivial scattering.

2.6.3 Trivial Scattering and the Normal Form

Theorem 2.6.1. Let the space dimensionality be $d \geq 2$ and let H_{int} be defined by the formulae (2.6.2); furthermore, let the corresponding system (2.6.3) have one more motion invariant (in addition to H and P) of the form

$$I_g = \sum_n g_n |a_n|^2 + \sum_{p \geq 2} \sum_{0 \ldots p} {}^g I_{01 \ldots p} a_0 a_1 \ldots a_p , \tag{2.6.5}$$

where $g_n \neq \text{const}$, and all coefficient functions ${}^g I_{01 \ldots p}$ (referred to in the following simply as "coefficients") are finite [7]. Then:

1) If the dispersion law ω_k is degenerative with respect to decays (2.3.1) (so that $d = 2$), then for any f_k satisfying (2.3.14) on (2.3.1) an integral of the motion I_f for the system (2.6.3) exists. The I_f can be obtained by substituting f for g in all terms of the series I_g (2.6.4), and all coefficients ${}^f I_{01 \ldots p}$ of I_f are finite. [We have learned that such f_k should have the form of (2.3.15)]. However the action-angle variables analytic in a_k, a_k^* do not exist in the periodic case.

2) If ω_k is nondegenerate and has a zero limit as $|k| \to 0$, then there exist integrals I_f with any continuous $f_k \to 0$ as $|k| \to 0$. If in addition $U_{nn0}^{-111} = 0$ is true, then there exists a canonical transformation,

$$\gamma_{n_0}^{s_0} = a_{n_0}^{s_0} + \sum_{p \geq 2} \sum_{1 \ldots p} \hat{\Gamma}_{-01 \ldots p} a_1 \ldots a_p , \tag{2.6.6a}$$

mapping the system (2.6.1) to the form

$$i s \dot{\gamma}_n^s = \Omega_n \gamma_n^s , \quad \Omega_n = \Omega_n \left[|\gamma|^2 \right] \tag{2.6.6b}$$

and its Hamiltonian to the normal form

$$H = \sum_n h_n \left[|\gamma|^2 \right] = \sum_n \omega_n |\gamma_n|^2 + \ldots , \tag{2.6.7}$$

and $\Omega_n = \partial H/\partial|\gamma_n|^2$. The quantities $\Gamma_{-01\ldots p}$ can be obtained by recurrence or with the aid of the diagram technique and are finite at any p so that the zero denominators are absent in the canonic transformation (2.6.6).

3) If ω_k has a singularity near $|k| \to 0$, then the results are the same when imposing additional constraints. For example, for the KP-2 equation (1.1.3) with $\alpha^2 = -i$, it is necessary to impose a condition $a_n = 0$ at $n = (0, \nu_2)$; this constraint is compartible with the equation.

4) Let $H_{int} = H^{(4)}$;

$$H^{(4)} = \frac{1}{4} \sum_{1,2,3,4} \hat{T}_{1234}^{+1+1-1-1} a_1 a_2 a_3^* a_4^* . \tag{2.6.8}$$

then the system (2.1.1) having an additional motion invariant I_g analytic in a, a^* also has the additional integrals I_f with any continuous f_k, under the assumption that there exist limits of ω_k as $|k| \to 0$ and of T_{1234} as $k_1 \to k_3$ or $k_1 \to k_4$. Under these conditions there exists a canonical transformation mapping this system to the form (2.6.6) and its Hamiltonian to the normal form (2.6.7). The canonical transformation can be constructed in full analogy with (2.6.6).

5) If T_{1234} does not have such a limit (as an example one can think of the Davey-Stewartson equation (1.1.4) [15]), then these singularities are to be analysed separately; for the Davey-Stewartson equation, all the results of statement 4) are true [7].

Proof. Statement 1) is actually proved in Sect. 2.6a; therefore let us go to the nondegenerative case and suppose for convenience that ω_k is nondecaying, non-singular at $|k| = 0$ and $\omega(0) = 0$. Then Theorem 2.2.1 does not imply any restriction in the first order on th quantity $U_{i,jl}$. Actually, in the space (k_1, k_2), the surface

$$\omega(k_1 + k_2) = \omega(k_1) + \omega(k_2) \tag{2.6.9}$$

consists of two planes $k_1 = 0$, $k_2 = 0$, and each function f_k with $f(0) = 0$ is degenerative on them.

We shall seek coefficients of the transformation (2.6.9) immediately from (2.6.6, 7). We see that $\gamma_k[a]$ differs from $a_k^-[a]$, given by (2.6.3), only due to the nonlinear frequency shift:

$$\Omega_k = \omega_k + \delta\Omega_k \left[|\gamma|^2\right]$$

$$\delta\Omega_{k_0} = \sum_1 \Omega_{k_0 k_1} |\gamma_{k_1}|^2 + \sum_{p \geq 2} \sum_{1\ldots p} \Omega_{01\ldots p} |\gamma_1|^2 \ldots |\gamma|^2 .$$

This means that in the first order, (2.6.3) and (2.6.6) coincide, i.e.,

$$E_{012} \hat{\Gamma}_{012} = E_{012} \hat{\Psi}_{012} = \frac{-s_0}{2} \hat{U}_{012} . \tag{2.6.10}$$

So, we see that for (2.6.6) to exist, we have to require that

$$U_{nn0}^{1-1\pm1} = 0 , \tag{2.6.11a}$$

and (2.6.2) becomes

$$H^{(3)} = \frac{1}{3!} \sum_{1,2,3}' \hat{U}_{123} a_1 a_2 a_3 , \tag{2.6.2a}$$

where the prime indicates that terms with zero n_1 or n_2 or n_3 are absent. If this is true, one can choose

$$\Gamma_{nn0}^{1-1\pm1} = 0 , \tag{2.6.11b}$$

and (2.6.6) does not have small denominators in the first order. In the second order we have

$$E_{0\,123}^{-1\,-111} \Gamma_{0\,123}^{-1\,-111} = S_{0\,123}^{-1\,-111} - \Omega_{n_0 n_1} \delta(T_2) , \tag{2.6.12}$$

where T_2 is a surface of trivial scattering of the second order:

$$T_2 : \quad \begin{matrix} n_0 = n_3 \\ n_1 = n_2 \end{matrix} \quad \text{or} \quad \begin{matrix} n_0 = n_2 \\ n_1 = n_3 \end{matrix} . \tag{2.6.13}$$

$\delta(T_2) = 1$ if $(n_0 \ldots n_3) \in T_2$, and $\delta(T_2) = 0$ otherwise, and ($\tilde{\sigma}$ is the symmetrization operator)

$$S_{0\,123}^{-1\,-111} = E_{0\,123}^{-1\,-111} \Psi_{0\,123}^{-1\,-111} = -\frac{1}{3} \tilde{\sigma}_{0123} \sum_{s'n'} \Psi_{-01n'}^{s'} \hat{U}_{n'n_2n_3}^{-s's_2s_3} .$$

We see that Γ_{0123} equals Ψ_{0123} outside T_2 and differs from it on T_2, where Ψ_{0123} does not exist. Outside T_2, Ψ_{0123} exists, due to Theorem 2.2.1. From (2.6.12) we find $\Omega_{n_0 n_1}$ as

$$\Omega_{n_0 n_1} = S_{0\,123}^{-1\,-111} \big|_{T_2} . \tag{2.6.14}$$

Now on T_2 the $\Gamma_{0\,123}^{-1\,-111}$ is undetermined and should be obtained from the canonicity conditions

$$\{\gamma_p, \gamma_q\} = 0 = \{\gamma_p^*, \gamma_q^*\} , \quad \{\gamma_p, \gamma_q^*\} = i\delta_{pq} , \tag{2.6.15}$$

where

$$\{\alpha, \beta\} = i \sum_n \left(\frac{\partial \alpha}{\partial a_n} \frac{\partial \beta}{\partial a_n^*} - \frac{\partial \alpha}{\partial a_n^*} \frac{\partial \beta}{\partial a_n} \right) .$$

This gives, for example,

$$\Gamma_{p\,qpq}^{-1\,-111} = \frac{1}{2} \left| \Gamma_{p\,q\,p+q}^{-1\,-11} \right|^2 . \tag{2.6.16}$$

We shall consider the higher orders by induction. It is more convenient to do this for the inverse of (2.6.6),

$$a_{n_0}^{s_0} = A_{n_0}^{s_0}[\gamma] = \gamma_{n_0}^{s_0} + \sum_{p \geq 2} \sum_{1 \dots p} \hat{A}_{-01 \dots p} \gamma_1 \cdots \gamma_p . \tag{2.6.17}$$

The recurrence condition for coefficients of A can be written down in more compact form if we introduce the notation

$$\hat{\Pi}_{01 \dots q} = \Pi^{(q)} , \quad \hat{U} = \hat{U}_{ijl} , \quad \Omega_{n_\alpha}^{(p)} = \Omega_{n_\alpha 1 \dots p}$$

and

$$\Pi^{(q)} * X^{(r)} = \sum_{n' s'} \hat{\Pi}_{01 \dots q-1} {}_{n'}^{s'} \hat{X}_{n' q+1 \dots q+r}^{-s'}$$

for any functions. Then we have

$$E^{(2p)} A^{(2p)} = - \hat{\sigma} \left\{ -s_0 \hat{U} * A^{(2p-1)} - \frac{s_0}{2} \sum_{q=2}^{2p-2} \hat{U} * A^{(q)} * A^{(2p-q)} \right.$$

$$\left. + \sum_{q=1}^{p-1} A^{(2p-2q)} \sum_{\alpha=1}^{2p-2q} s_\alpha \Omega_{n_\alpha}^{(q)} \delta \left(T_q \right) \right\} \tag{2.6.18a}$$

$$E^{(2p+1)} A^{(2p+1)} = - \hat{\sigma} \left\{ -s_0 \hat{U} * A^{(2p)} - \frac{s_0}{2} \sum_{q=2}^{2p-1} \hat{U} * A^{(q)} * A^{(2p-q+1)} \right.$$

$$\left. + \Omega_{n_0}^{(p)} \delta \left(T_{p+1} \right) + \sum_{q=2}^{p} A^{(2p-2q+1)} \sum_{\alpha=1}^{2p-2q+1} s_\alpha \Omega_{n_\alpha} \delta \left(T_q \right) \right\} ; \tag{2.6.18b}$$

$A^{(1)} = 1$, and outside the resonant surface $E_{ijl} = 0$, one has

$$A^{(2)} = \hat{A}_{ijl} = -\hat{\Gamma}_{ijl} = \frac{s_i}{2} \frac{\hat{U}_{ijl}}{E_{ijl}} . \tag{2.6.18c}$$

On the resonant surface, if $E_{ijl} = 0$, we put (as above for $\Gamma^{(2)}$)

$$\hat{A}_{ijl} = 0 .$$

After this, the right-hand side of (2.6.18b) for $A^{(3)}$ up to the sign coincides with the right-hand side of (2.6.12), and we obtain Ω_{01} in the $\delta\Omega$ of the form (2.6.14). On the resonant surface

$$E_{0123} = 0 , \quad P_{0123} = 0 .$$

Let us define $A^{(3)} \sim \delta(T_2)$; the coefficient of proportionality must be obtained from canonicity conditions (2.6.15).

When going to higher orders $p \geq 4$, we suppose that $A^{(2q)}$; $2q < p$ equals zero on the resonant surfaces

$$E^{(2q)} = 0, \quad P^{(2q)} = 0, \tag{2.6.19a}$$

while $A^{(2q+1)}$, $2q + 1 < p$, on the manifold

$$E^{(2q+1)} = 0, \quad P^{(2q+1)} = 0 \tag{2.6.19b}$$

may be nonzero only on the trivial scattering submanifolds T_q of (2.6.19b) indicated above. Their values on T_q should be defined from (2.6.15).

Now consider the inversion procedure used to obtain (2.6.17) from (2.6.6). If we write (2.6.6) symbolically as

$$\gamma = a + \Gamma^{(2)} aa + \Gamma^{(3)} aaa + \Gamma^{(4)} aaaa + \ldots, \tag{2.6.20}$$

then

$$\begin{aligned}
a =& \gamma - \Gamma^{(2)} \gamma\gamma - \Gamma^{(3)} \gamma\gamma\gamma + 2\Gamma^{(2)} * \Gamma^{(2)} \gamma\gamma\gamma - \big[\Gamma^{(4)} \\
&+ 3\Gamma^{(3)} * \Gamma^{(2)} + 2\Gamma^{(2)} \Gamma^{(3)} - \Gamma^{(2)} * \Gamma^{(2)} * \Gamma^{(2)} \\
&- 6\Gamma^{(2)} \Gamma^{(2)} * \Gamma^{(2)} \big] \gamma\gamma\gamma\gamma + \ldots .
\end{aligned} \tag{2.6.21}$$

So we see that

$$A^{(p)} = -\Gamma^{(p)} + \sum \alpha_q \Gamma^{(q)} * \Gamma^{(p-q+1)} + \ldots,$$

and that

$$E^{(p)} A^{(p)} \xrightarrow[E^{(p)} \to 0]{} - E^{(p)} \Gamma^{(p)}$$

in points of a general position on (2.6.19). Therefore we can apply the considerations in Sect. 2.6a based on Theorem 2.2.1, and prove the solvability of (2.6.18) in points of a general position on (2.6.19). According to the induction hypothesis, in other points (of a special position) on (2.6.19), the terms which do not contain Ω can be nonzero only on a submanifold T_q of (2.6.19b) with $q = p$, where Ω-containing terms are nonzero only.

We consider the term $U * A^{(q)} * A^{(r)}$ on the resonant surface $E^{(q+r)} = 0$ in special points. This means that $E^{(q)} = 0$ and $E^{(r)} = 0$. According to the induction hypothesis, $A^{(r)} \neq 0$ only on the trivial scattering submanifold if r is odd. As for $U * A^q$ on $E^{(q)} = 0$, we have already seen that due to Theorem 2.2.1, it can be nonzero only in special points of $E^{(q)} = 0$, i.e., on the trivial scattering submanifold of $E^{(q)} = 0$ only (if at all). The $\Omega^{(q)}$ with $q < p$ are already known from junior orders, while $\Omega^{(p)}$ in (2.6.18b) is not known and should be chosen so as to eliminate the right-hand side of (2.6.18b) on T_p.

Now (2.6.18) are solvable, but $A^{(2p+1)}$ are undetermined on corresponding resonant surfaces. We set $A^{(2p)}|_{E^{(2p)}=0} = 0$ while $A^{(2p+1)} \sim \delta(T_p)$, and the coefficient of proportionality is to be found from (2.6.15). We can come to the next order and Theorem 2.6.1 is proven. (For the self-contained proof of this theorem, which does not apply to the rapidly decreasing case, see [7].)

Consider now the singular dispersion laws. The typical example is the Kadomtsev-Petviashvili equation (1.1.3) with $\omega_k = P^3 + \alpha^2 Q^2/P$, $k = (P, Q)$. This equation is known to have an infinite number of motion invariants both at $\alpha^2 = 1$ (KP-1), when the dispersion law is degenerative, and at $\alpha^2 = -1$ (KP-2), when it is nondecaying and nondegenerative. In the Hamiltonian description it corresponds to the equation with $H_{int} = H^{(3)}$, and

$$V_{123}^{111} = 0 \, , \quad V_{123}^{-111} = \sqrt{P_1 P_2 P_3}\, \theta(P_1)\theta(P_2)\theta(P_3) \, , \tag{2.6.22}$$

where θ is a Heaviside function. From this form of H_{int} it follows that $(d/dt)|a_{0Q}|^2 = 0$. But $\omega_k \to \infty$ as $P \to 0$, $Q \neq 0$ so that the KP-1 equation is senseless with resepct to a_{0Q}. The complete determination of this equation for a_{0Q} in the case of rapidly decreasing initial conditions leads to infinite numbers of constraints [16] except in the periodic case, where there is only one constraint, which can be easily obtained from the description of the KP equation in the form of a system,

$$\begin{aligned} u_t + uu_x + u_{xxx} + 3\alpha w_y &= 0 \\ w_x &= \alpha u_y \end{aligned} \tag{2.6.23}$$

Let us consider the Fourier-image of (2.6.23) with boundary conditions periodic in x and y, and particularly the dynamics of components with $P = 0$. One can see that for the solvability of the second equation with respect to w, it is necessary to impose a constraint $u_n = 0$ at $n = (0, Q)$, and the requirement of its invariance means that w_{0Q}. The latter is equivalent to introducing an integration constant,

$$\frac{1}{l} \int_0^l dx \int_0^x u_{yy}\, dx' \, ,$$

when reducing (2.6.23) to a single equation (l is a period in x). This additional term was obtained in [17] from a consideration of the Hamiltonian structure of the KP equation as generated by the Lie-Berezin-Kirillov bracket on orbits of a coadjoint action of the gauge group. In the form of (2.6.3) this leads to the correct form of the periodic KP-1 equation [$n_j = (P_j, Q_j)$]:

$$\begin{aligned} \dot{a}_{0Q}^s &= 0 \\ is_0 \dot{a}_0 &= \omega_{n_0} a_0 + \sum_{1,2}' \hat{U}_{-012} a_1 a_2 \, , \quad P_0 \neq 0 \, , \end{aligned} \tag{2.6.24}$$

where the prime near the summation sign indicates the absence of terms with $P_1 = 0$, $P_2 = 0$.

Starting from (2.6.24) and imposing the constraint $a_{0Q} = 0$, one can perform all the above procedures and see that if $\alpha^2 = -1$, the canonical transformation (2.6.6) exists and thus the nondegenerative KP equation is completely integrable.

If $\alpha^2 = 1$, the canonical transformation (2.6.6) does not exist and actions analytic in field variables are absent.

It should be noted that if we consider the case of rapidly decreasing boundary conditions, the distinctions between degenerative and nondegenerative equations disappear (we have pointed out this fact in Sect. 2.5 already). This is the reason for thinking of both KP equations as completely integrable systems [18]. The analogical distinction between two KP equations was recently obtained by *Krichever*, using his algebrogeometric appraoch [13].

Consider now the singular four-particle interaction when T_{1234} in (2.6.8) has a singularity on T_2 [(2.6.20) and above]. The important example for physical applications is the Davey-Stewartson equation (1.1.4) having T_{1234} of the form [19]:

$$T_{1234} = \frac{(P_1 - P_3)^2 - (Q_1 - Q_3)^2}{(P_1 - P_3)^2 + (Q_1 - Q_3)^2}$$
$$+ \frac{(P_1 - P_4)^2 - (Q_1 - Q_4)^2}{(P_1 - P_4)^2 + (Q_1 - Q_4)^2} = \frac{\kappa_1^2 \kappa_2^2 - \mu_1^2 \mu_2^2}{(\kappa_1^2 + \mu_1^2)(\kappa_2^2 + \mu_2^2)} , \tag{2.6.25}$$

where κ_1, κ_2, μ_1, μ_2 and P_0, Q_0 are the coordinates parametrizing the resonant surface:

$$P_1 = P_0 + \tfrac{1}{2}(\kappa_1 + \kappa_2) , \quad P_2 = P_0 - \tfrac{1}{2}(\kappa_1 + \kappa_2)$$
$$P_3 = P_0 + \tfrac{1}{2}(\kappa_1 - \kappa_2) , \quad P_4 = P_1 - \tfrac{1}{2}(\kappa_1 - \kappa_2)$$
$$Q_1 = Q_0 + \tfrac{1}{2}(\mu_1 + \mu_2) , \quad Q_2 = Q_0 - \tfrac{1}{2}(\mu_1 - \mu_2) \tag{2.6.26}$$
$$Q_3 = Q_0 + \tfrac{1}{2}(\mu_1 - \mu_2) , \quad Q_4 = Q_0 + \tfrac{1}{2}(\mu_1 + \mu_2) ,$$

where the resonance condition $E_{1234} = 0$ takes the form

$$\kappa_1 \kappa_2 - \mu_1 \mu_2 = 0 . \tag{2.6.27}$$

In points of a general position we see that $T_{1234} = 0$, in accordance with Theorem 2.6.1. Those points are singular where

$$\kappa_1 = \mu_1 = 0 \quad \text{or} \quad \kappa_2 = \mu_2 = 0 \tag{2.6.28}$$

In points of a general position for the transformation (2.6.6), we have

$$\Gamma_{1234} = \frac{\kappa_1 \kappa_2 + \mu_1 \mu_2}{(\kappa_1^2 + \mu_1^2)(\kappa_2^2 + \mu_2^2)} .$$

In the periodic case we see from (1.4) that on the manifolds (2.6.28), say, $\kappa_2 = \mu_2 = 0$, one has to put

$$T = \frac{\kappa_1^2 - \mu_1^2}{\kappa_1^2 + \mu_1^2} \quad \text{if } \kappa_1^2 + \mu_1^2 \neq 0 , \quad \kappa_2 = \mu_2 = 0$$

$$T = 0 \quad \text{if } \kappa_1 = \kappa_2 = \mu_1 = \mu_2 = 0 .$$

As a result we come to the nonsingular vertex T_{1234}; furthermore, we may construct the transformation (2.6.6) and prove the absence of zero denominators in it (using the existence of an additional integral) via the scheme outlined above.

3. Applications to Particular Systems

3.1 The Derivation of Universal Models

The present volume contains a paper by F. Calogero devoted to the derivation of universal models for nonlinear wave interactions from rather general types of differential equations. Many of these models appear to be exactly solvable; in fact even the more interesting situations hold true.

Let us take as a starting point some particular physical wave system, any-where from solid state physics to astrophysics; sometimes this model can be stated in terms of differential equations. Then let us perform the asymptotic expansion procedure on it. In doing so we single out the essential kernel of the physical phenomenon under consideration. The resulting model will prove universal and applicable to many physical problems at once. It is very likely that it will appear to be exactly solvable. In that case the model itself represents an important mathematical object. In order to study it, we may have to use advanced mathematics, like Lie group theory or algebraic geometry.

The occurrence of such wonderful things seems incomprehensible and an explanation may lie in the field of philosophy rather than in science. Here it is worth recalling the well-known paper by E. Wiggner, "On the Incomprehensible Effectivity of Mathematics in Natural Science" [20]. All of the above concerns both conservative and dissipative systems. We do not have a sufficiently general language for describing dissipative systems, but for conservative ones, we do: it is the language of Hamiltonian mechanics. This language takes its most simple form in the case of translationally invariant systems; i.e., when considering phenomena occurring in homogeneous space. Then it is possible to introduce canonical variables (amplitudes of progressive waves) a_k and take a Hamiltonian of the system in the form of a functional power series in a_k, a_k^* as a starting point.

We have considered the form of a Hamiltonian in the beginning of the present paper; dealing with such a Hamiltonian, it is easy to construct different universal models. A detailed description of the procedure may be found in papers [21–23]; we now consider several particular examples.

Let the Hamiltonian of the system have the form (our notation is the same is in Chap. 2)

$$H = \int \omega_k |a_k|^2 \, dk + \frac{1}{3!} \int \hat{V}_{012} a_0 a_1 a_2 \, dk_0 dk_1 dk_2 \ . \tag{3.1.1}$$

Let the oscillations in the medium with the wave vectors lying near three values ξ_0, ξ_1 and ξ_2 be excited; then the resonance conditions are fulfilled:

$$\xi_0 = \xi_1 + \xi_2 \, , \quad \omega_{\xi_0} = \omega_{\xi_1} + \omega_{\xi_2} \, . \tag{3.1.2}$$

Suppose that the domains in k-space occupied by these three packets do not overlap. Then one can introduce three fields, $a_0(k)$, $a_1(k)$, $a_2(k)$, to describe the behaviour of the system at times shorter than the time of the next order interaction (when one can neglect the higher nonlinear processes). Thus

$$a_k = a_0(k) + a_1(k) + a_2(k) \, ,$$

and in the Hamiltonian (3.1.1), one can make a substitution:

$$V_{012} a_0 a_1 a_2 \cong V_{\xi_0 \xi_1 \xi_2}^{s_0 - s_0 - s_0} a_0^{s_0}(k_0) a_1^{-s_0}(k_1) a_2^{-s_0}(k_2)$$

$$H \approx \sum_{j=0} \int \left[\omega_{\xi_j} + (v_j, k - \xi_j) \right] |a_j(k)|^2 \, ,$$

where $v_j = \nabla_k w(k = \xi_j)$. Now coming to the envelope fields,

$$A_j(r,t) = \frac{\exp(i\omega_{\xi_j} t)}{(2\pi)^{d/2}} \int a_j(k) e^{ik \cdot r} dk \, ,$$

we obtain the well-known three-wave system, q is a constant:

$$\begin{aligned}
\dot{A}_0 + (v_0 \nabla) A_0 &= q A_1 A_2 \\
\dot{A}_1 + (v_1 \nabla) A_1 &= q^* A_0 A_2^* \\
\dot{A}_2 + (v_2 \nabla) A_2 &= q^* A_0 A_1^* \, .
\end{aligned} \tag{3.1.3}$$

Now we show how the Hamiltonian (3.1.3) arises in physics, using the example of waves in media with weak dispersion [23] in which

$$\omega_k^2 = c^2 k^2 \left(1 + \lambda k^2 \right) \, . \tag{3.1.4}$$

Such a dispersion law is characteristic for waves on the surface of shallow water or for ion-acoustic waves in plasma. Media with a weak dispersion are described by the hydrodynamiclike equations with an additional term [22]:

$$\frac{\partial \varrho}{\partial t} + \operatorname{div} \varrho \nabla \phi = 0 \tag{3.1.5}$$

$$\frac{\partial \phi}{\partial t} + \frac{(\Delta \phi)^2}{2} = -\frac{c^2}{\varrho_0} \left(\delta \varrho + \frac{3}{2} - g \frac{\delta \varrho^2}{\varrho_0} - 2\lambda \Delta \delta \varrho \right) \, .$$

Here, ϕ is a hydrodynamic potential, $\delta \varrho$ is a quantity canonically conjugated to it, which can be called the "denstiy"; g is a constant. Introducing new variables a_k by formulae $(d = 3)$

$$\delta\varrho = \frac{1}{(2\pi)^{3/2}} \int \frac{\varrho_0^{1/2}k^{1/2}}{\sqrt{2}c^{1/2}} \left(a_k + a_{-k}^*\right) e^{ik\cdot r} dk \tag{3.1.6}$$

$$\nabla\phi = \frac{-i}{(2\pi)^{3/2}} \int \frac{k}{k^{1/2}} \frac{c^{1/2}}{\sqrt{2}\varrho_0^{1/2}} \left(a_k - a_{-k}^*\right) e^{ik\cdot r} dk \, ,$$

we obtain an interaction Hamiltonian (3.1.1) with

$$V_{k_0k_1k_2}^{s_0-s_0-s_0} = V_{k_0k_1k_2}^{s_0s_0s_0} = \frac{c^{1/2}}{16(\pi^3\varrho_0)^{1/2}} \left\{ \frac{(k_0\cdot k_1)k_2^{1/2}}{k_0^{1/2}k_1^{1/2}} \right.$$

$$\left. + \frac{(k_1\cdot k_2)k_0^{1/2}}{k_1^{1/2}k_2^{1/2}} + \frac{(k_2\cdot k_0)k_1^{1/2}}{k_2^{1/2}k_0^{1/2}} + 3g\,(k_0k_1k_2)^{1/2} \right\} \, . \tag{3.1.7}$$

The corresponding equations are

$$\frac{\partial a_{k_0}}{\partial t} = -i\frac{\delta H}{\delta a_{k_0}^*} = -i\omega_{k_0}a_{k_0} - i\left\{2\int\left[2V_{k_0k_1k_2}^*a_1a_2^*\delta_{k_0+k_2-k_1}\right.\right.$$

$$+ V_{k_0k_1k_2}a_1a_2\delta_{k_0-k_1-k_2}$$

$$\left.\left. + V_{k_0k_1k_2}^*a_1^*a_2^*\delta_{k_0+k_1+k_2}\right]dk_1dk_2\right\} \, . \tag{3.1.8}$$

They describe weakly nonlinear waves which are close to sinusoidal if the nonlinear correction to the frequency is much less than the dispersion correction. In essence this is a validity condition for the approximation (3.1.7–8).

Now let us convince ourselves that weakly nonlinear and weakly dispersive waves in the system (3.1.7–8) are described by the KP equation. Remember that in the original equations (3.1.5) we supposed the long wave approximation. Because the nonlinearity and dependence of a transverse coordinate are small, in the interaction Hamiltonian this transverse coordinate can be omitted. It should be taken into account only in the linear part of (3.1.8). As a result we obtained from (3.1.6)

$$V_{k_0k_1k_2}^{-s_0s_1s_2} = \frac{c^{1/2}(3g+3)}{16(\pi^3\varrho_0)^{1/2}}(p_0p_1p_2)^{1/2} \, ,$$

where $k_i = (p_i, q_i)$, $i = 0, 1, 2$ and $q_i \ll p_i$. In the linear term one has to expand w_k in a power series of the small q at finite p to obtain ($\lambda p^2 \ll 1$, $q^2 \ll p^2$):

$$\omega_k = c\sqrt{p^2 + q^2 + \lambda(p^2 + q^2)^2} \approx cp\sqrt{1 + q^2/p^2 + \lambda p^2}$$

$$\approx c\left(p + \frac{\lambda}{2}p^3 + \frac{q^2}{2p}\right) \, .$$

Up to some coefficients which can be removed by scaling transformations and the term cp in ω_k, which can be removed by transformation to the movable reference system, we obtain the KP Hamiltonian in normal coordinates $a_k(t)$, related with the original variable $u(x, y, t)$ via a formula like (3.1.6a).

Now consider the general case of an interaction of high-frequency (short) and low-frequency (long) waves [23] in the conservative medium with the Hamiltonian H. We introduce normal coordinates a_k for short waves and b_k for long waves. In these coordinates the quadratic part of the Hamiltonian has the form

$$H_0 = \int \omega_k a_k a_k^* dk + \int \Omega_k b_k b_k^* dk , \tag{3.1.9}$$

where ω_k and Ω_k are the dispersion laws of high-frequency and low-frequency waves. The interaction Hamiltonian can be represented in the form

$$H_{int} = H_1 + H_2 + H_3 ,$$

where H_1 describes the mutual interaction of the short waves, H_2 describes their interaction with the long waves and H_3 describes the mutual interaction of long waves. The motion equations have the standard form

$$\dot{a}_k = -i\frac{\delta H}{\delta a_k^*} , \quad \dot{b}_k = -i\frac{\delta H}{\delta b_k^*} . \tag{3.1.10}$$

In what follows we shall suppose that the b-amplitudes are small ($b_k \ll a_k$), and neglect the H_3. In the H_3 we keep only terms linear in b and of the lowest order in a_k which do not disappear when averaging over the long-wave period. These requirements enable us to find the Hamiltonian

$$H_2 = \int \left[h_{k_0 k_1 k_2} b_{k_0} a_{k_1} a_{k_1} + (*) \right] \delta_{k_0 - k_1 - k_2} dk_0 dk_1 dk_2 .$$

Here, $(*)$ indicates the complex conjugated expression. The theory is valid when

$$H_2 \gg \int \omega_k |b_k|^2 dk$$

and the low-frequency waves are strongly rearranged by the action of the high-frequency waves. We choose the Hamiltonian H_1 as

$$H_1 = \tfrac{1}{2} \int W_{k_0 k_1 k_2 k_3} a_0^* a_1^* a_2 a_3 \delta_{k_0 + k_1 - k_2 - k_3} dk_0 \ldots dk_3 .$$

This structure of H_1 is characteristic for a medium with cubic nonlinearity and in some cases, for a medium with quadratic nonlinearity and in some cases, for a medium with quadratic nonlinearity when cubic terms in the Hamiltonian may be removed by canonical transformation [22].

The interaction Hamiltonian is greatly simplified when the high-frequency waves form a narrow packet in the k-space near $k \approx k_0$. Then one can put

$$W_{k k_1 k_2 k_3} \approx W_{k_0 k_0 k_0 k_0} = q \tag{3.1.11a}$$

$$h_{k k_1 k_2} \approx h_{k k_0 k_0} = f(k, k_0) \tag{3.1.11b}$$

$$\omega_k = \omega_{k_0} + \frac{\partial \omega}{\partial k} \delta k + \frac{1}{2} \frac{\partial^2 \omega}{\partial k_\alpha \partial k_\beta} \delta k_\alpha \delta k_\beta \ . \tag{3.1.11c}$$

If the low-frequency waves are acoustic waves, $\Omega = ck$, it is possible to calculate $f(k, k_0)$ and the Hamiltonian of the system explicitly. To do so, let us replace a_k by the new variable (envelope field),

$$\Psi(r, t) = \frac{1}{(2\pi)^{3/2}} \int a_k \exp \{i\omega(k_0)t + i(k - k_0) \cdot r\} \, dk \ , \tag{3.1.12}$$

and b_k by two scalar functions: the density variation $\delta\varrho$ and the medium velocity v defined by formula (3.1.6) (with b_k standing for a_k). The energy of the narrow packet in the k-space is $\omega(k_0)|\Psi(r, t)|^2$. In the presence of the sound wave the quantity $\omega(k_0)$ acquires a variation,

$$\delta\omega(k_0) = \frac{\partial\omega(k_0)}{\partial\varrho} \delta\varrho + \frac{\partial\omega(k_0)}{\partial v} v \ ,$$

and the corresponding variation of the high-frequency wave energy is

$$\delta\varepsilon = \int |\Psi|^2 \left(\frac{\partial\omega}{\partial\varrho} \delta\varrho + \frac{\partial\omega}{\partial v} v \right) dr \ . \tag{3.1.13}$$

The quantity $\delta\varepsilon$ obviusly coincides with H_2. In the isotropic medium,

$$\frac{\partial\omega(k_0)}{\partial v} = \alpha \frac{k_0}{k_0} \ .$$

Let us introduce the notation

$$\frac{\partial\omega(k_0)}{\partial\varrho} = \beta \ , \quad v = \nabla\Phi \ ,$$

where Φ is the hydrodynamic potential. As one can see from (3.1.12), the quantity Ψ is a canonical transformation of a_k and therefore

$$i\frac{\partial\Psi}{\partial t} = iv_g \frac{\partial\Psi}{\partial z} - \frac{1}{2}\omega_k'' \frac{\partial^2\Psi}{\partial z^2} - \frac{v_g}{2k_0} \Delta_\perp\Psi + \Psi \left(q|\Psi|^2 + \beta\delta\varrho \right.$$
$$\left. + \alpha\frac{\partial\Phi}{\partial z} \right) = \frac{\delta(H - \int \omega_0|\Psi|^2 dr)}{\delta\Psi^*} \ . \tag{3.1.14}$$

The variables Φ and $\delta\varrho$ are canonically conjugated and obey the equations

$$\frac{\partial\delta\varrho}{\partial t} = -\varrho_0\Delta\Phi - \alpha\frac{\partial}{\partial z}|\Psi|^2 = \frac{\delta H}{\delta\Phi} \tag{3.1.15}$$

$$\frac{\partial\Phi}{\partial t} = -c^2\frac{\delta\varrho}{\varrho_0} - \beta|\Psi|^2 = -\frac{\delta H}{\delta\varrho} \ . \tag{3.1.16}$$

Inserting (3.1.12) and formulae like (3.1.16) (expressing $\delta\varrho$ and v in terms of b_k) into (3.1.13), we find for $f(k, k_0)$ the expression

$$f(k, k_0) = \left(\frac{k}{16\pi c \varrho_0} \right)^{1/2} \left(\beta \varrho_0 + \alpha c \frac{(k, k_0)}{k k_0} \right) .$$

Equations (3.1.14–16) describe the interaction of high-frequency waves of any nature with waves of the acoustic type. In many physical problems the dependence of a high-frequency wave dispersion law on the medium velocity may be neglected, and one may put $\alpha = 0$. Then (3.1.15, 16) may be reduced to one equation:

$$\left(\frac{\partial^2}{\partial t^2} - c^2 \Delta \right) \delta \varrho = + \beta \varrho_0 \Delta |\Psi|^2 . \tag{3.1.17}$$

Let $v_g \neq c$ and the amplitude of high-frequency waves be sufficiently small. Then one may consider the low-frequency waves as purely forced and replace $\partial / \partial t$ by $v_g \partial / \partial z$. System (3.1.14–16) is reduced now to the form

$$i\Psi_l = i v_g \Psi_z - \frac{1}{2} \omega'' \Psi_{zz} - \frac{v_g}{2k} \Delta_\perp \Psi + u \Psi$$
$$L_1 u = L_2 |\Psi|^2 \tag{3.1.18}$$
$$u = q|\Psi|^2 + \beta \delta \varrho + \alpha \Phi_z ,$$

where L_1 and L_2 are second-order homogeneous partial differential operators:

$$L_n = C_{ij}^{(n)} \frac{\partial^2}{\partial \kappa_i \partial \kappa_j} . \tag{3.1.19}$$

The system (3.1.18) is universal for the description of small-amplitude, high-frequency waves with acoustic-type waves.

3.2 Kadomtsev-Petviashvili and Veselov-Novikov Equations

Let us apply the results obtained in the Chap. 2 to the KP equation. Let us begin with KP-2. The dispersion law of this equation,

$$\omega = p^3 - \frac{3q^2}{p} ,$$

is nondecaying. Therefore, from the results of Chap. 2 it follows that states asymptotic as $t \to \pm\infty$ coincide for KP-2 with rapidly decreasing boundary conditions [see (2.4.3) and [5], [6], [9] also]. It also follows that amplitudes of the classical scattering matrix become zero on the corresponding resonant surfaces in points of general position. In the first order this fact is trivial: the dispersion law for KP-2 is nondecaying, $V_{012} = (p_0 p_1 p_2)^{1/2}$, and (2.6.9) has in this case solutions $k_1 = 0$ or $k_2 = 0$ only. The analogous identity for second-order amplitudes was verified in [4].

The phenomenon of coincidence of asymptotic states for KP-2 was obtained independently in [18] via the inverse scattering technique. With periodic boundary conditions the KP-2 may be transformed to the normal form (2.6.6) and is completely integrable [7]. The hypothesis of complete integrability of periodic KP-2 was proposed for the first time in [5].

We have already mentioned in Chap. 2 that the dispersion law of the KP-1 equation is degenerative:

$$p = \xi_1 - \xi_2 , \quad q = \xi_1^2 - \xi_2^2 , \quad \omega_k = 4\left(\xi_1^3 - \xi_2^3\right) .$$

On the resonant surface the coefficient function in the Hamiltonian is

$$V_{kk_1k_2} = \sqrt{pp_1p_2} = [(\xi_1 - \xi_2)(\xi_2 - \xi_3)(\xi_1 - \xi_3)]^{1/2} \neq 0 .$$

Therefore the KP-1 equation describes a nontrivial scattering. The states asymptotic as $t \to \pm\infty$ do not coincide and are related by the formula

$$
\begin{aligned}
C^*_{\xi_1\xi_2} - C^-_{\xi_1\xi_2} =& \frac{\pi i}{12(\xi_1 - \xi_2)} \int_{\xi_2}^{\xi_1} [(\xi_1 - \xi_2)(\xi_1 - \xi_3)(\xi_3 - \xi_2)]^{1/2} \\
& \times \left(C^+_{\xi_1\xi_3} + C^-_{\xi_1\xi_3}\right)\left(C^+_{\xi_3\xi_2} + C^-_{\xi_3\xi_2}\right) d\xi_3 \\
& + \int_{-\infty}^{\xi_2} [(\xi_1 - \xi_2)(\xi_1 - \xi_3)(\xi_2 - \xi_3)]^{1/2} \\
& \times \left(C^{*+}_{\xi_2\xi_3} + C^{*-}_{\xi_2\xi_3}\right)\left(C^+_{\xi_1\xi_3} + C^-_{\xi_1\xi_3}\right) d\xi_3 ,
\end{aligned}
$$

where $C_{\xi_1\xi_j} = C(p, q)$, $p = \xi_i - \xi_j$, $q = \xi_i^2 - \xi_j^2$. In the periodic case (and in any case in which boundary conditions are vanishing the KP-1 equation proves to be a nonintegrable system.

Recently, the Veselov-Novikov equation [24]

$$v_t = \partial^3 v + \bar{\partial}^3 v + \partial(uv) + \bar{\partial}(\bar{u}v) \tag{3.2.1}$$

$$\bar{\partial} u = -3\partial v , \quad v = \bar{v}$$

has been considered. Here, $\partial = \partial_z = \partial_x - i\partial_y$, $z = x + iy$, and the bar indicates complex conjugation. The solutions independent of y for this equation are reduced to the solution of the Korteveg–de Vries equation. The Equation (3.2.1) can be solved via the inverse scattering transform method [24] and allows a L-A-B triad representation,

$$\frac{\partial L}{\partial t} + [L, A + \bar{A}] = fL ,$$

where

$$L = -\Delta + v(z, \bar{z}) , \quad A = \partial^3 + u\partial \quad \text{and} \quad f = \partial u + \bar{\partial} u .$$

The properties of (3.2.1) depend strongly on conditions asymptotic as $z \to \infty$. If $v \to 0$ as $z \to \infty$, then the Veselov-Novikov equation has a dispersion law,

$$\omega_k = 2\left(p^3 - 3q^2\right) , \quad k = (p, q) , \tag{3.2.2}$$

which is nondegenerative (because it is analytic in p and q), like the KP-2 ($\alpha^2 = -1$) dispersion law. Hence (3.2.2) must possess all the properties of the KP-2 equation. Since the dispersion law (3.2.2) is nondecaying, we have to verify that the first-order scattering amplitude becomes zero on the resonant surface. The resonant manifold is determined in the space (p_1, p_2, q_1, q_2) by the equation

$$(p_1 p_2 - q_1 q_2)(q_1 + q_2) + (p_1 q_2 + p_2 q_1)(p_1 + p_2)$$
$$= p_1 p_2(p_1 + p_2) + p_1 q_2^2 + p_1 q_2^2 + p_2 q_1^2 = 0 . \tag{3.2.3}$$

Now let us calculate the first-order scattering amplitude. For this we make the Fourier transformation in (3.2.1) via x and y :

$$v(x, y, t) = \frac{1}{2\pi} \int \left(v_k + v^*_{-k}\right) e^{i(px+qy)} dx\, dy .$$

Now the relation between u and v takes the form of $(k = (p, q))$:

$$u_k = -3\frac{\kappa^*}{\kappa} \left(v_k + v^*_{-k}\right) , \quad \kappa = p + iq . \tag{3.2.4}$$

Substituting (3.2.4) into (3.2.1), we obtain in the nonlinear term the expression (up to the coefficients unessential for us)

$$\int \left[\kappa \left(\frac{\kappa_1^*}{\kappa_1} + \frac{\kappa_2^*}{\kappa_2}\right) + \kappa^* \left(\frac{\kappa_1}{\kappa_1^*} + \frac{\kappa_2}{\kappa_2^*}\right)\right] v_{k_1} v_{k_2} \delta(k - k_1 - k_2) dk_1 dk_2 \tag{3.2.5}$$

and other terms containing v^*v, which we need not write down because we know beforehand that the (3.2.1) is a Hamiltonian equation. The squared bracket in (3.2.5), after making the substitution $\kappa = p + iq$, taking into account the δ-function and making some algebraic transformations, becomes

$$\frac{[p_1 p_2 + q_1 q_2][p_1 p_2(p_1 + p_2) + p_1 q_2^2 + p_2 q_1^2]}{(p_1^2 + q_1^2)(p_2^2 + q_2^2)} . \tag{3.2.6}$$

Now making the replacement

$$v_k = a_k(p^2 + q^2)^{1/2} ,$$

and rewriting (3.2.1) in terms of a_k, we have the interaction Hamiltonian

$$H_{\text{int}} = \int V_{k_1+k_2, k_1 k_2} \left(a^*_{k_1+k_2} a_{k_1} a_{k_2} + (^*)\right) dk_1 dk_2 ,$$

where

$$V_{k_1+k_2,k_1k_2} = \text{const} \frac{(p_1p_2 + q_1q_2)(p_1p_2(q_1+q_2) + p_1q_2^2 + p_2q_1^2)}{[[(p_1+p_2)^2 + (q_1+q_2)^2](p_1^2+q_1^2)(p_2^2+q_2^2)]^{1/2}} .$$

Comparing this expression with (3.2.3), we see that $V_{k_1+k_2,k_1k_2}$ contains the energy denominator $E = \omega_{k_1+k_2} - \omega_{k_1} - \omega_{k_2}$ as a factor and becomes zero simultaneously with it, in agreement with Theorem 2.2.1. For the Veselov-Novikov equation, other statements concerning $KP-2$ are also true, namely the coincidence of asymptotics as $t \to \pm\infty$, the triviality of scattering and the existence of a transformation to the normal form in the periodic case.

3.3 Davey-Stewartson-Type Equations.
The Universality of the Davey-Stewartson Equation in the Scope of Solvable Models

As we have seen in Sect. 3.1, the problem of the interaction of small-amplitude, quasimonochromatic wave packets with acoustic waves leads in an natural way to equations which we shall call Davey-Stewartson-type equations:

$$i\Psi_t + L_1\Psi + u\Psi = 0$$
$$L_2u = L_3|\Psi|^2 .$$

$$(3.3.1)$$

Here, $u(r,t)$ is a real function indicating a mean field while $\Psi(r,t)$ is a complex function representing the envelope, $r = (x_1, \ldots, x_d)$, $d = 2,3$ and

$$L_n = \sum_{i,j}^{d} C_{ij}^{(n)} \frac{\partial^2}{\partial x_i \partial x_j} , \quad n = 1,2,3 .$$

$$(3.3.2)$$

The Davey-Stewartson equation itself is written via operators (3.3.2) of the form

$$L_1 = L_3 = \frac{\partial}{\partial x^2} \pm \frac{\partial^2}{\partial y^2}$$

$$L_2 = \pm\left(\frac{\partial^2}{\partial x^2} \mp \frac{\partial}{\partial y^2}\right) .$$

It arises when applying the multiscale expansion technique to the KP equation [25, 26] and in the theory of two-dimensional long waves over finite depth liquids [15].

To study the system (3.3.1) it is convenient to rewrite it in the explicitly Hamiltonian form

$$i\dot{\Psi}_{k_0} + L_1(k_0)\Psi_{k_0} + \int \hat{T}_{0123}\Psi_{k_1}^* \Psi_{k_2}\Psi_{k_3} dk_1 dk_2 dk_3 = 0 ,$$

$$(3.3.3)$$

where $L(k)$ are symbols of the operators (3.3.2). The vertex $\hat{T}_{0123} = T_{0123}\delta(P_{0123})$ has the form

$$T_{0123} = \frac{L_3(k_0 - k_2)}{L_2(k_0 - k_2)} + \frac{L_3(k_0 - k_3)}{L_2(k_0 - k_3)} \qquad (3.3.4)$$

and is defined on the surface $P_{0123} \equiv k_0 + k_1 - k_2 - k_3$. The Hamiltonian of the equation (3.3.1) has the form

$$H = \int L_1(k)|\Psi|^2 dk$$

$$+ \frac{1}{2} \int \hat{T}_{0123} \Psi_{k_0}^* \Psi_{k_1}^* \Psi_{k_3} \Psi_{k_3} dk_0 \dots dk_3 . \qquad (3.3.5)$$

The quadratic form $L_1(k)$ may be transformed to the diagonal form via the non-degenerative map. After doing so, the new coefficients $\tilde{C}_{ij}^{(2)}$, $\tilde{C}_{ij}^{(3)}$ arise; we shall designate α_{ij}, β_{ij}, correspondingly. The dispersion law $w = L_1(k)$ is degenerative only when $L_1(k) = k_1^2$, $k = (k_1, \dots, k_d)$; we shall not take this case into account. In all other cases the system (3.3.1) may have additional motion invariants only if T becomes zero on the resonant surface (2.3.32). One should note that if T becomes zero at some L_2, L_3, then it also becomes zero upon interchanging L_2 and L_3. To distinguish between these systems, one has to analyse the second-order vertex [19], This analysis leads to the following results at $d = 2$, when

$$w_k = k_1^2 + \sigma k_2^2 , \quad \sigma = \pm 1 .$$

Let $\sigma = 1$. Then the solvable system is

$$\beta_{12} = 0 , \quad \beta_{11} = \beta_{22} = \beta , \quad \alpha_{11} = -\alpha_{22} = \alpha ,$$
$$i\Psi_t + \Delta\Psi + u\Psi = 0 \qquad (3.3.6)$$
$$\left[\alpha \left(\partial_x^2 - \partial_y^2 \right) + 2\alpha_{12}\partial_x\partial_y \right] u = \beta\Delta|\Psi|^2 .$$

By a change of variables the last equation could be transformed to the form

$$\left(\frac{\partial^2}{\partial x^2} - \frac{\partial^2}{\partial y^2} \right) u = \Delta|\Psi|^2 .$$

If $\sigma = -1$, we have the counterpart of (3.3.6): $\beta_{11} = \beta_{22} = \beta$, $\alpha_{11} = -\alpha_{22} = \alpha$, $\alpha_{12} = 0$. As in (3.3.6), in diagonal form we obtain:

$$i\Psi_t + \left(\partial_x^2 - \partial_y^2 \right) \Psi + u\Psi = 0$$
$$\alpha\Delta u = \beta \left(\partial_x^2 - \partial_y^2 \right) |\Psi|^2 \qquad (3.3.7)$$

and also the system

$$N \left(\partial_x \pm \partial_y \right) \left[\left(\partial_x \pm \partial_y \right) u + \left(\partial_x \mp \partial_y \right) |\Psi|^2 \right] = 0$$

$$i\Psi_t + \left(\partial_x^2 - \partial_y^2 \right) \Psi + u\Psi = 0 .$$

The latter in coordinates $x_1 - x_2 = \xi$, $x_1 + x_2 = \eta$ becomes

$$i\Psi_t + \Psi_{\xi\eta} + u\Psi = 0, \quad u_\xi = |\Psi|^2_\eta. \tag{3.3.8}$$

The system (3.3.8) and its L-A pair has been presented in [9]. Equations (3.3.6, 7) are also integrable via the inverse scattering; more detailed information can be found in [19]. At $d = 3$, analogous but much more extensive analysis shows that the system (3.3.1) does not have any additional invariants.

It is useful to keep in mind the following fact. If one takes some two-dimensional, exactly solvable model and considers the initial conditions, like rapidly oscillating waves with slowly varying amplitude, then after the averaging procedure (or multiple scale expansion), one obtains the envelope equation in the form of one of the Davey-Stewartson equations (with one of the two admissible combinations of L_2 and L_3), the so-called DS-1 and DS-2. Specific examples can be found in [26].

3.4 Applications to One-Dimensional Equations

The ideas developed above can be explored in the one-dimensional case using the results contained in Sect. 3.3a). We present here the results for: a) the two coupled nonlinear Schrödinger equation system [12]; b) the systems describing the long-acoustic and short-wave interaction (first neglecting [11] and then taking into account [27] the effects of eigen nonlinearity and the dispersion of long waves); and c) the system describing the interaction of two counter-directed wave packets in the cubic medium [28].

The system of two coupled nonlinear Schrödinger equations arises in nonlinear optics [29] and has the form

$$\begin{aligned}
i\Psi_{1t} &= C_1\Psi_{1xx} + 2\alpha|\Psi_1|^2\Psi_1 + 2\beta|\Psi_2|^2\Psi_1 \\
i\Psi_{2t} &= C_2\Psi_{2xx} + 2\gamma|\Psi_2|^2\Psi_2 + 2\beta|\Psi_1|^2\Psi_2.
\end{aligned} \tag{3.4.1}$$

It is a Hamiltonian system:

$$\begin{aligned}
H = \int \Big\{ &C_1|\Psi_{1x}|^2 + C_2|\Psi_{2x}|^2 + \alpha|\Psi_1|^2 \\
&+ 2\beta|\Psi_1|^2|\Psi_2|^2 + \gamma|\Psi_2|^2 \Big\}.
\end{aligned} \tag{3.4.2}$$

The exact solvability of (3.4.1) with $C_1 = C_2$, $\alpha = \beta = \gamma$ has been shown in [30]. To study the system (3.4.1) in the general case we have to first determine whether the set of dispersion laws

$$w_1(k) = C_1 k^2, \quad w_2(k) = C_2 k^2 \tag{3.4.3}$$

is degenerative to the process (2.3.8). As we have already seen, at $\varrho = C_1/C_2 \neq \pm 1$ the set (3.4.3) is nondegenerative to the process (2.3.8). Because the amplitude of the process (2.3.8) is a constant in all k-space and equal to $2\beta \neq 0$, the system (3.4.1) cannot have an additional integral at $\varrho \neq \pm 1$. At $\varrho = \pm 1$, one has to calculate the second-order amplitude corresponding to the next nonlinear process. One may calculate, for example, the amplitude of the process (2.3.10). The corresponding manifold in the space $(k_1, \ldots, k_0)$ is quadratic and has a rational parametrization [12]. Using it, one may show that the set (3.4.3) is nondegenerative to (2.3.10). The amplitude of the process (2.3.10) is rather complicated; it is important that this amplitude become zero in two cases:

$$\varrho = 1, , \quad \alpha = \beta \quad \text{and} \quad \varrho = -1, \quad \alpha = -\beta . \tag{3.4.4}$$

Analogously one may obtain $\beta = \gamma$ at $\varrho = 1$ and $\beta = -\gamma$ at $\varrho = -1$. Therefore except for the "vector Schrödinger equation" $\varrho = 1$, the equations (3.4.1) with

$$\varrho = -1 , \quad \alpha = -\beta = \gamma \tag{3.4.5}$$

may be integrable also. The system (3.4.1) with coefficients (3.4.5) is indeed integrable. In fact, in [31] it has been shown that the inverse scattering method is applicable to the system

$$i\Psi_t = \Psi_{xx} + \Psi X \Psi$$
$$- iX_t = X_{xx} + X\Psi X , \tag{3.4.6}$$

Where X and Ψ are matrices. We choose

$$\Psi = (\Psi_1, \ldots, \Psi_n) , \quad X = \begin{pmatrix} X_1 \\ \vdots \\ X_n \end{pmatrix}$$

and consider the reduction $X = A\Psi^+$, where A is a Hermitean matrix. Then we have

$$i\Psi_{mt} = \Psi_{mxx} + u\Psi_m , \quad m = 1, \ldots, n , \tag{3.4.7}$$

where $u = \Psi A \Psi^+$ is a real function. By a unitary transformation, the matrix A may be transformed to the diagonal form $A \to \alpha_i \delta_{ij}$. Therefore if $n = 2$, there are only two possibilities, namely the vector Schrödinger case [30] and the system with coefficients (3.4.5). This integrable system was obtained independently in [12] and [32].

The nonintegrability of the system describing the resonant interaction of long acoustic waves and short waves derived in [33] may be proved in an analogous way [11]:

$$i\Psi_t + \Psi_{xx} + u\Psi = 0$$
$$u_{tt} + c^2 u_{xx} = 2 \left(|\Psi|^2\right)_{xx} ; \tag{3.4.8}$$

as well as the nonintegrability of the system [27]

$$u_t + \left(u^2 + \alpha|\Psi|^2 + u_{xx}\right)_x = 0 \tag{3.4.9}$$

$$i\Psi_t + \Psi_{xx} + u\Psi = 0 ,$$

generalizing the system (3.4.8). In nonlinear optics a system also arises [34]:

$$\frac{\partial S^+}{\partial \xi} = S^+ \times I S^- + S^+ \times I^+ S^+$$
$$\frac{\partial S^+}{\partial \xi} = S^- \times I S^+ + S^- \times I^- S^- ,$$

(3.4.10)

where

$$\frac{\partial}{\partial \xi} = \frac{\partial}{\partial t} - v\frac{\partial}{\partial x} , \quad \frac{\partial}{\partial \eta} = \frac{\partial}{\partial t} + v\frac{\partial}{\partial x} ,$$

and I, I^+, I^- are diagonal matrices. If $I^+ = I^- = 0$, the system (3.4.10) coincides with the assymetric chiral field equations [35] and is integrable. In [28] it is shown that this case exhausts all the possibilities of integrability of (3.4.10). The proof uses Theorem 2.3.1 and the lemma from paper [36] concerning the system (3.4.10) with $\partial/\partial x = 0$. Let us discuss this point in more detail.

Lemma. For a reduced system (3.4.9) with $\partial/\partial x = 0$ to be integrable, it is necessary that the system

$$\frac{\partial S^+}{\partial t} = S^+ \times I^+ S^+$$
$$\frac{\partial S^-}{\partial t} = S^- \times I S^+$$

(3.4.11)

possess an additional integral to $I = (S^+ J^+ S^+)$ linear in S^- and of the degree 1 in S^+, l is an integer.

From this lemma it follows that if matrices

$$J^+ = \text{diag} \left(J_1^+, J_2^+, J_3^+ \right) , \quad J_1^+ \neq J_2^+ \neq J_3^+$$

are nondegenerative, then the equality

$$\left(J_1^+ - J_2^+ \right) \left(J_3^+ \right)^2 + \left(J_2^+ - J_3^+ \right) \left(J_1^+ \right)^2 + \left(J_3^+ - J_1^+ \right) \left(J_2^+ \right)^2$$
$$+ k^2 \left(J_1^+ - J_2^+ \right) \left(J_2^+ - J_3^+ \right) \left(J_3^+ - J_1^+ \right) = 0 , \quad k \in \mathbb{N} , \quad k \neq 0$$

is a necessary condition for the system (3.4.11) to be integrable. Even if at $\partial/\partial \xi = \partial/\partial \eta = \partial/\partial t$ this condition is fulfilled, under other reductions,

$$\frac{\partial}{\partial \xi} \rightarrow m\frac{\partial}{\partial t} , \quad \frac{\partial}{\partial \eta} \rightarrow m^2\frac{\partial}{\partial t} ,$$

it is not fulfilled. This means that for (3.4.10) to be integrable, it is necessary that two entries J_j^+, $j = 1, 2, 3$ coincide. Becuase we may add to J^+ any diagonal matrix, one can set $J_1^+ = J_2^+ = 0$, $J_3^+ \neq 0$. Then the first equation in (3.4.11) can be solved easily and

$$S^+ = \begin{pmatrix} M_1 \cos \left(J_3^+ M_0 t + \varphi_0 \right) \\ -M_1 \sin \left(J_3^+ M_0 t + \varphi_0 \right) \\ M_0 \end{pmatrix} .$$

Here M_0, M_1, φ_0 are arbitrary constants. Further, let an integral exist,

$$Z = \sum S_i^- P_{l_i}^-(S^+) , \tag{3.4.12}$$

where $P_l^i(S^+)$ are polynomials of S^+ of degree l, $i = 1, 2, 3$. As has been shown in [36], from the existence of an additional integral of the reduced system (3.4.11), it follows that $J_1 = J_2$.

Because all the aforesaid also applies to J^-, we conclude that there are only two possibilities for the system (3.4.10) to be integrable:

$$J = \text{diag} \left(J_1, J_1, J_3 \right) , \quad J^{\pm} = \text{diag} \left(0, 0, J_3^{\pm} \right)$$

and

$$J = \text{diag} \left(J_1, J_1, J_1 \right) , \quad J^+ = \text{diag} \left(0, J_2^+, 0 \right) , \quad J^- = \text{diag} \left(0, 0, J_3^+ \right) .$$

Now one has to use the Holdstein-Primakov variables,

$$S_1^+ + i S_2^+ = a \sqrt{2M^+ - |a|^2} , \quad M^+ = |S^+| ,$$

$$S_1^- + i S_2^- = b \sqrt{2M^- - |b|^2} , \quad M^- = |S^-| ,$$

by which the system (3.4.10) acquires the standard form (2.1.1) with $\alpha = 2$, $a^{(1)} = a$, $a^{(2)} = b$ and dispersion laws

$$\omega_k^{(1),(2)} = \omega_k^{\pm} = c_1 \pm c_2 \sqrt{c_3 k^2 + 1} , \quad c_i = \text{const} .$$

$$c_i = \text{const} .$$

The set $\{\omega^+\}$ is nondegenerative to the six-particle processes. By calculating the second-order vertex and checking that it is nonzero on the resonant surface at $J^{\pm} \neq 0$, we obtain the required statement.

The system generalizing (3.4.1) with the Hamiltonian has also been studied [37]

$$II = \int \left\{ c_1 |\Psi_{1x}|^2 + c_2 |\Psi_{2x}|^2 + \alpha |\Psi_1|^4 + 2\beta |\Psi_1|^2 |\Psi_2|^2 + \gamma |\Psi_2|^4 \right.$$
$$\left. + \delta \left(\Psi_1^2 \Psi_2^{*2} + \Psi_1^{*2} \Psi_2^2 \right) \right\} dx .$$

Quite analogously to (3.4.2), when $\beta \neq 0$, one obtains $c_1 = \pm c_2$. At $\delta \neq 0$, only the possibility $c_1 = c_2$ remains (one has to consider the process $\Psi_1 + \Psi_1 + \Psi_2 \rightarrow \Psi_2 + \Psi_2 + \Psi_2$). The result is that the integrable cases are already known and can be found in [38].

Appendix I [6]

Proofs of the Local Theorems
(of Uniqueness and Others from Sect. 2.3)

We are seeking the functions $\omega(p, q)$ and $f(p, q)$ determined parametrically in the form

$$p = \xi_1 - \xi_2 \quad q = a(\xi_1) - a(\xi_2)$$

$$\omega = b(\xi_1) - b(\xi_2) + \sum_{n=1}^{\infty} \varepsilon^n \omega_n(\xi_1, \xi_2)$$

$$f = c(\xi_1) - c(\xi_2) + \sum_{n=1}^{\infty} \varepsilon^n f_n(\xi_1, \xi_2) .$$

(A.1.1)

Here, ε is a small denominator. It is convenient to set the three-dimensional resonance manifold parametrically in the form

$$p_1 = \xi_1 - \xi_2 , \quad q_1 = a(\xi_1 + \eta) - a(\xi_2 + \eta)$$
$$p_2 = \xi_3 - \xi_2 , \quad q_2 = a(\xi_3 + \nu) - a(\xi_2 + \nu) ,$$

(A.1.2)

requiring additionally that

$$q = q_1 + q_2 = a(\xi_1) - a(\xi_2)$$
$$= a(\xi_1 + \eta) - a(\xi_3 + \eta) + a(\xi_3 + \nu) - a(\xi_2 + \nu) .$$

(A.1.3)

Now conditions (2.3.13, 14) together with (A.1.3) will impose three equations upon five parameters $\xi_1, \xi_2, \xi_3, \eta, \nu$. This system of equations must define η and ν in the form of a series in ε:

$$\eta = \sum_{n=1}^{\infty} \varepsilon^n \eta_n(\xi_1, \xi_2, \xi_3) , \quad \nu = \sum_{n=1}^{\infty} \varepsilon^n \nu_n(\xi_1, \xi_2, \xi_3) .$$

We have a linear overdetermined system in the first order in ε :

$$\left[a'(\xi_1) - a'(\xi_3)\right] \eta_1 + \left[a'(\xi_3) - a'(\xi_2)\right] \nu_1 = 0$$
$$\left[b'(\xi_1) - b'(\xi_3)\right] \eta_1 + \left[b'(\xi_3) - b'(\xi_2)\right] \nu_2 = \Omega_1$$
$$\left[c'(\xi_1) - c'(\xi_3)\right] \eta_1 + \left[c'(\xi_3) - c'(\xi_2)\right] \nu_3 = F_1 .$$

(A.1.4)

Here

$$\Omega_1 = \omega_1(\xi_1, \xi_2) - \omega_1(\xi_1, \xi_3) - \omega_1(\xi_3, \xi_2)$$
$$F_1 = f_1(\xi_1, \xi_2) - f_1(\xi_1, \xi_3) - f_1(\xi_3, \xi_2) .$$

(A.1.5)

The consistency condition of the system (A.1.4) has the form

$$\Omega_1 B = F_1 A ,$$

(A.1.6)

where

$$A(\xi_1, \xi_2, \xi_3) = \Delta_{ab} = \alpha(\xi_1, \xi_2) + \alpha(\xi_2, \xi_3) + \alpha(\xi_3, \xi_1)$$
$$B(\xi_1, \xi_2, \xi_3) = \Delta_{ac} = \beta(\xi_1, \xi_2) + \beta(\xi_2, \xi_3) + \beta(\xi_3, \xi_1)$$

(A.1.7)

$$\alpha(\xi_1, \xi_2) = b'(\xi_1)a'(\xi_2) - b'(\xi_2)a'(\xi_1)$$
$$\beta(\xi_1, \xi_2) = c'(\xi_1)a'(\xi_2) - c'(\xi_2)a'(\xi_1) \ .$$

(A.1.8)

Functions $A(\xi_1, \xi_2, \xi_3)$ and $B(\xi_1, \xi_2, \xi_3)$ are antisymmetric relative to all argument permutations. By interchanging ξ_1 and ξ_3 in (A.1.6) and summing up the results, we can convince ourselves that functions $\omega_1(\xi_1, \xi)$ and $f_1(f_1, f_2)$ are antisymmetric:

$$\omega_1(\xi_1, \xi_2) = -\omega_1(\xi_2, \xi_1) \quad f_1(\xi_1, \xi_2) = -f_1(\xi_2, \xi_1) \ .$$

Thus, we may put

$$\Omega_1 = \omega_1(\xi_1, \xi_2) + \omega_1(\xi_2, \xi_3) + \omega_1(\xi_3, \xi_1)$$
$$F_1 = f_1(\xi_1, \xi_2) + f_1(\xi_2, \xi_3) + f_1(\xi_3, \xi_1) \ .$$

(A.1.9)

So our problem is to solve the functional equation (A.1.6). It is easy to check that (A.1.6) has the following solution:

$$\omega_1(\xi_1, \xi_2) = \frac{b'(\xi_1) - b'(\xi_2)}{a'(\xi_1) - a'(\xi_2)} \, (l(\xi_1) - l(\xi_2))$$

(A.1.10)

$$f_1(\xi_1, \xi_2) = \frac{c'(\xi_1) - c'(\xi_2)}{a'(\xi_1) - a'(\xi_2)} \, (l(\xi_1) - l(\xi_2)) \ .$$

(A.1.11)

Here, $l(\xi)$ is any function. This solution does not result in a new dispersion law, but represents the result of reparametrization in (A.1.1).

Let us put

$$\xi_1 - \xi_2 = \eta_1 - \eta_2$$

$$a(\xi_1) - a(\xi_2) = a(\eta_1) - a(\eta_2) + \varepsilon \, [l(\eta_1) - l(\eta_2)]$$

(A.1.12)

$$b(\xi_1) - b(\xi_2) = b(\eta_1) - b(\eta_2) + \varepsilon\omega(\eta_1, \eta_2) \ .$$

The $\omega(\eta_1, \eta_2)$ represents in itself a series in powers of ε, the first term of which is given by formulae (A.1.10, 11). One more trivial solution of (A.1.12) is

$$\omega_1 = p(\xi_1) - p(\xi_2) \ , \quad f_1 = q(\xi_1) - q(\xi_2) \ ;$$

($p(\xi)$ and $q(\xi)$ are any functions, representing variations of $b(\xi)$ and $c(\xi)$).

It is important to note that (A.1.6) possesses one more solution as well. Let us assume

$$\omega_1(\xi_1, \xi_2) = \alpha(\xi_1, \xi_2)S(\xi_1, \xi_2)$$
$$f_1(\xi_1, \xi_2) = \beta(\xi_1, \xi_2)S(\xi_1, \xi_2) \ .$$

(A.1.13)

After substitution of (A.1.13) into (A.1.6) we can be convinced that $S(\xi_1, \xi_2)$ satisfies the surprisingly simple equation,

$$S(\xi_1, \xi_2) \left[a'(\xi_1) - a'(\xi_2) \right] + S(\xi_2, \xi_3) \left[a'(\xi_2) \right.$$
$$\left. - a'(\xi_3) \right] + S(\xi_3, \xi_1) \left[a'(\xi_3) - a'(\xi_1) \right] = 0 \tag{A.1.14}$$

$$S(\xi_1, \xi_2) = \frac{r(\xi_1) - r(\xi_2)}{a'(\xi_1) - a'(\xi_2)} . \tag{A.1.15}$$

Here, $r(\xi)$ is an arbitrary function again. The solution (A.1.15) is also a trivial one and results from reparametrization of a dispersion law of the form

$$p = \xi_1 - \xi_2 + \varepsilon \left[r(\xi_1) - r(\xi_2) \right] ,$$

$$q = a(\xi_1) - a(\xi_2) , \quad \omega = b(\xi_1) - b(\xi_2) ,$$

which is to the first order in ε equivalent to (2.3.15) with a modified function $a(\xi)$. To obtain given $a(\xi)$, one needs to make a change of variables of the form

$$\xi_1 = \eta_1 + \varepsilon a'(\eta_2) \frac{r(\eta_1) - r(\eta_2)}{a'(\eta_1) - a'(\eta_2)} ;$$

$$\xi_2 = \eta_2 + \varepsilon a'(\eta_1) \frac{r(\eta_1) - r(\eta_2)}{a'(\eta_1) - a'(\eta_2)} .$$

Substituting new variables into the expression for ω, and expanding in ε, we go to expression (A.1.1) with the term linear in ε being of the form (A.1.13, 15).

We shall consider (A.1.6) as a system of linear algebraic equations relative to the unknown functions $\omega(\xi_1, \xi_2)$ and $f(\xi_1, \xi_2)$. Let variable ξ_3 take two arbitrary values $\xi_3 = \sigma_1$ and $\xi_3 = \sigma_2$. Let us write:

$$A_{1,2} = A_{1,2}(\xi_1, \xi_2) = A|_{\xi_3 = \sigma_{1,2}}$$
$$B_{1,2} = B_{1,2}(\xi_1, \xi_2) = B|_{\xi_3 = \sigma_{1,2}} \tag{A.1.16}$$

$$f(\xi, \alpha_i) = g_i(\xi) , \quad \omega(\xi, \alpha_i) = h_i(\xi) , \quad i = 1, 2 . \tag{A.1.17}$$

We can see from (A.1.17) that in the most general case, the solution of (A.1.6) may depend on not more than four functions of one variable $g_{1,2}(\xi)$ and $h_{1,2}(\xi)$.

Our solution depends upon these very four functions, $l(\xi)$, $p(\xi)$, $q(\xi)$ and $r(\xi)$. Solving (A.1.6) at $\xi_3 = \sigma_{1,2}$ and making an elementary analysis of the solution, we can be convinced that we have constructed a general solution of the functional equation (A.1.6). The result obtained can be considered as the local uniqueness theorem for degenerative dispersion laws. This theorem without a complete proof was presented in [9]. The global uniqueness theorem appears in Appendix II.

Let $\omega(p, q)$ be a differentiable function, and $\omega(0, 0) = 0$. Let $\omega(p, q)$ satisfy one more condition,

$$\frac{|\omega(p, q)|}{R} \xrightarrow[R \to 0]{} 0 \quad R = |p^2 = q^2|^{1/2} . \tag{A.1.18}$$

Then the dispersion law $\omega(p, q)$ is decaying. There is a manifold $\Gamma^{1,2}$, because it contains a two-dimensional plane $p_2 = q_2 = 0$ and a vicinity of this plane, given by the following equation

$$\frac{\partial \omega}{\partial p}(p_1, q_1)p_2 + \frac{\partial \omega}{\partial q}(p_1, q_1)q_2 = 0 \qquad (A.1.19)$$

Putting $p_2 = q_2 = 0$ in (2.3.14), we get $f(0,0) = 0$; moreover,

$$\lim_{R \to 0} \left[\frac{f(R, \vartheta)}{R} \right] = f_0(\theta) < \infty \quad \text{at all } \theta \ .$$

Here, $\vartheta = \text{arctg}\,(q_2/p_2)$.

Thus, in the vicinity of zero, $f(p, q)$ may tend asymptotically to the homogeneous function of the first order. But we assumed that this function is analytic. Thus, $f_0(\vartheta) = 0$ and function f also submit to condition (A.1.19). Now in the vicinity of $p_2 = q_2 = 0$ we have, from (2.3.13):

$$\frac{\partial f}{\partial p}(p_1, q_1)p_2 + \frac{\partial f}{\partial q}(q_1, p_1)q_2 = 0 \ .$$

This means that the Jacobian between functions f and ω is equal to zero, and the latter are functionally dependent,

$$f(p, q) = F[\omega(p, q)] \ .$$

Now we have from (2.3.13, 14):

$$F[\omega(p_1, q_1) + \omega(p_2, q_2)] = F[\omega(p_1, q_1)] + F[\omega(p_2, q_2)] \ ,$$

from which we conclude $F(\xi) = \lambda\xi$, λ is a constant. The important consequence of this result is Theorem 2.3.3.

Let us designate a wave number corresponding to a new space dimension via "r", and consider the dispersion law, which becomes (2.3.15) at $r = 0$. The proof of the theorem 2.3.5 [6]:

Let the degenerative law $\omega(p, q, r)$ be parametrized in the vicinity of $r = 0$ as follows:

$$p = \xi_1 - \xi_2 \quad q = a(\xi_1) - a(\xi_2)$$

$$\omega(p, q, r) = b(\xi_1) - b(\xi_2) + r \sum_{n=0}^{\infty} r^n \omega_n(\xi_1, \xi_2) , \qquad (A.1.20)$$

and let manifold $\Gamma^{1,2}$ have dimensionality 5. Then $\omega_0 = \text{const}$, $\omega_n = 0$, $n > 0$. Then the resonance manifold $\Gamma^{1,2}$ for the dispersion law (A.1.20) may be given in the form

$$a(\xi_1) - a(\xi_2) = a(\xi_1 + \eta) - a(\xi_3 - \eta) + a(\xi_3 + \nu) - a(\xi_2 + \nu)$$

$$\sum_{k=0}^{\infty} (r_1 + r_2)^{k+1} \omega_k(\xi_1, \xi_2) = -b(\xi_1) + b(\xi_2) + b(\xi_1 + \eta)$$

$$+ b(\xi_3 + \eta) + b(\xi_3 + \nu) - b(\xi_2 + \nu) + \sum_{n=0}^{\infty} \left[r_1^{n+1} \omega_n(\xi_1 + \eta, \xi_3 \right.$$

$$\left. + \eta) + r_2^{n+1} \omega_n(\xi_3 + \nu, \xi_2 + \nu) \right] . \tag{A.1.21}$$

Let us choose ξ_1, ξ_2, ξ_3, r_1 and r_2 as independent variables and then consider ν and η as their functions, analytical in r_1 and r_2.

The degeneracy condition can be written in its usual form,

$$f(p, q, r_1 + r_2) = f(p_1, q_1, r_1) + f(p_2, q_2, r_2) . \tag{A.1.22}$$

The solution of (A.1.22) may be found in the form

$$f(p, q, r) = c(\xi_1) - c(\xi_2) + r \sum_{n=0}^{\infty} r^n f_n(\xi_1, \xi_2)$$

$$\eta = \sum_{m+n=1}^{\infty} \eta_{mn} r_1^m r_2^n , \quad \nu = \sum_{m+n=1}^{\infty} \nu_{mn} r_1^m r_2^n . \tag{A.1.23}$$

Considering terms linear in r_1 and r_2 in (A.1.21, 22), and marking $\eta_0 = \eta_{10} r_1 + \eta_{01} r_2$, $\nu_0 = \nu_{10} r_1 + \nu_{01} r_2$, we obtain

$$\eta_0 \left[a'(\xi_1) - a'(\xi_3) \right] + \nu_0 \left[a'(\xi_3) - a'(\xi_2) \right] = 0$$

$$(r_1 + r_2) \omega_0(\xi_1, \xi_2) = r_1 \omega_0(\xi_1, \xi_3) + r_2 \omega_0(\xi_3, \xi_2)$$
$$+ \eta_0 \left[b'(\xi_1) - b'(\xi_3) \right] + \nu_0 \left[b'(\xi_3) - b'(\xi_2) \right] \tag{A.1.24}$$

$$(r_1 + r_2) f_0(\xi_1, \xi_2) = r_1 f_0(\xi_1, \xi_3) + r_2 f_0(\xi_3, \xi_2)$$
$$+ \eta_0 \left[c'(\xi_1) - c'(\xi_3) \right] + \nu_0 \left[c'(\xi_3) - c'(\xi_2) \right] .$$

Setting coefficients equal in (A.1.24) at r_1, r_2 separately, we obtain overdetermined the system of equations for η_{10}, ν_{10}, η_{01} and ν_{01}. Their consistency conditions are

$$[\omega_0(\xi_1, \xi_2) - \omega_0(\xi_1, \xi_3)] B$$
$$= [f_0(\xi_1, \xi_2) - f_0(\xi_1, \xi_3)] A \tag{A.1.25}$$
$$[\omega_0(\xi_1, \xi_2) - \omega_0(\xi_2, \xi_2)] B$$
$$= [f_0(\xi_1, \xi_2) - f_0(\xi_3, \xi_2)] A . \tag{A.1.26}$$

Here, A and B are given by formulae (A.1.7, 8).

In contrast to (A.1.6), (A.1.25, 26) do not possess nontrivial solutions. To convince ourselves of this, let us differentiate (A.1.25) in ξ_3 and then apply operator $\partial^3/\partial\xi_3^3 - \partial^3/\partial\xi_3^2\partial\xi_2$ for the same equation, further putting $\xi_3 = \xi_2$. We obtain the system of the two homogeneous equations for $\partial\omega/\partial\xi_2$, $\partial f/\partial\xi_2$, having a nonzero determinant. So, $\partial\omega/\partial\xi_2 = 0$, $\partial f/\partial\xi_2 = 0$. Similarly, we get $\partial\omega/\partial\xi_1 = 0$, $\partial f/\partial\xi_1 = 0$ from (A.1.26). Thus, the unique solution of (A.1.25) is

$\omega_0 = $ const, $f_0 = $ const, $\nu_0 = \eta_0 = 0$. We can further prove this fact via induction. Let ν_k, η_k, be the sums of the sequence terms in (A.1.23), for which $m + n = k$. Let $\nu_q = \eta_q = 0$ at $q < k$. Collecting in (A.1.21, 22) terms of degree k, we have

$$(r_1 + r_2)^k \omega_{k-1}(\xi_1, \xi_2) = r_1^k \omega_{k-1}(\xi_1, \xi_3) + r_2^k \omega_{k-1}(\xi_3, \xi_2)$$
$$+ \eta_k \left[b'(\xi_1) - b'(\xi_3) \right] + \nu_k \left[b'(\xi_3) - b'(\xi_2) \right] = 0 \qquad \text{(A.1.27)}$$

$$\eta_k \left[a'(\xi_1) - a'(\xi_3) \right] + \nu_k \left[a'(\xi_3) - a'(\xi_2) \right] = 0 ,$$

and an analogous equation for f. Taking the mixed derivative in r_1, r_2 of the k-th order $\partial^k / \partial r_1^{k-1} \partial r_2$, we get

$$k! \omega_k(\xi_1, \xi_2) = \left[b'(\xi_1) - b'(\xi_3) \right] \frac{\partial^k \eta_k}{\partial r_1^{k-1} \partial r_2}$$

$$+ \left[b'(\xi_3) - b'(\xi_2) \right] \frac{\partial^k \nu_k}{\partial r_1^{k-1} \partial r_2}$$

$$k! f_k(\xi_1, \xi_2) = \left[c'(\xi_1) - c'(\xi_3) \right] \frac{\partial^k \eta_k}{\partial r_1^{k-1} \partial r_2}$$

$$+ \left[c'(\xi_3) - c'(\xi_2) \right] \frac{\partial^k \nu_k}{\partial r_1^{k-1} \partial r_2} .$$

Consistency of these equations with (A.1.27) results in the equation of the form (A.1.6),

$$\omega_k(\xi_1, \xi_2) \Delta_{ac} = f_k(\xi_1, \xi_2) \Delta_{ba} ,$$

whihc is not fulfilled, as $\Delta_{ac} / \Delta_{ba}$ is a function of ξ_1, ξ_2, ξ_3.

Actually, Δ_{ac} and Δ_{ba} are totally antisymmetric functions, so their ratio is a totally symmetric function of ξ_1, ξ_2 and ξ_3, and is not equal to the constant, as b and c are different functions. The theorem is proven.

On the basis of this theorem, one may suggest the hypothesis that at $d > 2$ and under the condition of maximal dimensionality of $\Gamma^{1,2}$, no dispersion laws exist which are degenerate with respect to the process $1 \leftrightarrow 2$. Requirement of maximum dimensionality of $\Gamma^{1,2}$ is essential, indeed, at any $d \geq 2$, the linear dispersion law $\omega = |k| \varphi(k/|k|)$ is degenerative. However, manifold $\Gamma^{1,2}$ is given by the parallelism condition on k_1, k_2 and k and so has dimensionality 4, less than maximum.

Let us now consider the scattering process of two interacting waves. The manifold $\Gamma^{2,2}$ is given by the equations (2.3.32). The dispersion law $\omega(k)$ is nondegenerate relative to this process, if in some region of the manifold $\Gamma^{2,2}$, functional equation (2.3.33) has a nontrivial solution. Apparently, manifold $\Gamma^{2,2}$ includes two hypersurfaces, set by conditions

$$k = k_2 , \quad k_1 = k_3 \quad \text{or} \quad k = k_3 , \quad k_1 = k_2 ,$$

crossing each other via $k = k_1 = k_2 = k_3$. On this submanifold, the $\bar{\Gamma}^{2,2}$ equation is fulfilled at any $f(k)$. At $d = 1$, $\Gamma^{2,2} = \bar{\Gamma}^{2,2}$, and any dispersion law is denenerative.

Theorem 2.3.4 is the evident consequence of the following lemma:

Lemma 1. The quadratic dispersion law with any signature is nondegenerative with respect to (2.3.32) at $d \geq 2$.

Proof. Let us reduce the quadratic form (2.3.31) to a diagonal form via coordinate system rotation; then $(k = (k^{(1)}, \ldots, k^{(d)}))$:

$$\omega(k) = k^{(1)^2} + \sigma_2 k^{(2)^2} + \ldots + \sigma_d k^{(d)^2}$$

$$\sigma_i = \pm 1, \quad i = 2, \ldots, d. \tag{A.1.28}$$

All signs in (A.1.28) are independent. With the dispersion law (A.1.28) the manifold $\Gamma^{2,2}$ has a rational parametrization,

$$\begin{aligned}
k^{(1)} &= P_1 + \tfrac{1}{2}\mu(1 - Q) & k_1^{(1)} &= P_1 - \tfrac{1}{2}\mu(1 - Q) \\
k_2^{(1)} &= P_1 - \tfrac{1}{2}\mu(1 + Q) & k_3^{(1)} &= P_1 + \tfrac{1}{2}\mu(1 + Q) \\
k^{(i)} &= P_i + \tfrac{1}{2}\mu(\tau_i + s_i) & k_3^{(i)} &= P_i - \tfrac{1}{2}\mu(\tau_i + s_i) \\
k_2^{(i)} &= P_i + \tfrac{1}{2}\mu(\tau_i - s_i) & k_3^{(i)} &= P_i - \tfrac{1}{2}\mu(\tau_i - s_i)
\end{aligned} \tag{A.1.29}$$

$$i = 2, \ldots, d,$$

where

$$Q = \sum_{n=2}^{d} \sigma_n \tau_n s_n,$$

and $P_1, \ldots, P_d$, μ, τ_i, s_i are independent coordinates on resonance surface (2.3.32). Let us put parametrization (A.1.29) into the functional equation

$$\begin{aligned}
f\left(P_1 + \tfrac{1}{2}\mu(1 - Q),\ P_2 + \tfrac{1}{2}\mu(\tau_2 - s_2),\ \ldots\right) & \\
+ f\left(P_1 - \tfrac{1}{2}\mu(1 - Q),\ P_2 - \tfrac{1}{2}\mu(\tau_2 + s_2),\ \ldots\right) & \\
= f\left(P_1 - \tfrac{1}{2}\mu(1 + Q),\ P_2 + \tfrac{1}{2}\mu(\tau_2 - s_2),\ \ldots\right) & \\
+ f\left(P_1 + \tfrac{1}{2}\mu(1 + Q),\ P_2 - \tfrac{1}{2}\mu(\tau_2 - s_2),\ \ldots\right) &.
\end{aligned} \tag{A.1.30}$$

Differentiating (A.1.30) in τ_i, s_i, supposing $\tau_i = s_i$, subtracting one from the other, differentiating in τ and supposing $\mu = 0$, we find

$$\partial^2 f(P_1, \ldots, P_d)/\partial P_1 \partial P_i = 0, \quad i = 2, \ldots, d,$$

from which

$$f = F_1\left(k^{(1)}\right) + \Phi\left(k^{(2)}, \ldots, k^{(d)}\right). \tag{A.1.31}$$

Substituting (A.1.31) into (A.1.30), writing down the equations obtained via differentiation in τ_i, τ_j, s_i, s_j and supposing all τ, s to be equal to zero, after simple transformations, we obtain that $\partial^2\Phi/\partial P_i\partial P_j = 0$ or

$$f = F_1\left(k^{(1)}\right) + \ldots + F_d\left(k^{(d)}\right) . \tag{A.1.32}$$

Let us substitute (A.1.32) into (A.1.30) and differentiate in P_1. We obtain

$$F_1'\left(P_1 + \tfrac{1}{2}\mu(1-Q)\right) + F_1'\left(P_1 - \tfrac{1}{2}\mu(1-Q)\right)$$
$$= F_1'\left(P_1 - \tfrac{1}{2}\mu(1+Q)\right) + F_1'\left(P_1 + \tfrac{1}{2}\mu(1-Q)\right) .$$

Differentiating in Q and μ, we get two equations on F_1'', whose consistency condition is written in the form of the equation (at $Q = 0$)

$$F_1''\left(P_1 - \mu/2\right) = F_1''\left(P_1 + \mu/2\right) .$$

On account of the arbitrariness of P_1 and μ we obtain that $F'' = \text{const}$. Exactly in the same way, differentiating (A.1.30) in P_i and then in τ_i, s_i, subtracting one from the other and supposing $\tau_i = -s_i$, we obtain

$$F_i''\left(P_i + \mu\tau_i\right) = F_i''\left(P_i - \mu\tau_i\right) ,$$

from which, on account of the arbitrariness of P_i, μ, τ_i, we conclude that $F_i'' = \text{const.}$

Thus, $F_i = c_i k^{(i)^2} + B_i k^{(i)} + D_i$. It is easy to see that $c_i = \sigma_i c$ from (A.1.30) that proves nondegeneracy. It follows from this that dispersion laws which are completely degenerative relative to process (2.3.32) do not exist. Besides theorem 2.3.4, the statement which follows suggests that it is doubtful that even partially degenerative dispersion laws exist relative to process (2.3.32).

Let the dispersion law $\omega(k)$ be decaying. Then manifold $\Gamma^{2,2}$ of codimensionality one is given by the system of equations

$$k + k_1 = k_2 + k_3 = q$$
$$\omega(k) + \omega(k_1) = \omega(k_2) + \omega(k_3) = \omega(q) . \tag{A.1.33}$$

If the dispersion law is degenerative relative to the process "one into two", then on manifold $\Gamma_M^{2,2}$, function $f(k)$ is sure to satisfy the following equation:

$$f(k) + f(k_1) = f(k_2) + f(k_3) = f(q) , \tag{A.1.34}$$

which, of course, does not mean even partial degeneracy of the dispersion law $\omega(k)$. For degeneracy to occur, it is necessary to fulfill (2.3.33) on $\Gamma^{2,2}$ in the vicinity of just one point of manifold (2.3.32).

Let us study this possibility in the simplest case $d = 2$, when the dispersion law belongs to the class (2.3.15) we are considering. Now manifold $\Gamma_M^{2,2}$ (A.1.33) is parametrized as follows (at $d = 2$ its dimensionality is equal to four):

$$p = \xi_1 - \xi_2, \quad p_1 = \xi_2 - \xi_3, \quad p_2 = \xi_1 - \xi_4, \quad p_3 = \xi_4 - \xi_3$$
$$q = a(\xi_1) - a(\xi_2), \qquad\qquad q_1 = a(\xi_2) - a(\xi_3),$$
$$q_2 = a(\xi_1) - a(\xi_4), \qquad\qquad q_3 = a(\xi_4) - a(\xi_3).$$

(A.1.35)

Let us consider the vicinity of a point on $\Gamma^{2,2}$, given on $\Gamma_M^{2,2}$ via coordinates ξ_1, ξ_2, ξ_3, ξ_4. We may set it, having retained expression (A.1.35) for p_i and defined

$$q = a(\xi_1) - a(\xi_2), \qquad\qquad q_1 = a(\xi_2 + \nu_1) - a(\xi_3 + \nu_1),$$
$$q_2 = a(\xi_1 + \nu_2) - a(\xi_4 + \nu_2), \quad q_3 = a(\xi_4 + \nu_3) - a(\xi_3 + \nu_3).$$

Similarly we can define ω_i. Resonance conditions impose two conditions upon ν_i:

$$\left[a'(\xi_2) - a'(\xi_3)\right] \nu_1 = \left[a'(\xi_1) - a'(\xi_3)\right] \nu_2$$
$$+ \left[a'(\xi_4) - a'(\xi_3)\right] \nu_3$$

$$\left[b'(\xi_2) - b'(\xi_3)\right] \nu_1 = \left[b'(\xi_1) - b'(\xi_3)\right] \nu_2$$
$$+ \left[b'(\xi_4) - b'(\xi_3)\right] \nu_3.$$

Degeneracy condition yields one more equation:

$$\left[c'(\xi_2) - c'(\xi_3)\right] \nu_1 = \left[c'(\xi_1) - c'(\xi_3)\right] \nu_2$$
$$+ \left[c'(\xi_4) - c'(\xi_3)\right] \nu_3.$$

If functions a, b, c are linearly independent, these equations possess zero solutions only. Thus, submanifold $\Gamma_M^{2,2}$ cannot be locally enlarged while retaining degeneracy.

Appendix II

Proof of the Global Theorem
for Degenerative Dispersion Laws [40]

Consider the $d = 2$ case. Our goal is to find the resonant manifold Γ itself instead of the dispersion law $\omega(p, q)$. the latter is defined by

$$\omega(p_1 + p_2, q_1 + q_2) = \omega(p_1, q_1) + \omega(p_2, q_3).$$

(A.2.1)

Due to the degeneracy of $\omega(p, q)$, functions $f_i(p, q)$ $(i = 1, 2, 3)$ exist, satisfying the same equation on Γ:

$$f_i(p_1 + p_2, q_1 + q_2) = f_i(p_1, q_1) + f_i(p_2, q_3).$$

Consider the function

$$\tilde{\omega}(p,q) = \omega(p,q) + ap + bq + \sum_{i=1}^{3} c_i f_i(p,q) \, .$$

Here a, b, c_i are some constants. The function $\tilde{\omega}$ satisfies the same equation (A.2.1) on Γ, and we shall think of ω in (A.2.1) as containing five arbitrary constants.

Let ξ_1, ξ_2, ξ_3, be the coordinates on Γ described by functions $p_1(\xi_i)$, $p_2(\xi_i)$, $q_1(\xi_i)$, $q_2(\xi_i)$. Let us fix a point ξ on Γ and differentiate (A.2.1) via ξ_i. In what follows we designate

$$\omega_1 = \frac{\partial \omega}{\partial p} \, , \quad \omega_2 = \frac{\partial \omega}{\partial q} \, , \quad \omega_{20} = \frac{\partial^2 \omega}{\partial p^2} \, , \quad \omega_{11} = \frac{\partial^2 \omega}{\partial p \partial q} \, , \quad \omega_{02} = \frac{\partial^2 \omega}{\partial q^2} \, , \cdots \, .$$

We have

$$\tilde{\omega}_1(p_1, q_1)\frac{\partial p_1}{\partial \xi_i} + \tilde{\omega}_2(p_1, q_1)\frac{\partial q_1}{\partial \xi_i} + \tilde{\omega}_1(p_2, q_2)\frac{\partial p_2}{\partial \xi_i}$$

$$+ \tilde{\omega}_2(p_2, q_2)\frac{\partial q_2}{\partial \xi_i} = F_i \equiv \tilde{\omega}_1(p_1 + p_2, q_1 + q_2) \qquad (A.2.2)$$

$$\times \left(\frac{\partial p_1}{\partial \xi_i} + \frac{\partial p_2}{\partial \xi_i} \right) + \tilde{\omega}_2(p_1 + p_2, q_1 + q_2) \left(\frac{\partial q_1}{\partial \xi_i} + \frac{\partial q_2}{\partial \xi_i} \right) \, .$$

We also adopt the following notation for the Jacobi determinant of three functions $A(\xi)$, $B(\xi)$, $C(\xi)$:

$$\{A, B, C\} = \begin{vmatrix} \frac{\partial A}{\partial \xi_1} & \frac{\partial B}{\partial \xi_1} & \frac{\partial C}{\partial \xi_1} \\ \frac{\partial A}{\partial \xi_2} & \frac{\partial B}{\partial \xi_2} & \frac{\partial C}{\partial \xi_2} \\ \frac{\partial A}{\partial \xi_3} & \frac{\partial B}{\partial \xi_3} & \frac{\partial C}{\partial \xi_3} \end{vmatrix} \, ,$$

and also set

$$w_1 = \{p_1, p_2, q_1\} \, ; \quad w_2 = \{p_1, p_2, q_2\}$$

$$v_1 = \{q_1, q_2, p_1\} \, ; \quad v_2 = \{q_1, q_2, p_2\} \, .$$

From (A.2.2) we obtain

$$w_1 \omega_1(p_2, q_2) + v_1 \omega_2(p_2, q_2) = R \, , \qquad (A.2.3)$$

where

$$R = \begin{vmatrix} \frac{\partial p_1}{\partial \xi_1} & \frac{\partial q_1}{\partial \xi_1} & F_1 \\ \frac{\partial p_1}{\partial \xi_2} & \frac{\partial q_1}{\partial \xi_2} & F_2 \\ \frac{\partial p_1}{\partial \xi_3} & \frac{\partial q_1}{\partial \xi_3} & F_3 \end{vmatrix} \, .$$

Differentiating (A.2.3) in ξ_i one obtains

$$w_1 \frac{\partial p_2}{\partial \xi_i} w_{20}(p_2, q_2) + \left(w_1 \frac{\partial q_2}{\partial \xi_i} + v_1 \frac{\partial p_2}{\partial \xi_i} \right) w_{11}(p_2, q_2)$$

$$+ v_1 \frac{\partial q_2}{\partial \xi_i} w_{02}(p_2, q_2) + \frac{\partial w_1}{\partial \xi_i} w_1(p_2, q_2) \qquad \text{(A.2.4)}$$

$$+ \frac{\partial v_1}{\partial \xi_i} w_2(p_2, q_2) = \frac{\partial R}{\partial \xi_i} .$$

From (A.2.4) we have

$$\{p_2, q_2, w_1\} w_1(p_2, q_2) + \{p_2, q_2, v_1\} w_2(p_2, q_2) = \{p_2, q_2, R\} .$$

Now by choosing special values of a, b, c_i let us achieve that, in the given point ξ,

$$w_1(p_1 + p_2, q_1 + q_2) = 0 , \quad w_2(p_1 + p_2, q_1 + q_2) = 0 , \quad w_{ij}(p_1 + p_2, q_1 + q_2) = 0$$

so that $R = 0$, and also $\{p_2, q_2, R\} = 0$. From the compartibility condition of (A.2.3) ($R = 0$) and the latter equality

$$\{p_2, q_2, w_1\} w_1(p_2, q_2) + \{p_2, q_2, v_1\} w_2(p_2, q_2) = 0$$

we obtain

$$v_1 \{p_2, q_2, w_1\} = w_1 \{p_2, q_2, v_1\} . \qquad \text{(A.2.5a)}$$

Because all of the expression is symmetric with respect to the permutation of indices 1 and 2, we also have

$$v_2 \{p_1, q_1, w_2\} = w_2 \{p_1, q_1, v_2\} . \qquad \text{(A.2.5b)}$$

Analogously choosing the constants a, b, c in another way, it is easy to obtain the relations

$$\begin{aligned}
(v_1 + v_2) \{p_2, q_2, w_1 + w_2\} &= (w_1 + w_2) \{p_2, q_2, v_1 + v_2\} \\
v_2 \{p_1 + p_2, q_1 + q_2, w_2\} &= w_2 \{p_1 + p_2, q_1 + q_2, v_2\} \\
(v_1 + v_2) \{p_1, q_1, w_1 + w_2\} &= (w_1 + w_2) \{p_1, q_1, v_1 + v_2\} \\
v_1 \{p_1 + p_2, q_1 + q_2, w_1\} &= w_1 \{p_1 + p_2, q_1 + q_2, v_1\} .
\end{aligned} \qquad \text{(A.2.6)}$$

Now we consider new functions α, β, γ such that

$$v_1 = \alpha w_1 , \quad v_2 = \beta w_2 , \quad v_1 + v_2 = \gamma(w_1 + w_2) .$$

From (A.2.5, 6) we have

$$\{p_2, q_2, \alpha\} = 0 ; \quad \{p_1, q_1, \beta\} = 0 ;$$

$$\{p_2, q_2, \gamma\} = 0 ; \quad \{p_1, q_1, \gamma\} = 0 ;$$

$$\{p_1 + p_2, (q_1 + q_2), \beta\} = 0 , \quad \{p_1 + p_2, (q_1 + q_2), \alpha\} = 0 . \qquad \text{(A.2.7)}$$

It follows from (A.2.7) that

$$p_2 = P_2(\alpha, \gamma) , \quad q_2 = Q_2(\alpha, \gamma) , \quad p_1 + p_2 = A(\alpha, \beta)$$

$$p_1 = P_1(\beta, \gamma) , \quad q_1 = Q_1(\beta, \gamma) , \quad q_1 + q_2 = B(\alpha, \beta) .$$

Then we obviously obtain

$$P_1(\beta, \gamma) + P_2(\alpha, \gamma) = A(\alpha, \beta)$$
$$Q_1(\beta, \gamma) + Q_2(\alpha, \gamma) = B(\alpha, \beta) .$$

(A.2.8)

Functional equations (A.2.8) can be solved easily, leading to

$$P_1 = a_1(\beta) - a_2(\gamma) , \quad P_2 = a_2(\gamma) - a_3(\alpha)$$

$$A = a_1(\beta) - a_3(\alpha) , \quad Q_1 = b_1(\beta) - b_2(\gamma)$$

$$Q_2 = b_2(\gamma) - b_3(\alpha) , \quad B = b_1(\beta) - b_3(\alpha) .$$

Here, a_i, b_i, $i = 1, 2, 3$ are arbitrary functions of one variable. The result obtained leads to dispersion laws of the form (2.3.15, 26). In the above it has been supposed that functions α, β, γ are functionally independent. This is really true in the general case. Special cases should be obtained by some limiting procedure. Obviously the unique possibility is to obtain the homogeneous functional of degree one.

Conclusion

Let us summarize. In the present paper we have aimed at showing that a method like Poincaré's analysis of the integrability of dynamical systems, based on the study of the perturbation theory series, proves to be very effective. Earlier, an analogous method proved the nonexistence of a strong recursive operator for multidimensional systems [41]; we can only hope that this does not exhaust its capacities. However, it has recently been shown [42, 43] that, by generalizing the recursion operator concept, it is possible to construct both recursion operators and bi-Hamiltonian structures for multidimensional solvable equations. Interesting examples include the KP and DS systems. One can not exclude a priori the possibility that only essentially nonlinear integrals exist for some systems (2.1.1).

With regard to the systems considered in this paper, i.e., those containing integrals which are quadratic in the main part, certain questions have been answered since our paper [6] was published in 1987: namely, the question of action-angle variables in nondegenerative systems with periodic boundary conditions ([7]; Sect. 2.6) and that of a global description of the degenerative dispersion laws ([40]; Appendix II).

Nevertheless some questions remain unanswered; for example: Can the resonant manifold for decays $1 \to 2$ always be described via only one parametrization (i.e. consisting of two parts) corresponding to the replacements $\xi_i \to \xi_i$,

$a(\xi_i) \to -a(-\xi_i)$, $b(\xi_i) \to -b(-\xi_i)$, as in the KP-1 equation? KP-1-like equations with degenerative dispersion laws are especially interesting. Although they are exactly solvable by the inverse scattering technique, current methods still cannot provide solutions which are not rapidly decreasing and are in general position. In contrast to soliton and finite gap solutions which in the space of all solutions of such equations are not dense, these types of solutions of a general position possess stochastic properties and must be studied statistically. The study of these solutions (which are generally not weakly nonlinear) is rather important from the viewpoint of understanding the turbulent nature of dynamical systems. A weakly nonlinear solution of these equations may be studied by the kinetic equation technique (see [20]), which is particularly interesting and was first considered in [14].

Finally, we wish to point out that the integrals of the two-dimensional systems we have considered do not exhaust the algebra of integrals; and it is only its commutative subalgebra. It corresponds to commutative symmetries. Symmetries and integrals, explicitly dependent of space-time variables exist, which comprise a noncommutative algebra. Corresponding equations are also solvable; see, for example [44, 45].

Note added in proof. The existence condition for the *three* additional functions f_i in Theorem 2.3.4 cannot be relaxed. Let us consider the equations

$$\omega(p, q) + \omega(p_1, q_1) + \omega(p_2, q_2) = 0$$

$$p + p_1 + p_2 = 0$$

$$q + q_1 + q_2 = 0 .$$

They are satisfied on the manifold

$$2p_1 p_2(q_1 + q_2) + p_1^2 q_2 + p_2^2 q_1 = 0 \tag{A}$$

for three linearly independent functions

$$\omega_1(p, q) = qp^2 \quad \omega_2(p, q) = \frac{q^3}{p^2} \quad \omega_3(p, q) = \frac{q}{p} . \tag{B}$$

This fact, which is easily directly verified, is important for the weakly turbulent theory of drift waves in plasmas and Rossby waves in geophysics. It was established by *Balk, Nazarenko* and *Zakharov* [46] who also found that the number of functions $\omega_i(p, q)$ can not be increased. The two functions $\omega_1(p, q)$ and $\omega_2(p, q)$ are odd and also satisfy the relations

$$\omega(p, q) = \omega(p_1, q_1) + \omega(p_2, q_2)$$

$$p = p_1 + p_2 \tag{C}$$

$$q = q_1 + q_2 .$$

The function $\omega(p, q)$ is analytic. In accordance with Theorem 2.3.2, the function $\omega_2(p, q)$ is not analytic, but it is unique. This fact is generic for any analytic dispersion law (*Schulman, Tsakaya,* [47]). It is interesting that the function

$$\omega(p, q) = qp^2 + \frac{q^3}{p^2} = \omega_1(p, q) + \omega_2(p, q)$$

is a degenerative dispersion law belonging to the class (2.3.15). In this case one has

$$a(\xi) = \xi^2 \qquad b(\xi) = \tfrac{1}{4}\xi^4 \; .$$

This is nothing but the dispersion law in the "KP-hierarchy" which follows after KP-1. In this case the resonant manifold C is a sum of three disconnected parts. One of them is given by (A), the two others by (2.3.20).

References

1   C.S. Gardner, J.M. Green, M.D. Kruskal, R.M. Miura: Phys. Lett. **19**, 1095–1097 (1967)
2   N.J. Zabusky, M.D. Kruskal: Phys. Rev. Lett. **15**, 240–243 (1965)
3   A. Poincaré: *New methods of celestial mechanics,* in *Selected Works,* Vol. I, II (Nauka, Moscow 1971), pp. 1–358
4   V.E. Zakharov, E.I. Schulman: Physica **D1**, 191–202 (1980)
5   V.E. Zakharov, E.I. Schulman: Dokl. Akad. Nauk **283**, 1325–1328 (1985)
6   V.E. Zakharov, E.I. Schulman: Physica **D29**, 283–320 (1988)
7   E.I. Schulman: Teor. Mat. Fizika **76**, 88–99 (1988) , in Russian
8   H.H. Chen, Y.C. Lee, J.E. Lin: Physica Scripta **20**, 490–492 (1979); Physica **D26**, 165–170 (1987)
9   V.E. Zakharov: "Integrable Systems in Multidimensional Spaces", in *Lect. Notes Phys.* Vol. 153 (Springer, Berlin–Heidelberg 1983) pp. 190–216
10   H. Umezava: *The Quantum Field Theory* (Inostr. Literatura, Moscow 1985), p. 380
11   E.I. Schulman: Dokl. Acad. Nauk, **259**, 579–781 (1981)
12   V.E. Zakharov, E.I. Schulman: Physica **D4**, 270–274 (1982)
13   I.M. Krichever: Dokl. Akad. Nauk in press
14   V.E. Zakharov: Integrable turbulence. Talk presented at the International Workshop "Nonlinear and Turbulent Processes in Physics", Kiev, 1983
15   A. Daveay, K. Stewartson: Proc. Roy. Soc. Lond. **A338**, 101–110 (1974)
16   H.H. Chen, J.E. Lin: "Constraints in the Kadomtsev-Petviashvili equation" – preprint N82–112, University of Maryland (1981)
17   A. Reiman, M. Semionov-Tian Shanskii: Proc. of LOMI Scientific Seminars **133**, 212–227 (1984)
18   R.K. Bullogh, S.V. Manakov, Z.J. Jiang: Physica **D18**, 305–307 (1988)
19   E.I. Schulman: Teor. Mat. Fizika **56**, 131–136 (1983)
20   E.P. Wigner: *Symmetries and Reflections* (Indiana University Press, Bloomington 1970)
21   V.E. Zakharov: "Kolmogorov Spectra in the Theory of Weak Turbulence", in *Basic Plasma Physics,* ed. by M.N. Rosenbluth, R.Z. Sagdeev (North-Holland, Amsterdam 1984)
22   V.E. Zakharov: Izv. Vyssh. Uchebn. Zaved. Radiofiz. **17**, 431–453 (1974)
23   V.E. Zakharov, A.M. Rubenchik: Prikl. Mat. Techn. Fiz. **5**, 84–98 (1972)
24   A.P. Veselov, S.P. Novkov: Dokl. Akad. Nauk **279**, 784–788 (1984)
25   V.I. Shrira: J. Nnl. Mech. **16**, 129–138 (1982)
26   V.E. Zakharov, E.A. Kuznetsov: Physica **D18**, 455–463 (1986)
27   E.S. Benilov, S.P. Burtsev: Phys. Lett. **A98**, 256–258 (1983)

28   D.D. Tskhakaia: Teor. Math. Fizica **77**, in press
29   A.L. Berkhoer, V.E. Zakharov: Zh. Eksp. Teor. Fiz. **58**, 903–911 (1970)
30   S.M. Manakov: Zh. Eksp. Teor. Fiz. **65**, 505–516 (1973)
31   V.E. Zakharov, A.B. Shabat: Funct. Anal. Appl. **8**, 43–53 (1974)
32   V.G. Makhankov, O.K. Pashaev: "On Properties of the Nonlinear Schrödinger Equation with
      $U(p, q)$ Symmetry", JINR-Dubna preprint E2-81-70 (Dubna, USSR 1981)
33   V.E. Zakharov: Zh. Eksp. Teor. Fiz. **62**, 1745–1759 (1972)
34   V.E. Zakharov, A.V. Mikhailov: Pisma Zh. Eksp. Teor. Fiz. **45**, 279–282 (1987)
35   I.V. Cherednic: Teor. Mat. Fizika **47**, 537–542 (1981)
36   A.P. Veselov: Dokl. Acad. Nauk **270**, 1298–1300 (1983)
37   V.Z. Khukhunashvili: Teor. Mat. Fizika **79**, 180–184 (1989)
38   A.P. Fordy, P.P. Kulish: Comm. Pure Appl. Math. **89**, 427–443 (1983)
39   S.V. Manakov: Physica **D3**, 420–427 (1981)
40   V.E. Zakharov: Fund. Anal. Appl. **23** (1989), in press
41   V.E. Zakharov, B.G. Konopelchenko: Comm. Math. Phys. **93**, 483–509 (1984)
42   A.S. Fokas, P.M. Santini: Stud. Appl. Math., **75**, 179 (1986); Comm. Math. Phys., **116**,
      449–474 (1988)
43   P.M. Santini, A.S. Fokas: Comm. Math. Phys., **115**, 375–419 (1988)
44   A.Yu. Orlov, E.I. Schulman: "Additional Symmetries of Two-Dimensional Integrable Sys-
      tems", preprint Inst. Autom. Electrometry N217 (1985), Teor. Mat. Fizika **64**, 323–327 (1985);
      Lett. Math. Phys. **12**, 171–179 (1986)
45   A.Yu. Orlov: "Vertex Operator, $\bar{\partial}$-Problem, Symmetries, Variational Identifies and Hamil-
      tonian Formalism for (2+1) Integrable Systems", in *Proc. Int. Workshop on Nonlinear and
      Turbulent Processes in Physics* (1987), ed. by V.G. Bar'akhtar, V.M. Chernousenko, N.S.
      Erokhin, V.E. Zakharov (World Scientific, Singapore 1988), p. 116–134
46   A.M. Balk, S.B. Nazarenko, V.E. Zakharov: Zh. Exp. Teor. Fiz. **97** (1990), in press
47   E.I. Schulman, D.D. Tzakaya: Funkt. Anal. Pril. **25** (1991), in press

What Is an Integrable Mapping?

A.P. Veselov

Introduction

Rational mappings of $\mathbb{C}P^1$ and dynamic properties of their iterations once again attract the attention of mathematicians. The dynamic theory of such mappings has been developed in the classical works of G. Julia and P. Fatou. The recent investigations of Sullivan, Thurston, Douady and Hubbard throw new light upon this problem and uncover deep connections with the theory of Kleinian groups and Teichmüller space [1]. It is a very surprising fact that the notion of the integrability for such mappings is not discussed in these papers.

The first part of the present paper is devoted to such discussion. As the basis of the definition of the integrability, we place the existence of commuting mapping with suitable properties. Such a definition is motivated by the classical results of *Julia, Fatou* and *Ritt* [2–4] and by modern soliton theory, more precisely, the theory of finite-gap operators [5] and the theory of symmetries of the partial differential equations (PDE) [6] (see also the paper by Mikhailov, Shabat and Sokolov in this book). The most interesting result which we propose is the intriguing connection of such integrable polynomial mappings of $\mathbb{C}^n$ with the theory of Lie algebras. The construction, discovered in [7], allows us to match every simple complex Lie algebra of rank n to the family of the integrable polynomial (rational) mappings of $\mathbb{C}^n(\mathbb{C}P^n)$. We discuss also the analogous construction for the correspondences in $\mathbb{C}^n \times \mathbb{C}^n$ (or $\mathbb{C}P^n \times \mathbb{C}P^n$) and its relation with the Yang-Baxter equation. A separate section is devoted to the polynomial Cremona mappings of $\mathbb{C}^2$.

In the second part we consider the discrete analogs of the integrable systems of classical mechanics, following in the main the author's paper [8]. The corresponding class of mappings contains the following Lagrangean systems with discrete time.

Let M^n be any smooth manifold, $\mathcal{L}$ be the function on $M^n \times M^n$. Let us consider the problem of the extremum of the functional $S(q)$, $q = (q_i)$, $q_i \in M^n$, $i \in \mathbb{Z}$:

$$S(q) = \sum_{k \in \mathbb{Z}} \mathcal{L}(q_k, q_{k+1}) . \qquad (1)$$

In a coordinate system (x^i, y^i) on $Q = M^n \times M^n$, which is induced by the coordinates u^i on M^n, we have

$$\delta S = 0 \iff \frac{\partial \mathcal{L}}{\partial x^i}(q_k, q_{k+1}) + \frac{\partial \mathcal{L}}{\partial y^i}(q_{k-1}, q_k) = 0 . \qquad (2)$$

The equation (2) is the natural discrete analog of the equations of classical mechanics: $\mathcal{L}$ is the "Lagrangean function", S is "the action functional", k is "time". Such equations occur in recent papers on solid state physics and dynamical theory (see [9–13]). The simplest example is the Frenkel-Kontorova model [12]; it corresponds to $M^n = \mathbf{R}^1$. The problem when M^n is the unit sphere in $\mathbf{R}^3$ has been considered in connection with the ground state of a one-dimensional classical spin chain [9–11]. In this interpretation $\mathcal{L}$ is the energy of interaction of neighbouring spins, S is the full energy of the configuration. The problem about certain wave functions of the quantum Heisenberg model can be reduced to the problem for $\mathcal{L} = (Jq_k, q_{k+1})$. It has been investigated for this special case in [10, 11, 14].

The first paragraph of the second part is devoted to the Hamiltonian theory of the Lagrangean system with discrete time. The corresponding analogs of the Liouville and Noether theorems are proposed. In the remaining paragraphs, we consider some interesting examples of the integrable (in Liouville sense) systems of such type: the one-dimensional Heisenberg chain with classical spins and the discrete analog of the C. Neumann problem, the billiard in quadrics and the discrete version of the top's dynamics. For the first two problems we give the interpretation of the solutions in terms of the eigen-functions of some difference operators. We propose the discrete analog of Moser-Trubowitz isomorphism [15], using the algebraic-geometrical approach to the spectral theory of difference operators, begun by Novikov, Date and Tanaka and developed by Krichever (see reviews [5], [16]). As a consequence, we give explicit formulas for the general solution in θ-functions.

1. Integrable Polynomial and Rational Mappings

1.1 Polynomial Mapping of $\mathbb{C}$: What Is Its Integrability?

Iterations of any linear function $f(z) = az + b$ can be found easily in explicit form, so the first nontrivial case corresponds to degree 2. Every quadratic mapping can be reduced by a linear change of variable to the form $f(z) = z^2 + c$. The dynamic defined by this mapping in general is very complicated and was investigated by Douady and Hubbard [17]. In particular cases, for example for $c = 0$, this dynamic has a simple description and can be natrually considered as an integrable. But what is the integrability for such mappings? The starting point for the author was the formal analogy of this problem with the spectral theory of the Hill's operator $L = -d^2/dx^2 + u(x)$. As was found by S.P. Novikov [18] and developed in [5] (see also [19]) such an operator L has in its spectrum a finite number of gaps if there exist a differential operator A of odd order, which commutes with L: $[L, A] = 0$.

Definition. A polynomial mapping $f : \mathbb{C} \to \mathbb{C}$ is called *integrable*, if there exists a polynomial map $g : \mathbb{C} \to \mathbb{C}$, which commutes with f, such that the sets of the iterations of g and f are disjoint.

In this definition the degrees of f and g are supposed to be greater than 1. In suggesting such a definition, the author was not aware of the remarkable papers of G. Julia, P. Fatou and J. Ritt [2–4], in which the problem of the description of all such mappings was solved. Their results can be formulated as follows: to within a linear change of variable, there exist only two series of the integrable polynomial maps: $f = z^k$ and $f = \pm T_k(z)$, where $T_k(z) = \cos k \arccos z$ – the Tchebycheff polynomials. In the quadratic case $f = z^2 + c$ only $c = 0$ and $c = -2$ correspond to the integrable maps. The last result in a strange way coincides with the coefficient in θ-functional formulas for the potential of finite-gap operator [5]. The analogy with the spectral theory can be prolonged if we try to construct the analogue of the Bloch eigenfunction ψ [5] in the following natural way [7]. Let's determine a formal series $\psi = 1/z + \sum_{s=0}^{\infty} \zeta_s z^s$ for the polynomial $f(z) = z^k + a_{k-1} z^{k-1} + \ldots + a_0$ by the relation $f \circ \psi = \psi \circ z^k$. The coefficients ζ_s are calculated uniquely by recursion formulas and determine a series converging in some neighbourhood of the zero. Such a function – more precisely its inversion – has been considered in the beginning of our century by F. Böttcher [20], so we call it the Böttcher function. It gives a mapping of the unit disk into the domain of attraction of infinity for the mapping f.

Theorem 1. A polynomial f defines an integrable mapping if and only if the Böttcher function ψ is rational. In this case, every commuting mapping g satisfies the equality $g \circ \psi = \psi \circ \lambda z^m$, $\lambda^{k-1} = 1$ [7].

Only two variants of such functions are possible: $\psi = 1/z + a$ or $\psi = 1/z + a + bz$, which lead to the cases of integrability mentioned above. One part of this theorem follows from the results of Julia, Fatou and Ritt. Another part can be proven as follows. It's easy to show that a rational Böttcher function cannot have poles different from zero and infinity. Thus ψ must have the form $\psi = 1/z + \zeta_0 + \ldots + \zeta_N z^N$.

Notice that the function $\varphi = g \circ \psi$ also satisfies the equalities $f \circ \varphi = \varphi \circ z^n$ $(f \circ g) \circ \psi = g \circ (f \circ \psi) = (g \circ \psi) \circ z^m$. the same is valid for $\psi \circ 1/z$.

Lemma. All formal solutions $\varphi = \alpha/z^p + \ldots$ $(p > 0)$ of the functional equation $f \circ \varphi = \varphi \circ z^m$ have the form $\varphi = \psi \circ (\lambda z^p)$ for suitable λ.

In particular, $\psi \circ (1/z) = \zeta_N/z^N + \ldots + \zeta_0 + z$ must coincide with $\psi(\lambda z^N)$, which is possible only for $N = 1$. The case $N = 0$ is evident. The relation $g \circ \psi = \psi \circ \lambda z^m$ follows from the lemma. The theorem is proved.

If we compare the proposed proof with the considerations of Krichever [21], we see that the analogy between mappings and differential operators is much deeper than it first appeared.

Our definition of integrability is in good agreement with the symmetry approach to PDE, which allows classification of all integrable systems of a certain form (see [6] and this book). In fact, the existence of the commuting mapping g

determines the symmetry of the dynamical system, defined by $f : x_{k+1} = f(x_k)$ because g transforms the solution x_k of it into the new solution $x'_k = g(x_k)$. The absence of the common iterations permits repetition of this process infinitely often in the general case.

But the best motivation for our approach to integrability are the remarkable papers of Julia, Fatou and Ritt, who first realized the importance of the problem about the commuting mappings in dynamic theory.[1]

1.2 Commuting Polynomial Mappings of $\mathbf{C}^N$ and Simple Lie Algebras

Let now $f : \mathbf{C}^n \to \mathbf{C}^n$, $f = (f_1(z), \ldots, f_n(z))$, $z = (z_1, \ldots, z_n)$ be a polynomial mapping: $f_i \in \mathbf{C}[z_1, \ldots, z_n]$. The degree of this mapping is defined to be the number of preimages of the generic point. In this paragraph the degrees of all mappings will be assumed greater than one. The case of invertible polynomial mappings will be considered separately (see §4). How can we define the integrability of such a mapping f? How many commuting mappings must we demand? The author thinks that for the following class of irreducible mappings the integrability can be defined in the same manner as in the one-dimensional case. We say that mapping f is *reducible* if after some invertable polynomial change of coordinates the functions $f_1, \ldots, f_k$ $(k < n)$ depend only on $z_1, \ldots, z_k$. In the contrary case f is said to be *irreducible*.

Definition. An irreducible mapping f will be called *integrable* if there corresponds to it an irreducible mapping g, which commutes with f, such that the sets of iterates of g and f are disjoint [7].

This definition is again motivated by the analogy with the theory of symmetries of the certain class of PDEs [6], where the existence of one symmetry implies infinite sets of such symmetries and the complete integrability of corresponding PDEs.

Problem. Describe to within a change of variables all integrable polynomial mappings.

For $n = 1$ this problem is solved by Juilia, Fatou and Ritt (see §1). For $n \geq 2$ there exists only the conjecture, due to the author, which says that all integrable mappings are given by the following construction [7], generalizing the construction of Tchebycheff mappings.

Let G be a simple complex Lie algebra of rank n, H its Cartan subalgebra, H^* its dual space, $\mathcal{L}$ a lattice of weights in H^* generated by the fundamental weights $\omega_1, \ldots, \omega_n$ and L the dual lattice in H (see [23]). We define the mapping $\phi_G : H/L \to \mathbf{C}^n$, $\phi_G = (\varphi_1, \ldots, \varphi_n)$, $\varphi_k = \sum_{w \in W} \exp[2\pi i w(\omega_k)]$, where W is the Weyl group, acting on the space H^*.

---

[1] Note that the problem about the commuting ordinary differential operators was first considered at the same time by *Burchnall* and *Chaundy* [22].

Theorem 2. With each simple complex Lie algebra G of rank n is associated an infinite series of integrable polynomial mappings P_G^k, $k = 2, 3, \ldots$, determined from the condition:

$$\phi_G(kx) = P_G^k(\phi_G(x)) \; ,$$

where

$$P_G^k \circ P_G^l = P_G^{kl} = P_G^l \circ P_G^k \; .$$

All coefficients of the polynomials, determining this mapping, are integers.

The proof is based on the theorem of Chevalley asserting that the algebra of exponential invariants of W is freely generated by $\varphi_1, \ldots, \varphi_n$ [23]. In place of φ_k we may take the characters of the fundamental representation of the Lie algebra G.

For $n = 1$ there is a unique simple algebra A_1. Here $\phi_{A_1} = 2\cos(2\pi x)$ and the $P_{A_1}^k$, are, to within a linear substitution, Tchebycheff polynomials. For arbitrary n the explicit form of the mapping P_G^k may be found from the generating mapping $\xi_G = \sum_{k=0}^\infty P_G^k t^k$: $\xi_G = (\xi_1, \ldots, \xi_n)$, $\xi_i = \sum_{w \in W}(1 - t \exp[2\pi i w(\omega_i)])^{-1}$.

The proposed series may be extended in the following manner. Let the point $a \in H$ be such that $w(a) \equiv a(\mathrm{mod}\, L)$ for all $w \in W$, σ be the automorphism of the root system. Then the polynomial mappings $P_{G,a,\sigma}^k$ are determined from the condition

$$\phi_G(k\sigma(x) + a) = P_{G,a,\sigma}^k(\phi_G(x)) \; .$$

For $n = 1$ this leads to the minus sign before the Tchebycheff polynomials: for the polynomial of the odd degree this sign is essential. The list of all inequivalent polynomial mappings $P_{G,a,\sigma}^k$ connected with the simple Lie algebras of rank 2 can be found in Appendix A.

Dynamical properties of the mapping $f = P_{G,a,\sigma}^k$ follow from their definition.

An analogue of the Julia set J may be defined in the given case as the set of points whose images, under all iterations of f, remain in a bounded domain in $\mathbb{C}^n$.

Theorem 3. The set J is a singular n-dimensional simplex σ^n, the image of the real Weyl alcove under the mapping ϕ_G. The mappings $f = P_{G,a,\sigma}^k$ preserve the measure on J with density $\mu = (-1)^{N/4} j^{-1}(z)$, where N is the total number of roots of G,

$$j(z) = \sum_{w \in W}(\det w) \exp[2\pi i w(\varrho)] \; , \quad \varrho = \omega_1 + \ldots + \omega_n \; ,$$

and are ergodic on J.

We remark that the function $j(z)$ is not invariant with respect to W, but $j^2(z)$ already is and may, therefore, be expressed in the form of a polynomial in the $z_k = \varphi_k(x)$. For the Tchebycheff's polynomials, J is the segment $[-1, 1]$, $\mu = (1 - x^2)^{-1/2}$. The points outside of J tend to infinity under the iterations.

The important information about mapping give the eigenvalues of the derived mapping $f_*(z_0)$ in the fixed point z_0 (the "spectrum" of z_0). The mappings P_G^k have a unique common fixed point $z_0 = \phi_G(0)$, so the spectrum of z_0 is of interest.

Theorem 4. The spectrum of z_0 has the form $k^{\alpha_1+1}, \ldots, k^{\alpha_n+1}$, where $\alpha_1, \ldots, \alpha_n$ are the exponents of the algebra G [23].

Corollary. The series of the integrable mappings P_G^k, corresponding to Lie algebras with nonisomorphic Weyl groups are inequivalent.

The proof follows from the independence of the spectrum from the coordinate system.

Chalykh has computed the spectrum of all fixed points and proved that $P_{B_n}^k$ and $P_{C_n}^k$ are also inequivalent when $n > 2$ [24]. It is interesting that they are semiconjugate, i.e. conjugate with the help of a noninvertible polynomial mapping of degree two.

The natural conjecture is that the mappings $P_{G,\sigma,a}^k$ and $P_{G',\sigma',a'}^k$ are equivalent under the polynomial change of variables if and only if $G \approx G'$ and endomorphisms of $H/L : x \to k\sigma(x) + a$ and $x \to k\sigma'(x) + a'$ are conjugated by some linear automorphism. This follows for the Lie algebras G of rank 2 from the results of Chalykh [24].

Another interesting problem is concerned with the compactifications of $\mathbb{C}^n$ and the dynamics of P_G^k on them. Let us consider the "weighted" $\mathbb{C}P^n$, which are defined as the quotients of $\mathbb{C}^{n+1}$ by the action of $\mathbb{C}^*$: $(z_0, \ldots, z_n) \sim (tz_0, t^{k_1}z_1, \ldots, t^{k_n}z_n)$, $t \in \mathbb{C}^* = \mathbb{C}\backslash\{0\}$, where $k_1, \ldots, k_n$ are the natural numbers (weights).

Theorem 5. The mapping P_G^k can be extended to the mapping of $\mathbb{C}P_n^n$ with the weights $k_1, \ldots, k_n$ if and only if the vector $v = \sum_{i=1}^m k_i\alpha_i$, where $\alpha_1, \ldots, \alpha_n$ be the basis of the simple roots of G, satisfies the conditions $(\alpha_i, v) \geq 0$ for all i, i.e. belongs to the corresponding Weyl chamber [24] .

For the description of dynamics it is very convenient to use the polyhedron corresponding to this variety. In this case it is the simplex, which is cut off from the Weyl chamber by the plane, orthogonal to the vector v . In these terms the dynamics has a nice geometric description [25].

Our results relate also to the real case, because the mappings P_G^k have integer coefficients. The dynamic of P_G^k for the p-adic case and the finite field case also can be of interest.

But what about the analogy with commuting differential operators? The author supposes that the analog of our family of integrable mappings P_G^k is the commuting ring of quantum integrals of the Calogero system and its generalization, found by *Olshanetsky* and *Perelomov* [26] and connecting with the simple Lie algebras. The new work of the author and *Chalykh* [27, 55] is devoted to this question.

1.3 Commuting Rational Mappings of $\mathbb{C}P^n$

We consider again the case of degree greater than 1. The integrability of such mappings can be defined in the same manner as for polynomial mappings.

For $n = 1$ the problem of describing integrable rational mappings was solved by *J. Ritt* [4]. He has proved that all such mappings $f(z)$ must satisfy the functional equations

$$\varphi(\alpha x + \beta) = f(\varphi(x))$$

for a certain meromorphic periodic function $\varphi(x)$. To within the transformation $\varphi \to (a\varphi + b)(c\varphi + d)^{-1}$, only the following possibilities exist:

1) $\varphi(x) = \cos x$
2) $\varphi(x) = \mathcal{P}(x)$, $\mathcal{P}(x)$-Weierstrass elliptic function,
3) $\varphi(x) = \mathcal{P}^2(x)$, where $\mathcal{P}(x)$ corresponds to the lemniscatic curve with parameter $\tau = i$,
4) $\varphi(x) = \mathcal{P}'(x)$ for the curve with parameter $\tau = 1/2 + i\sqrt{3}/2$,
5) $\varphi(x) = \mathcal{P}^3(x)$ for the same curve.

In all the elliptic cases the transformation $x \to \alpha x$, is any endomorphism of the corresponding elliptic curve, i.e., the complex multiplication or the ordinary isogeny $x \to kx$. The list of all inequivalent integrable rational mappings of degree two can be found in Appendix B.

The construction of the previous paragraph can be applied also for the rational case. Let G, H, H^*, L be the same as in §2. Consider an Abelian variety $M_{G,\tau} = H/L + \tau L$ which is a product of elliptic curves $\mathcal{R}$ with parameter τ. There exists a corresponding analog of the Chevalley theorem [28] [29], which says that the quotients of this variety by the natural action of the Weyl group W is the weighted projective space $\mathbb{C}P^n$. So we can associate with every endomorphism of $\mathcal{R}: z \to \lambda z$ the rational mapping $R^\lambda_{G,\tau} : \mathbb{C}P^n \to \mathbb{C}P^n$.

Theorem 6. To each simple complex Lie algebra G and to each elliptic curve $\mathcal{R}$ is associated a commuting semigroup of rational mappings of $\mathbb{C}P^n$ into itself isomorphic to the semigroup of endomorphism of $\mathcal{R}$ [7].

As well as for the polynomial case this family can be extended with the help of a shift and automorphism of the root system.

Unfortunately, the basis of invariant functions on $M_{G,\tau}$ is defined in an ineffective way [29], therefore explicit formulas for the mappings $R^\lambda_{G,\tau}$ can be written only for particular cases. One of the examples has the following form:

$$(x : y : z) \longrightarrow \left(\frac{1}{2i}x(y - z) : -\frac{1}{4}\left[(y + z)^2 - x^2\right] : yz \right).$$

Notice that already for $n = 1$ it is a non-trivial thing to find the explicit formula for $\mathcal{P}(nx)$ through $\mathcal{P}(x)$. The following beautiful forms for such expressions have been found by Frobenius and Stickelberger [30]. Let Δ_n be a Wronskian of $\mathcal{P}'(x), \ldots, \mathcal{P}^{(n-1)}(x)$:

$$\Delta_n = \det \begin{vmatrix} \mathcal{P}'(x) \ldots & \mathcal{P}^{(n-1)}(x) \\ \mathcal{P}^{(n-1)}(x) \ldots & \mathcal{P}^{(2n-3)}(x) \end{vmatrix} .$$

Then,

$$\mathcal{P}(nx) = \mathcal{P}(x) + \frac{1}{n^2} \frac{d^2}{dx^2} \ln \Delta_n(x) .$$

This formula again surprisingly resembles the formulas in soliton theory (see [5], [31]).

The dynamics of the proposed rational integrable mappings is more complicated than in the polynomial case. In particular, the Julia set J for such mappings coincides with the whole space. Notice that this property does not yet determine the class of integrable mappings even for $n = 1$ (see [1]).

1.4 Commuting Cremona Mappings of $\mathbb{C}^2$

In this paragraph we consider the invertible polynomial mappings of $\mathbb{C}^2$, forming the so called affine Cremona group $GA_2(\mathbb{C})$.

Let $f = (P(x, y), Q(x, y))$ be a such polynomial mapping. If f as an element of $GA_2(\mathbb{C})$ has a finite order, then the description of the dynamics of f as evident. The mapping f of infinite order we call *integrable* if there exists a commuting mapping g, generating with f a subgroup $\mathbb{Z} \oplus \mathbb{Z} \subset GA_2(\mathbb{C})$.

For the description of all such integrable mapping we use the algebraic results about the structure of Cremona group $GA_2(\mathbb{C})$. As was proven by *Jung* [32] this group is the amalgamated product of the groups of affine transformations of the plane and the triangular transformations of the form $(x, y) \rightarrow (ax + R(y), by + c)$, where $R(y)$ is an arbitrary polynomial.

Based on this result, *Wright* [33] gives a description of all Abelian subgroups in $GA_2(\mathbb{C})$, which is sufficient for our goals.

Theorem 7. A mapping f from $GA_2(\mathbb{C})$ is integrable if there exists a change of coordinate from $GA_2(\mathbb{C})$, which transforms f to the triangular form, or equivalently, if the degrees of all polynomials, defining the iterations of f, are bounded [25, 34].

For the real case, the integrable mapping $f \in GA_2(\mathbb{R})$ can be transformed to the linear or triangular form. In all these cases all iterations can be found easily in explicit form.

Polynomial Cremona mappings have a constant Jacobian; the inverse statement is the famous Jacobian conjecture. If this constant is equal to 1, this mapping defines the symplectic transformation. For such mapping one can formulate the following the problem.

A function I on $\mathbb{C}^2$ is called as the integral of symplectic mapping f, if $I(f(p)) = I(p)$ for every $p \in \mathbb{C}^2$.

The problem is: describe all symplectic f from GA_2 which has a polynomial integral I.

Using the previous theorem we can get the following result.

Theorem 8. The integrable (in our sense) symplectic mapping f from $GA_2(\mathbb{C}; \mathbb{R})$ always has a nontrivial polynomial integral [34].

Is the inverse statement also valid? – is an open question.
Let us consider the examples.

Example 1. *The Invertible Quadratic Mappings.* It is easy to show that every mapping f, determined by polynomials of degree two, can be transformed to the following form by a linear change of coordinate:

$$f(x, y) = \left(\alpha_0 + \alpha_1 x + \alpha_2 y + \alpha x^2 , \ \beta_0 + \beta_1 x + \beta_2 y + \beta x^2 \right) , \tag{3}$$

where

$$\alpha_1 \beta_2 - \alpha_2 \beta_1 \neq 0 , \quad \alpha \beta_2 - \alpha_2 \beta = 0 .$$

Proposition 1. The mapping f of the form (3) is integrable if and only if $\alpha = \alpha_2 = 0$ or $\alpha = \beta = 0$.

If $\alpha \neq 0$ then the degrees of the polynomials, determining the iterations of f, increase to infinity. According to Theorem 7, this contradicts the integrability and proves the proposition.

In particular, the *Henon mapping* $f(x, y) = (1+y-ax^2, bx)$, $b \neq 0$ is integrable in our sense only for $a = 0$.

Example 2. *Moser Mapping*

$$f(x, y) = \left((x + y^3) \cos \alpha - y \sin \alpha , \ (x + y^3) \sin \alpha + y \cos \alpha \right) , \quad \sin \alpha \neq 0$$

also is nonintegrable because of the growth of degrees of iterations [35]. This is in good agreement with the results of Moser, which state that the Birkhoff series for that mapping diverge.

Moser also discussed the problem of including the mapping into the flow. If the flow is polynomial, this problem was investigated in detail by *Bass* and *Meisters* [36]. It is easy to see that the last property implies the integrability. As follows from [36] the inverse statement is false: the mapping $f(x, y) = (x + y^2 + 1, -y)$ has the commuting mapping $g(x, y) = (x + y^2, y)$, but can be included in any polynomial flow.

Example 3. The mapping, corresponding to the discrete equations of the form

$$x_{n+1} - 2x_n + x_{n-1} = \varphi(x_n) . \tag{4}$$

This type of equation is very popular in mathematical and physical literature. For $\varphi(x) = \sin x$ this equation describes the famous Frenkel-Kontorova model in solid state physics. We will discuss in detail the general theory of such systems in the second part and now consider the case when φ is a polynomial.

In variables $x = x_{n-1}$, $y = x_n$ the mapping, corresponding to (4), has a form $f(x,y) = (y, 2y - x + \varphi(y))$ and belongs to the group GA_2.

Proposition 2. The mapping f, describing the discrete equation (4) is integrable only in the linear case, i.e. for $\varphi(x) = ax + b$.

The reasoning is the same as in previous cases.

The investigation of the rational Cremona transformation of a plane is much more difficult because of the complexity of the group structure. Some examples of such Cremona mappings have appeared in physical literature devoted to the Chew-Low equation.

The problem of integrability of this example is very important but still unsolved [37].

1.5 Euler–Chasles Correspondences and the Yang–Baxter Equation

Let us consider the biquadratic relation of the following form

$$\phi(x,y) = a_0 x^2 y^2 + a_1 xy(x+y) + a_2(x^2 + y^2) + a_3 xy + a_4(x+y) + a_5 = 0 . \quad (5)$$

This relation determines the many valued mapping or $2 - 2$ correspondence. At first this relation was considered by Euler, who used it in the proof of the addition law for the elliptic integrals. Chasles has noticed that this correspondence describes Poncelet mapping for two quadrics in the plane (Fig. 1).

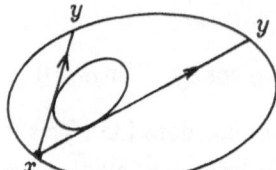

Fig. 1

Theorem 9. For the general correspondence (5) there exists an even elliptic function of the second order $\varphi(z)$ such that if $x = \varphi(z)$, then $y = \varphi(z \pm a)$ for some shift a (Euler et al).

In degenerate cases a trigonometric function appears instead of an elliptic function.

Our Lie-algebraic approach can be applied to the construction of the commuting correspondences of $\mathbb{C}^n$ and $\mathbb{C}P^n$. Indeed, again let G be a simple complex Lie algebra, ϕ_G be defined as in §2, A be a finite set, which is invariant under the Weyl group W. Then the manyvalued mapping $x = \phi_G(z) \rightarrow y = \phi_G(z \pm a)$ for all $a \in A$ defines some algebraic correspondence of $\mathbb{C}^n$: $\mathcal{E}_{G,A}(x,y) = 0$, according to the Chevalley theorem. Considering the Abelian variety $M_{G,\tau}$ (see §3) we come to the algebraic correspondences of $\mathbb{C}P^n$: $\mathcal{E}_{G,A,\tau}(x,y) = 0$. For $G = A_1$, $A = \pm a$ it is the Euler–Chasles correspondence (5). So we have

Theorem 10. To every simple complex Lie algebra and elliptic curve $\mathcal{R}$ corresponds the family of commuting symmetric correspondences $\mathcal{E}_{G,A,\tau}(x,y) = 0$ in $\mathbb{C}P^n \times \mathbb{C}P^n$ [25].

The importance of the problem about the commuting correspondences was discoverd by *Krichever* [38]. He has shown that some important particular cases of the Yang–Baxter equation can be reduced to this problem. The Euler–Chasles correspondences correspond to the famous Baxter solution of this equation. Our construction can possibly supply new solutions of the Yang–Baxter equation. This possibility is now under consideration.

Notice that the Felderhof solution corresponds to the transformation $z \rightarrow (az + b)(cz + d)^{-1}$, which is subjected to the conjugation by the mapping $f = z^2$. The last mapping is noninvertible and therefore leads to a manyvalued mapping. The natural generalization of such correspondences can be got in an analogous manner from projective transformations in every dimension.

2. Integrable Lagrangean Mappings with Discrete Time

2.1 Hamiltonian Theory

Definition. The mapping $\varphi : U \rightarrow M^n$ for some open domain $U \subset M^n \times M^n = N$ defines a *dynamic system with discrete time*, with a corresponding "Lagrangean" $\mathcal{L} : N \rightarrow \mathbb{R}^1$ if every sequence $q = (q_k)$, $k \in \mathbb{Z}$, $q_k \in M^n$ such that

$$q_{k+1} = \varphi(q_{k-1}, q_k) \tag{1}$$

is the stationary point for the functional

$$S(q) = \sum_{k \in \mathbb{Z}} \mathcal{L}(q_k, q_{k+1}) .$$

This means that equation (1) is a suitable solution of the algebraic equation $\delta S = 0$ or

$$\frac{\partial \mathcal{L}}{\partial y}(q_{k-1}, q_k) + \frac{\partial \mathcal{L}}{\partial x}(q_k, q_{k+1}) = 0 . \tag{2}$$

We will assume that the domain U is invariant under the following mapping $\Phi : U \rightarrow N$, $\phi(x,y) = (y, \varphi(x,y))$.

Let us consider the decomposition of the differntial $d\mathcal{L} = \alpha + \beta$ according to the natural isomorphism $T^*N = T^*M^n \oplus T^*M^n$. In coordinate system x^i, y^i on M^n, generated by some *coordinate system on* M^k, this decomposition has the form

$$dL = \frac{\partial \mathcal{L}}{\partial x^i} dx^i + \frac{\partial \mathcal{L}}{\partial y^i} dy^i \ , \quad \alpha = \frac{\partial \mathcal{L}}{\partial x^i} dx^i \ , \quad \beta = \frac{\partial \mathcal{L}}{\partial y^i} dy^i \ .$$

Let us define the closed 2-form ω on N as

$$\omega = d\beta = -d\alpha \ ,$$

or, in the coordinates

$$\omega = \frac{\partial^2 \mathcal{L}}{\partial x^i \partial y^j} dx^i \wedge dy^j \ .$$

This form is the image of the standard symplectic structure on T^*M^n under the mapping $f_{\mathcal{L}} : N \rightarrow T^*M^n$, $f_{\mathcal{L}}(x,y) = (y,\beta) = (y,(\partial \mathcal{L}/\partial y)(x,y))$.

Theorem 1. The mapping Φ presevers the form ω, and, what is more, $\mathcal{L}$ is the generating function of the canonical transformation in the domain V, where ω is nondegenerate.

The domain V is invariant under Φ, so we can consider the dynamics in $W = V \cap U$. The function F on V is called the integral of the corresponding system with discrete time if $F(\phi(P)) = F(P)$. The form ω determines the Poisson bracket on W and, therefore, the notion of the integrals in involutuion.

Theorem 2. If the dynamic system with discrete time (1) has n independent integrals in involution $F_1, \dots, F_n$, then every connected compact nonsingular level $M_c = \{P \in W : F_i(P) = c_i\}$ is diffeomorphic to the torus T^n, on which the mapping Φ corresponds to the ordinary translation. This translation can be considered as the translation for unit time along the trajectory of the flow with the Hamiltonian H,

$$H = -\mathcal{L}(P) + \int \beta$$

where the integration can be made along any curve on M_c, connecting the pointss P and $\phi(P)$.

The proof is based on the fact that the mapping commutes with all Hamiltonian flows, corresponding to $F_1, \dots, F_n$ and therefore must be the translation on Liouville's torus. Notice that the formula for H resembles the ordinary Legendre transformation, but such a formula is possible only in the integrable case. Already in the simplest situation, when $M^n = \mathbf{R}^1$, $\mathcal{L} = (x-y)^2 + f(x)$, there exist no integrals for the general function f, and therefore, no "Hamiltonian" [12].

This discrete analogue of the Liouville theorem [39] reduces the problem of solution of the system with discrete time, having sufficiently many integrals in involution, to the integration of the ordinary Hamiltonian system. Such discrete systems we will call an integrable.

How can we find the integrals? A possibility is presented by the following modification of the Noether theorem.

Let $\dot{x} = v(x)$ be the flow on M^n such that $(d/dt)\mathcal{L}(x, y) = Q(x) - Q(y)$ for some function Q on M^n. Then this flow preserves the set of stationary points of the functional S and it is possible to restrict it to this set.

Theorem 3. The flow $\dot{x} = v(x)$, $\dot{y} = v(y)$ on V is Hamiltonian, with the Hamiltonian

$$H(x, y) = \frac{\partial \mathcal{L}}{\partial y}(x, y)v(y) + Q(y) = -\frac{\partial \mathcal{L}}{\partial x}(x, y)v(x) + Q(x) ,$$

which is the integral of the corresponding system Φ.

This theorem can be considered as the simplest discrete variant of the Bogoyavlensky–Novikov theorem [40] (see also [41]) and its generalization, found by *Mokhov* [42]. An analogous result is valid for the case $\mathcal{L} = \mathcal{L}(q_k, q_{k+1}, \dots q_{k+l})$ with every l.

Let us consider now examples of integrable systems with discrete time.

2.2 Heisenberg Chain with Classical Spins and the Discrete Analog of the C. Neumann System

Consider a one-dimensional chain with classical spins S_k, which interact with the energy $\mathcal{L} = (S_k, JS_{k+1})$, $J = \text{diag}(J_1, J_2, J_3)$. It is the classical variant of the XYZ Heisenberg chain. The stationary solution of such models were investigated in many papers [9–14]. In particular, the authors of [9] have conjectured the "integrability" of this problem. *Granovsky* and *Zhedanov* [11] have found 2 algebraic integrals and some particular solutions.

Corresponding equations have the following form:

$$S_{k+1} + S_{k-1} = \lambda_k J^{-1} S_k \tag{3}$$

where the multiplier λ_k is determined from the condition $|S_{k+1}| = 1 : \lambda_k = 0$ or

$$\lambda_k = 2 \left(S_{k-1}, J^{-1} S_k \right) \left| J^{-1} S_k \right|^{-2} . \tag{4}$$

We will consider only the second case. After simple algebraic manipulations we come to the integrals of Granovsky and Zhedanov:

$$H_1 = \left(x, J^{-1} y \right) , \quad H_2 = |Jx|^2 + |Jy|^2 - (x, Jy)^2 .$$

So we must only prove that these integrals are in involution. For this purpose, the following variables are very useful (compare with [43]):

$$M = [x, Jy] , \quad l = x ; (M, l) = 0 , \quad l^2 = 1 .$$

Proposition 1. The symplectic structure ω in the variables M, l coincides with the Kostant–Kirilov structure on the corresponding coadjoint orbit of the group $E(3)$ of the motions of $\mathbf{R}^3$.

In other words the Poisson bracket of M_i, l_j are as follows

$$\{M_i, M_j\} = \varepsilon_{ijk}M_k , \quad \{M_i, l_j\} = \varepsilon_{ijk}l_k , \quad \{l_i, l_j\} = 0 .$$

The integrals H_1, H_2 have a form

$$\det J^2 - H_1^2 = (J^2 M, M) - \det J^2 (J^2 l, l) , \quad H_2 = M^2 + (J^2 l, l) .$$

The integrals in the Clebsch integrable case of the motion of a solid body in an ideal fluid, which can be realized on the same phase space [44] have the same form. On our orbit, where $(M, l) = 0$, $l^2 = 1$ this case is equivalent to the famous Neumann system, describing the motion of a point on the unit sphere under a force with potential $U(x) = (J^2 x, x)/2$ [43].

So we come to

Theorem 4. The system (3), (4), describing the stationary states of the classical anisotropic Heisenberg chain, is integrable and corresponds to the translation on the tori of the classical problems of Clebsch and Neumann.

This system (3, 4) is the discrete analog of the Neumann system for the sphere of any dimension, because of the following result [8].

Proposition 2. The system (3, 4) has the following integrals in involution:

$$F_\alpha(x, y) = x_\alpha^2 + \sum_{\beta \neq \alpha} \frac{(x \wedge Jy)_{\alpha\beta}^2}{J_\alpha^2 - J_\beta^2} , \quad \sum_{\alpha=0}^{n} F_\alpha \equiv 1$$

where

$$(x \wedge y)_{\alpha\beta} = x_\alpha y_\beta - x_\beta y_\alpha .$$

One can see the resemblance of the last formulae with the formulae for the Neumann system [15].

It can be shown also that the Neumann system is the continuous limit of the system (3, 4).

One can find explicit formulas for the solutions in θ-functions in [8], see also the last paragraph.

2.3 The Billiard in Quadrics

Billiards in convex domains $\Omega \subset \mathbf{R}^{n+1}$ are the most natural examples of the systems with discrete time. They correspond to the function $\mathcal{L} = |x - y|$, $x, y \in M^n = \partial\Omega$. Let us consider now the case of the ellipsoid

$$M^n = \{q \in \mathbf{R}^{n+1} : (Aq, q) = 1\} , \quad A_{\alpha\beta} = A_\alpha \delta_{\alpha\beta} , \quad \alpha, \beta = 0, \dots, n .$$

The corresponding equations have the following form:

$$\frac{q_{k+1} - q_k}{|q_{k+1} - q_k|} - \frac{q_k - q_{k-1}}{|q_k - q_{k-1}|} = \lambda_k A q_k \tag{5}$$

where q_k is the sequence of the blow's points. Introducing $p_k = (q_{k+1} - q_k)/|q_{k+1} - q_k|$ we can rewrite it as

$$p_k - p_{k+1} = \lambda_k A q_k \, , \quad q_k - q_{k-1} = \mu_k p_{k-1} \tag{6}$$

where λ_k and μ_k must be determined from the condition $|p| = 1$, $(Aq, q) = 1$.

Proposition 3. The symplectic form ω coincides with the restriction of standard form $dp \wedge dq$ on the phase manifold $p^2 = 1$, $(Aq, q) = 1$.

The integrals for this problem coincide with those for the Jacobi problem about the geodesics on ellipsoid, because the billiard is its limiting case when one of the axes of ellipsoid in $\mathbb{C}^{n+2}$ tends to zero. The formula for the integrals have a form [15]

$$F_\alpha = p_\alpha^2 + \sum_{\beta \neq \alpha} \frac{(p \wedge q)_{\alpha\beta}^2}{B_\beta - B_\alpha} \, , \quad B_\alpha = A_\alpha^{-1} \, . \tag{7}$$

The conservation of F_α means that the trajectory of point is tangent to the fixed confocal quadrics as well as in the Jacobi problem [15]. Notice that geodesic flow is in its turn the limiting case of the billiard when the distance between the initial data q_0, q_1 tends to zero.

Using the results of [15] and Proposition 3, we come to the following theorem.

Theorem 3. The billiard in an ellipsoid $(Aq, q) = 1$ has a complete set of involutive integrals F_α (7) and corresponds to the translation on the tori of the Neumann problem with potential $U(x) = (Bx, x)/2$, $B = A^{-1}$.

The explicit formulas for the solution in θ-functions are found in Sect. 2.5.

Notice that analogous conclusions are valid for the billiard in any quadric in $\mathbb{R}^{n+1}$.

One can consider also the billiard in the domain in $\mathbb{R}^{n+1}$, bounded by several quadrics, belonging to one confocal family

$$\sum_{\alpha=0}^{n} \frac{q_\alpha^2}{B_\alpha - z} = 1 \, . \tag{8}$$

This problem has the same integrals F_α and also is integrable (see [8]).

Another geometrical application of our results is a new proof of the classical *Poncelet porism* [45], concerning the polygons inscribed in one quadric and circumscribed about another. It states that the condition that the polygon be closed on n sides is independent of the initial point and its satisfaction imposes one condition on the pair of quadrics. This statement is a consequence of Theorem 5. In fact, the corresponding pair of quadrics can be transformed by a suitable projective transformation into the pair of confocal quadrics of the form (8).

The corresponding polygons transform into the trajectories of a certain billiard belonging to the fixed torus. According to Theorem 5 this situation is described by the translation on the elliptic curve and therefore the properties of trajectories of polygons do not depend on the initial point.

These geometrical considerations are closely connected with the Euler–Chasles correspondences (see above) and the addition law on an elliptic curve. For details we refer the reader to the book by *Darboux* [46] and the very interesting paper of *Griffiths* and *Harris* [47].

2.4 The Discrete Analog of the Dynamics of the Top

This example was first considered in [8]. Let $\mathcal{L}$ be the function $\mathcal{L} = \text{tr}\,(xIy^{\text{T}})$, where $x, y \in \text{SO}(N)$, $I = \text{diag}(I_1, \ldots, I_N)$. This function is invariant under the transformation $x \rightarrow gx$, $y \rightarrow gy$, $g \in \text{SO}(N)$ and therefore Theorem 3 can be applied. The corresponding "moment in the space" [39] has a form $m = yIx^{\text{T}} - xIy^{\text{T}}$. Introducing "moment in the body", $M = x^{\text{T}}mx = \omega^{\text{T}}I - I\omega$, where $\omega = y^{\text{T}}x$, we come to the discrete analog of the Euler equation

$$M_{k+1} = \omega_k M_k \omega_k^{-1}, \quad M_k = \omega_k^{\text{T}} I - I\omega_k, \quad \omega_k \in \text{SO}(N). \tag{9}$$

In the continuous limit, these equations transform into the ordinary Euler equations of the free dynamics of a solid body:

$$\dot{M} = [M, \omega], \quad M = I\omega + \omega I, \quad \omega \in \text{SO}(N). \tag{9'}$$

For the last equations Mischenko [48] has found the quadratic integrals

$$F_l = \sum_{p+q+r=l} \text{tr}\left(I^{2p} M I^{2q} M I^{2r}\right) \tag{10}$$

which proved to be in involution [49] and to be sufficient for the integrability of Euler equations for $\text{SO}(N)$, $N = 3, 4$.

Theorem 6. The discrete analogue of the Euler equations (9) has the involutive integrals, determined by the same formulas (10) [8]. For the groups SO (3) and SO (4) these discrete systems are integrable and correspond to the translation on the tori of the Euler equations (9') of an N-dimensional solid body.

For the equations (9') *Manakov* [50] has found the Lax representation $L' = [L, A]$ with $L = M + \lambda I^2$, $A = \omega + \lambda I$, which ensured the integrability of the system. The discrete analog of such a representation was first considered by Novikov (see [5]) and has the from $L_{k+1} = A_k L_k A_k^{-1}$. As follows from the previous theorem the matrices L_{k+1} and L_k, where $L_k = M_k + \lambda I^2$, are conjugate, but the explicit opportunate formula for A_k is unknown.

There is no doubt of the integrability of the discrete Euler equations for all N, but meanwhile a good proof has not been found.[2]

---

[2] *Note added in proof:* Such a proof and the discrete analogs of the Lax representations for this and the two previous systems are found in [56].

2.5 Connection with the Spectral Theory of the Difference Operators: A Discrete Analogue of the Moser–Trubowitz Isomorphism

As was found by *Moser* and *Trubowitz* [15] the solutions of the Neumann system describing the motion of a point on the unit sphere under a linear force, can be expressed in terms of eigenfunctions of finite-gap Schrödinger operators (the full generality was shown in [51, 52]). We present here a discrete version of this connection. We will use the algebraic-geometrical approach to the spectral theory of the difference operators (for detail we refer to [16]).

Consider the curve Γ of the form

$$y^2 = R(E), \quad R(E) = \prod_{i=0}^{2n} \left(E^2 - E_i^2 \right)$$

where E_i are different real numbers, $E_i \neq 0$. Let us define meromorphic functions $\psi_k(P)$ on it with the following properties: 1) ψ_k has on Γ n independent of k simple poles $P_1, \ldots, P_n$; 2) in the neighbourhood of the infinities $P \pm \psi_k$ has asymptotes $\psi_k^\pm(E) = \alpha_k^\pm E^{\pm k}(1 + O(1/E))$, where the sign indicates the sheet to which the point $P\pm$ belongs. The divisor $\mathcal{D} = P_1 + P_2 + \ldots + P_n$ is supposed to be invariant under the involution $\sigma : \Gamma \to \Gamma$, $\sigma(y, E) = (-y, -E)$.

Theorem 7. Functions ψ_k with the previous properties do exist, are determined up to the sign and satisfy the difference equation [5, 16]

$$c_{k+1}\psi_{k+1} + c_k\psi_{k-1} = E\psi_k,$$

where

$$c_k = \frac{\alpha_{k-1}}{\alpha_k}.$$

For c_k to be real, we must demand that the projections P_j on the E-plane belong to the "forbidden zones" – the segments of real line, where $R(E) \geq 0$.

The explicit formulae for ψ_k and c_k can be found in [16].

Let us consider the following class of such operators $(L\varphi)_k = c_{k+1}\varphi_{k+1} + c_k\varphi_{k-1}$, corresponding to the curves of the form

$$y^2 = \prod_{i=1}^{h} \left(E^2 - e_i^2 \right) \prod_{i=0}^{n} \left(E^2 - I_\alpha^2 \right).$$

Here $I_\alpha = J_\alpha^{-1}$, J_α are the same as in 2.2 for the opportunity we suppose that $0 < I_0 < I_1 < \ldots < I_n$. The real numbers e_i must be different and such that only one forbidden zone is placed between I_α and $I_{\alpha+1}$ (compare with [52]).

Let us define the sequence of the points q_k in $\mathbf{R}^{n+1}$, whose coordinates are the normalized eigenfunctions of the operators L:

$$q_k^\alpha = \beta_\alpha \psi_k(I_\alpha), \quad \beta_\alpha = I_\alpha \sqrt{P(I_\alpha) \prod_{\beta \neq \alpha} (I_\alpha^2 - I_\beta^2)},$$

where

$$P(E) = \prod_{i=1}^{n} (E^2 - \gamma_i^2) \ , \quad \text{and} \quad \pm \gamma_i$$

are the projections of the poles P_m on the plane E. The previous condition on the position of e_i ensures the expression under the radical is positive.

Lemma. The vectors q_k belong to the ellipsoid $(I^2 q, q) = 1$ and satisfy the equation [8]

$$c_{k+1} q_{k+1} + c_k q_{k-1} = I q_k \ .$$

Let us consider the Gauss mapping of this ellipsoid into the unit sphere: $S = g(q) = J^2 q / |J^2 q|$.

Theorem 8. Let q_k be the sequence of the vectors in $\mathbf{R}^{n+1}$ with the coordinates defined by the normalized eigenfunction of the difference operator L from the class described above [8]. Then $S_k = g(q_k)$ is the general solution of the discrete Neumann system (3) (4) (for $n = 2$ – the general stationary solution for XYZ Heisenberg chain).

The proofs can be found in [8].

Comparing with the Moser–Trubowitz consideration one can notice one new feature – the presence of intermediate Gauss mapping.

An analogous construction is possible for the interpretation of the dynamics in an ellipsoidal billiard (see [8]). As a consequence of this result we can find the explicit formulae for the general solution of this discrete system. For example, for the billiard they have the form:

$$q_k^\alpha = \varepsilon_\alpha \frac{\theta_\alpha(kU + \zeta)\theta(\zeta)}{\theta_\alpha(\zeta)\theta(kU + \zeta)} \ ,$$

where θ_α is a θ-function corresponding to the n-dimensional hyperelliptic Jacobian with certain characteristics depending on α (for the details, see [8]).

Note added in proof. In a recent paper [56] J. Moser and this author proposed an approach to the integration of discrete systems, based on the factorization of matrix polynomials. The idea is to represent the discrete system as the isospectral transformation $L(\lambda) \rightarrow L'(\lambda)$, where $L(\lambda)$ and $L'(\lambda)$ are some matrix polynomials, such that $L(\lambda) = A(\lambda)B(\lambda)$, $L'(\lambda) = B(\lambda)A(\lambda)$ are factorizations of a certain type. The form of the polynomials $L(\lambda)$ can be guessed from the Lax representation for the corresponding continuous system.

This approach turns out to be fruitful for all classical integrable systems discussed above and gives explicit formulas for the discrete versions in terms of θ-functions. The same idea works also for some billiard problems in spaces with constant curvature [57].

The fact that in all these cases we arrive at *Lagrangian* discrete systems (compare with [53, 54]) seems to be very important and calls for better understanding.

Appendix A

Polynomial Mappings of the Plane $P^2_{G,\sigma,a}$ Connected with the Simple Lie Algebras G of Rank 2

Lie algebra G	Φ_G	$P^2_{G,\sigma,a}$
$G = A_2$	$\Phi_G(x,y) = \begin{pmatrix} e^x + e^y + e^{-(x+y)} \\ e^{-x} + e^{-y} + e^{(x+y)} \end{pmatrix}$	$P^2_{A_2}\begin{pmatrix} u \\ v \end{pmatrix} = \begin{pmatrix} u^2 - 2v \\ v^2 - 2u \end{pmatrix}$ $P^2_{A_2,\sigma}\begin{pmatrix} u \\ v \end{pmatrix} = \begin{pmatrix} v^2 - 2u \\ u^2 - 2v \end{pmatrix}$
$G = B_2 \simeq C_2$	$\Phi_G\begin{pmatrix} x \\ y \end{pmatrix} = \begin{pmatrix} e^x + e^{-x} + e^y + e^{-y} \\ e^{x+y} + e^{x-y} + e^{-x-y} + e^{-x+y} \end{pmatrix}$	$P^2_{B_2}\begin{pmatrix} u \\ v \end{pmatrix} = \begin{pmatrix} u^2 - 2v - 4 \\ v^2 - 2u^2 + 4v + 4 \end{pmatrix}$
$G = G_2$	$\Phi_G\begin{pmatrix} x \\ y \end{pmatrix} = \begin{pmatrix} \cos x + \cos y + \cos(x+y) \\ \cos(x-y) + \cos(2x+y) + \cos(2y+x) \end{pmatrix}$	$P^2_{G_2}\begin{pmatrix} u \\ v \end{pmatrix} = \begin{pmatrix} u^2 - 2v - 2u - 6 \\ v^2 - 2u^3 - 6uv + 18u + 10v + 18 \end{pmatrix}$

Appendix B

Integrable Rational Mappings of $\mathbb{C}P^1$ of Degree 2

As follows from Ritt's theorem all such mappings f satisfiy a functional equation of the form $\varphi(\lambda x + \beta) = f(\varphi(x))$, where φ must be exp x, cos x or a suitable elliptic function. The condition that the degree of the mapping must be equal to 2 leads to a limitation on the corresponding elliptic curve. Namely, this curve must have the complex multiplicity of degree 2. There are only three such curves, with paramter τ equal to i, $i\sqrt{2}$ or $(1 + i\sqrt{7})/2$. Thus, we have the following 10 possibilities:

	$\varphi(x)$	λ	β	$f(z)$
1)	exp x	2	0	$f(z) = z^2$
2)	cos x	2	0	$2z^2 - 1$
3)	$\mathcal{P}(x), \tau = i$	$1 + i$	0	$\frac{1}{2i}(z - 1/z)$
4)	$\mathcal{P}^2(x), \tau = i$	$1 + i$	0	$-\frac{1}{4}(z - 2 + 1/z)$
5)	$\mathcal{P}(x), \tau = i\sqrt{2}$	$i\sqrt{2}$	0	$-\frac{1}{2}(z + \frac{1}{z - 2\sqrt{2}/3})$
6)	$\mathcal{P}(x), \tau = i\sqrt{2}$	$i\sqrt{2}$	$\frac{1}{2}$	$\sqrt{2} + 3 + \frac{18(\sqrt{2}-1)}{z+(z-2\sqrt{2}/3)^{-1}+2\sqrt{2}+6}$
7)	$\mathcal{P}(x), \tau = i\sqrt{2}$	$i\sqrt{2}$	$\frac{i}{2}$	$-2\sqrt{2} + \frac{18}{z+(z-2\sqrt{2}/3)^{-1}-4\sqrt{2}}$
8)	$\mathcal{P}(x), \tau = \frac{1+i\sqrt{7}}{2}$	$\frac{1+i\sqrt{7}}{2}$	0	$\frac{2}{-3+i\sqrt{7}}(z + \frac{1}{z - \sqrt{2}(5+i\sqrt{7})/4})$
9)	$\mathcal{P}(x), \tau = \frac{1+i\sqrt{7}}{2}$	$\frac{1+i\sqrt{7}}{2}$	$\frac{1}{2}$	$-\frac{\sqrt{7}}{2} + \frac{2(11-7\sqrt{2})}{-(3+\sqrt{7})(z+(z-\sqrt{2}(5+i\sqrt{7})/4)^{-1})+4\sqrt{7}}$
10)	$\mathcal{P}(x), \tau = \frac{1+i\sqrt{7}}{2}$	$\frac{1+i\sqrt{7}}{2}$	$\frac{3+i\sqrt{7}}{4}$	$\frac{i-\sqrt{7}}{2\sqrt{2}} + \frac{2(-1+i(\sqrt{2}-1))}{(-3+\sqrt{7})(z+(z-\sqrt{2}(5+i\sqrt{7})/4)^{-1})+4(\sqrt{7}-i)/\sqrt{2}}$

References

1   M. Ljubich: "Dynamics of the rational transformations: topological picture", Uspekhi Mat. Nauk, **41**, 35–95 (1986) [Russian]

2   G. Julia: "Memoire sur la permutabilité des fractions rationelles", Ann. Scient. Ec. Norm. (Sup.) **39**, 131–215 (1922)

3   P. Fatou: "Sur l'iteration analytique et les substitutions permutables", J. Math. Pure Appl., **3**, 1–49 (1924)

4   J. Ritt: "Permutable rational functions", Trans. Amer. Math. Soc., **25**, 399-448 (1923)

5   B.A. Dubrovin, V.B. Matveev, S.P. Novikov: "Nonlinear equations of Korteveg–de Vries type, finite zone linear operators and Abelian varieties", Russ. Math. Surveys, **31**, 59–164 (1976)

6   A.V. Mikhailov, A.B. Shabat, R.I. Yamilov: "Symmetry approach to classification of nonlinear equations. The full list of the integrable systems", Uspekhi Mat. Nauk, **42**, 3–53 (1987) [Russian]

7   A.P. Veselov: "Integrable mappings and Lie algebras", Sov. Math. Dokl., **35**, 211–213 (1987)

8   A.P. Veselov: "Integrable systems with discrete time and difference operators", Funct. Anal. Pril., **22**, 1–13 (1988) [Russian]

9   P.I. Belobrov et al: "Order and chaos in classical models of spin chains", J. Expt. Theor. Phys. **89**, 310–322 (1984)

10   Ya.I. Granovsky, A.S. Zhedanov: "Periodic structures on quantum spin chain", J. Expt. Theor. Phys. **89**, 2156–2163 (1985)

11   Ya.I. Granovsky, A.S. Zhedanov: "The solution of domain type in magnetic chain", Theor. Mat. Phys., **71**, 145–153 (1987) [Russian]

12   S. Aubry, Le Daeron: "The discrete Frenkel–Kontorova model and its extension. I. Exact result for the ground states", Physica **8D**, 381–422 (1983)

13   V.L. Pokrovsky, S.B. Khohlachev: "Inhomogeneous stationary states in the Heisenberg model", J. Expt. Theor. Phys. Lett. **22**, 371–373 (1975)

14   A.P. Veselov: "The integration of the stationary problem for classical spin chains", Theor. Mat. Phys., **71**, 154–159 (1987) [Russian]

15   J. Moser: "Various aspects of integrable Hamiltonian systems" in *Dynamical Systems* (C.E.M.E.) 1978

16   B.A. Dubrovin, I.M. Krichever, S.P. Novikov: "Integrable systems, I" in *Itogi nauki i tehniki*, Ser. Sovr. Probl. Mat., V.4, 1985 (Moscow), 179–285 [Russian]

17   A. Douady, J. Hubbard: "Iteration des polynomes quadratiques complexes", C.R. Acad. Sci. Paris, **294**, 123–126 (1982)

18   S.P. Novikov: "Periodic problem for the Korteweg–de Vriees equation, I", Funkt. Anal. Pril., **8**, 54–66 (1974)

19   H.P. McKean, P. van Moerbeke: "The spectrum of Hill's equation", Invent. Math. **30**, 217–274 (1975)

20   F. Böttcher: "The main laws of convergency of iterations and applications to analysis", Izv. Kazanskogo ph.-mat. ob., Ser.2, **14**, 155–200 (1904) [Russian]

21   I.M. Krichever: "Commutative rings of ordinary linear differential operators", Funct. Anal. Pril., **12**, 20–31 (1978)

22   J.L. Burchnall, T.W. Chaundy: "Commutative ordinary differential operators I, II", Proc. London. Math. Soc., **21**, 420–440 (1922); Proc. Royal. Soc. London. **118**, 557-583 (1928)

23   N. Bourbaki: "Groupes et algebras de Lie", Actualités Sci. Indust., No.1337 (Hermann, Paris 1968) Chaps.4–6

24   O.A. Chalykh: "On some properties of polynomial mappings connected with Lie algebras", Vestn. Moscow State University Ser. Mat., Mech., **24**, (1988)

25   A.P. Veselov: "The dynamic of mappings of toroidal varieties connected with Lie algebras", in *Proc. Int. Topological Conf.*, Baku, (1987)

26   M. Olshanetsky, A. Perelomov: "Quantum integrable systems related to Lie algebras", Phys. Rep., **94** 313–404 (1983)

27   A. Veselov, O. Chalykh: "Quantum Calogero Problem and commutative rings of many-dimensional differential operators", Usp. Mat. Nauk, **43** No.4 (1988)

28   E. Looijenga: "Root systems and elliptic curves", Invent. Math., **38**, 17–32 (1976)

29   I.N. Bernshtein, O.V. Shvartsman: "Chevalley theorem for the complex crystallographic Coxeter groups", Funct. Anal. Pril., **12**, 79–80 (1978) [Russian]

30   E. Whittaker, G. Watson: *A Course of Modern Analysis*, (Cambridge, University Press 1927)

31   S.P. Novikov (ed.): *The Theory of Solitons* (Nauka, Moscow 1980), [Russian]

32   H. Jung: "Über ganze birationale Transformationen der Ebene", J. Reine Angew. Math., **184**, 161–174 (1942)

33   D. Wright: "Abelian subgroups of Aut $(k[x, y])$ and applications to actions of affine plane", Illinois J. Math., **23**, 579–634 (1979)

34   A. Veselov: "Cremona Group and Dynamical Systems" Math. Notes, **45**, No.3 (1989)

35   J. Moser: "Lectures on Hamiltonian Systems", New York 1968

36   H. Bass, G. Meisters: "Polynomial flows in the plane", Adv. Math., **55**, 173–208 (1985)

37   K.V. Rerikh: "Chew–Low equations as Cremona transformations. Structure of general integrals.", Theor. Mat. Phys. **50**, 251–260 (1982) [Russian]

38   I.M. Krichever: "Baxter equation and algebraic geometry", Funct. Anal. Pril. **15**, 22–35 (1981)

39   V.I. Arnold: *Mathematical Methods of Classical Mechanics* (Springer, Berlin-Heidelberg 1978)

40   O.I. Bogoyavlensky, S.P. Novikov: "On the connection of the Hamiltonian formalisms of stationary and nonstationary problems", Funct. Anal. Pril. **10**, 9–12 (1976)

41   I.M. Gelfand, L.A. Dickey: "Fractional degrees of operators and Hamiltonian systems", Funct. Anal. Pril. **10**, 13–29 (1976)

42   O.I. Mokhov: "Hamiltonity of evolutional flow on the set of stationary points of its integral", Usp. Mat. Nauk, **39**, 173–174 (1984) [Russian]

43   A.P. Veselov: "Landau–Lifschitz equation and the integrable systems of classical mechanics", Sov. Math. Dokl., **270**, 1094–1097 (1983)

44   S.P. Novikov: "Hamiltonian formalism and the many-valued analog of Morse theory", Usp. Mat. Nauk., **37**, 3–49 (1982)

45   M. Berger: *Geometrie* (Paris 1977)

46   G. Darboux: *Principes de Geometrie Analytique* (Gauthier-Villars, Paris 1917)

47   P. Griffiths, J. Harris: "On Cayley's explicit solution to Poncelet porism", Enseign. Math., Ser. 2, **24**, N1–2, 31–40 (1978)

48   A.S. Mischenko: "The integrals of geodesic flows on Lie groups", Funct. Anal. Pril. **4**, 73–78 (1970)

49   L.A. Dickey: "Remark on the Hamiltonian systems connected with the group of rotations", Funct. Anal. Pril. **6**, 83–84 (1972)

50   S.V. Manakov: "Remarks on the integrals of the Euler equations of the n-dimensional heavy top", Funkt. Analy. Appl. **10**, 93–94 (1976)

51   A.P. Veselov: "Finite-zone potentials and integrable systems on sphere with quadratic potential" Funct. Anal. Pril., **14**, 48–50 (1980)

52   A.P. Veselov: *Geometry of the Hamiltonian systems connected with nonlinear PDEs*, Ph. D. Thesis, Moscow State University (1981)

53   M. Ablowitz, J. Ladic: "A non-linear difference scheme and inverse scattering", Stud. Appl. Math. **55**, 213 (1976)

54   R. Hirota: "Nonlinear partial difference equations I–IV", J. Phys. Soc. Japan **43**, 1424–1433, 2074–2084 (1977); **45**, 321–332 (1978)

55   O.A. Chalykh, A.P. Veselov: "Commutative rings of partial differential operators and Lie algebras", Comm. Math. Phys. **126**, 597–611 (1990)

56   J. Moser, A.P. Veselov: "Discrete versions of some classical integrable systems and factorization of matrix polynomials" Reprint of ETH, Zürich, 1989. (To appear in Comm. Math. Phys.)

57   A.P. Veselov: "Confocal surfaces and integrable billiards on the sphere and in the Lobachevsky space", J. Geometry and Physics (in press)

The Cauchy Problem for the KdV Equation With Non-Decreasing Initial Data

V.A. Marchenko

At the present time there exist numerous modifications of the inverse problem method which can be used to obtain wide classes of solutions of nonlinear equations. However, with the exception of rapidly decreasing and periodic initial data, the applicability of these methods for the solution of the Cauchy problem has not, in fact, been studied. For instance, the approach developed in [4] for the KdV equation leads to integral equations of the type

$$y(z) + e^{-2z(x-4z^2t)}p(z)\left[\nu(z)y(-z) + \int \frac{y(z')}{z'+z}d\mu(z') - 1\right] = 0 \qquad (0.1)$$

whose solution leads to the solutions of the KdV equation $u_t = \sigma u u_x - u_{xxx}$ by the formula

$$u(x,t) = 2\frac{d}{dx}\int y(z)d\mu(z) .$$

The functions $p(z)$, $\nu(z)$ and the measure $d\mu(z')$ play the role of parameters these solutions depend upon. But it has not been found yet whether it is possible (and if possible, then by what means) to determine these parameters so as to obtain the solutions satisfying the given initial data $u(x,0) = q(x)$, i.e., to solve the Cauchy problem. The main goal of the present work is to study this problem. The initial data should necessarily be somehow bounded to ensure, at least, solvability of the Cauchy problem. The most general class of initial data for which the Cauchy problem is undoubtedly solvable is obtained in [3]. The description of this class is given in the first two sections which also contain the refinements necessary for further discussion. The parameters of the first equation above corresponding to the initial data are determined in the third section.

1. Reflectionless Potentials

The one-dimensional Schrödinger operator

$$H = -\frac{d^2}{dx^2} + q(x) \quad (-\infty < x < \infty) \tag{1.1}$$

with a real infinitely differentiable potential $q(x)$ satisfying the inequalities

$$\int_{-\infty}^{\infty} (1 + |x|) \, |q^{(k)}(x)| \, dx < \infty \quad (k = 0, 1, 2, \ldots)$$

is selfadjoint, and its spectrum consists of two parts: an absolutely continuous part which fills the whole positive half-exis and a finite number of simple eigenvalues $\mu_n < \mu_{n-1} \cdots < \cdots < \mu_2 < \mu_1 < 0$. For all λ from the closed upper (lower) half-plane the equation

$$-y'' + q(x)y = \lambda^2 y$$

has the solutions $e^+(\lambda, x) \ (e^-(\lambda, x))$, which can be presented in the form

$$e^+(\lambda, x) = e^{i\lambda x} + \int_x^{\infty} K^+(x, t) e^{i\lambda t} dt \tag{1.2}$$

$$e^-(\lambda, x) = e^{i\lambda x} + \int_{-\infty}^{x} K^-(x, t) e^{i\lambda t} dt . \tag{1.2'}$$

These solutions for large $|\lambda|$ can be also represented as

$$e^+(\lambda, x) = \exp \left\{ i\lambda x - \int_x^{\infty} \sigma^+(\lambda, t) dt \right\} \quad (\text{Im } \lambda \geq 0)$$

$$e^-(\lambda, x) = \exp \left\{ i\lambda x + \int_{-\infty}^{x} \sigma^-(\lambda, t) dt \right\} \quad (\text{Im } \lambda \leq 0) \tag{1.3}$$

where the functions $\sigma^{\pm}(\lambda, x)$ satisfy the equation

$$\sigma'(\lambda, x) + 2i\lambda\sigma(\lambda, x) + \sigma^2(\lambda, x) - q(x) = 0$$

and in the respective half-planes can be expanded in the same asymptotic series when $|\lambda| \to \infty$

$$\sigma^{\pm}(\lambda, x) \sim \sum_{k=1}^{\infty} \frac{\sigma_k(x)}{(2i\lambda)^k} \tag{1.4}$$

where

$$\sigma_1(x) = q(x), \quad \sigma_2(x) = -q'(x), \quad \sigma_3(x) = q''(x) - q(x)^2 ,$$

$$\sigma_{k+1}(x) = -\sigma_k'(x) - \sum_{j=1}^{k-1}\sigma_{k-j}(x)\sigma_j(x) \quad (k = 2, 3, \dots) . \tag{1.5}$$

For real λ the solutions $e^+(\lambda, x)$, as well as the solutions $e^-(\lambda, x)$, form a complete set of eigenfunctions of a continuous spectrum which correspond to the eigenvalues $\mu = \lambda^2$. They are related by the equations

$$e^+(\lambda, x) = b(\lambda)e^-(-\lambda, x) + a(\lambda)e^-(\lambda, x)$$

$$e^-(-\lambda, x) = -b(-\lambda)e^+(\lambda, x) + a(\lambda)e^+(\lambda, x)$$

where

$$2i\lambda a(\lambda) = \left\{e^+(\lambda, x), e^-(\lambda, x)\right\} , \quad 2i\lambda b(\lambda) = \left\{e^-(\lambda, x), e^+(\lambda, x)\right\}$$

and $\{f, g\} = f'(x)g(x) - f(x)g'(x)$ denotes the Wronskian of the functions $f(x)$, $g(x)$. The function $2i\lambda a(\lambda)$ is holomorphic in the upper half-plane and is continuous in the closed upper half-plane. In this half-plane it has a finite number of simple roots $i\kappa_k$ ($\kappa_n > \kappa_{n-1} > \dots > \kappa_1 > 0$), the squares of which are negative eigenvalues $\mu_k = (i\kappa_k)^2$ of the operator H.

The solution $e^+(i\kappa_k, x)$, as well as the solution $e^-(-i\kappa_k, x)$, is the eigenfunction of the operator H which corresponds to the eigenvalue $\mu_k = (-i\kappa_k)^2$. They are related by the equalities

$$e^+(i\kappa_k, x) = C_k^+ e^-(-i\kappa_k, x) , \quad e^-(-i\kappa_k, x) = C_k^- e^+(i\kappa_k, x) \tag{1.6}$$

$(C_k^+ C_k^- = 1)$, and their normalizing coefficients

$$m_k^{\pm} \overset{\text{def}}{=} \left[\int_{-\infty}^{\infty} \left|e^\pm(\pm i\kappa_k, x)\right|^2 dx\right]^{-1/2}$$

are expressed in terms of C_k^+, C_k^-, $a(\lambda)$ follows:

$$\left(m_k^\pm\right)^{-2} = iC_k^\pm \dot{a}(i\kappa_k) . \tag{1.7}$$

(The point means differentiation with respect to λ). The functions $r^-(\lambda) = b(\lambda)a(\lambda)^{-1}$, $r^+(\lambda) = -b(-\lambda)a(\lambda)^{-1}$ are called the left and right reflection coefficients and the sets $\{r^-(\lambda), i\kappa_k, m_k^-\}$, $\{r^+(\lambda), i\kappa_k, m_k^+\}$ are called the left and right scattering data of the operator H, respectively. The potential $q(x)$ can be uniquely recovered either from the left or the right scattering data. For this, it is necessary to solve the equation

$$F^+(x + y) + K^+(x, y) + \int_x^\infty F^+(y + t)K^+(x, t)dt = 0 \quad (x \le y < \infty) \tag{1.8}$$

$$F^+(\xi) = \sum_k \left(m_k^+\right)^2 e^{-\kappa_k \xi} + \frac{1}{2\pi}\int_{-\infty}^\infty r^+(\lambda)e^{i\lambda\xi}d\lambda$$

(or the equation

$$F^-(x+y) + K^-(x,y) + \int_\infty^x F^-(y+t)K^-(x,t)dt = 0 \quad (-\infty < y \le x)$$

$$F^-(\xi) = \sum_k \left(m_k^-\right)^2 e^{\kappa_k \xi} + \frac{1}{2\pi} \int_{-\infty}^\infty r^-(\lambda)e^{i\lambda\xi} d\lambda \right) \tag{1.8'}$$

with respect to $K^+(x,y)$ (or $K^-(x,y)$). The potential $q(x)$ can be found then in terms of $K^+(x,y)$ $(K^-(x,y))$ as

$$\int_x^\infty q(t)dt = 2K^+(x,x) , \quad \int_{-\infty}^x q(t)dt = 2K^-(x,x) .$$

The necessary and sufficient conditions for the scattering data are known. In particular, for any $\kappa_x > 0$, $m_k^+ > 0$ the set $\{0, i\kappa_k, m_k^+\}$ $(k = 1, 2, \ldots n)$ is the right scattering data of a certain operator H of the type in question. The corresponding potentials $q(x)$ are called reflectionless because the reflection coefficients $r^+(\lambda)$, $r^-(\lambda)$ are identically equal to zero. Note that in the reflectionless case

$$a(\lambda) = \prod_{k=1}^n \frac{\lambda - i\kappa_k}{\lambda + i\kappa_k} , \quad e^+(\lambda, x) = a(\lambda)e^-(\lambda, x) . \tag{1.9}$$

The existence of non-trivial reflectionless potentials was first discovered by V. Bargman [1], who also found their explicit expressions. We will denote by $B(-\mu^2)$ the set of those reflectionless potentials for which the spectra of corresponding operators (1.1) are located to the right of the point $-\mu^2$ (i.e. $H + \mu^2 I \ge 0$), and by B the set of all reflectionless potentials $(B = \cup_{\mu \ge 0} B(-\mu^2))$.

Lemma 1.1. If $q(x) \in B$, then the solutions $e^\pm(\lambda, x)$ assume the following form:

$$e^\pm(\lambda, x) = e^{i\lambda x} \prod_{k=1}^n \frac{\lambda - i\lambda_k(x)}{\lambda \pm i\kappa_k}$$

where $\lambda_k(x)$ are real infinitely differentiable functions, and if $k \ne l$, then $\lambda_k(x) \ne \lambda_l(x)$ for all $x \in (-\infty, \infty)$.

Proof. since in our case $r^\pm(\lambda) \equiv 0$, we have

$$F^\pm(\xi) = \sum_{k=1}^n \left(m_k^\pm\right)^2 e^{\mp\kappa_k \xi}$$

then (1.8, 8') give

$$K^+(x,y) = - \sum_{k=1}^n \left(m_k^+\right)^2 e^{-\kappa_k y} \left\{ e^{-\kappa_k x} + \int_x^\infty K^+(x,t)e^{-\kappa_k t}dt \right\}$$

$$= - \sum_{k=1}^n \left(m_k^+\right)^2 e^{-\kappa_k y} e^+(i\kappa_k, x)$$

$$K^-(x,y) = -\sum_{k=1}^{n} \left(m_k^-\right)^2 e^{\kappa_k y} e^-\left(-i\kappa_k, x\right) .$$

Substituting for $K^\pm(x,y)$ in the formulas (1.2, 2') the right parts of these equalities, we find that

$$e^+(\lambda, x) = e^{i\lambda x}\left[1 + \sum_{k=1}^{n} \frac{(m_k^+)^2 e^{-\kappa_k x} e^+(i\kappa_k, x)}{i(\lambda + i\kappa_k)}\right]$$

$$e^-(\lambda, x) = e^{i\lambda x}\left[1 - \sum_{k=1}^{n} \frac{(m_k^-)^2 e^{\kappa_k x} e^-(-i\kappa_k, x)}{i(\lambda - i\kappa_k)}\right] ,$$

that is

$$e^\pm(\lambda, x) = e^{i\lambda x}\frac{P^\pm(\lambda, x)}{\Pi^\pm(\lambda)}$$

where $\Pi^\pm(\lambda) = \prod_{k=1}^{n}(\lambda \pm i\kappa_k)$ and $P^\pm(\lambda, x) = \lambda^n + \sum_{k=0}^{n-1} p_k^\pm(x)\lambda^k$ are certain polynomials in λ with coefficients which are infinitely differentiable with respect to x. Hence, according to (1.9), it follows that

$$e^{i\lambda x}\frac{P^+(\lambda, x)}{\Pi^+(\lambda)} = e^{i\lambda x}\frac{P^-(\lambda, x)\Pi^-(\lambda)}{\Pi^-(\lambda)\Pi^+(\lambda)} \quad (-\infty < \lambda < \infty)$$

and, therefore, $P^+(\lambda, x) = P^-(\lambda, x) = P(\lambda, x)$. Denoting the roots of this polynomial by $i\lambda_k(x)$ we obtain

$$e^\pm(\lambda, x) = e^{i\lambda x}\prod_{k=1}^{n}\frac{\lambda - i\lambda_k(x)}{\lambda \pm i\kappa_k} .$$

Since for Im $\lambda > 0$ (Im $\lambda < 0$) the solutions $e^+(\lambda, x)$ $(e^-(\lambda, x))$ belong to the space $L_2(a, \infty)$ $(L_2(-\infty, a))$ for any $a \in (-\infty, \infty)$, the numbers $i\lambda_k(x_0)$, placed in the upper (lower) half-plane have the following spectral meaning: their squares are eigenvalues of the selfadjoint boundary value problem

$$-y'' + q(x)y = \mu y , \quad y(x_0) = 0 , \quad x_0 < x < \infty$$

$$(-y'' + q(x)y = \mu y , \quad y(x_0) = 0 , \quad -\infty < x < x_0) .$$

Therefore, for all $x \in (-\infty, \infty)$ the roots $i\lambda_k(x)$ $(k = 1, 2, \ldots, n)$ lie on the imaginary axis and do not coincide, i.e., they are simple roots of the polynomial $P(\lambda, x)$ with the coefficients infinitely differentiable with respect to x. Hence, the functions $\lambda_k(x)$ are real, infinitely differentiable and, if $k \neq l$, then $\lambda_k(x) \neq \lambda_l(x)$ for all $x \in (-\infty, \infty)$.

Lemma 1.2. For all $x \in (-\infty, \infty)$ the moduli of the roots $i\lambda_k(x)$ of the polynomial $P(\lambda, x)$ alternate with the roots of the function $\lambda a(i\lambda)$, i.e., the following inequalities hold:

$$0 = \kappa_0 \leq |\lambda_{k_1}(x)| \leq \kappa_1 \leq |\lambda_{k_2}(x)| \leq \kappa_2 \leq \ldots \leq \kappa_{n-1} \leq |\lambda_{k_n}(x)| \leq \kappa_n .$$

$$(1.10)$$

Proof. The function $\lambda a(i\lambda)$ has on the half-axis $0 \leq \lambda \leq \infty$ $n+1$ roots $0 = \kappa_0 < \kappa_1 < \ldots \kappa_n < \infty$. Let us denote by K the set of all $x \in (-\infty, \infty)$, for which at least one of equalities $e^+(i\kappa_k, x) \, e^-(-i\kappa_k, x) = 0$ $(k = 1, 2, \ldots, n)$ is true. The set K is obviously finite, and if $x \notin K$, the function

$$f(\lambda) = \frac{e^+(i\lambda, x)e^-(-i\lambda, x)}{\lambda a(i\lambda)}$$

is continuous on each interval $(0, \kappa_1), (\kappa_1, \kappa_2) \ldots (\kappa_{n-1}, \kappa_n), (\kappa_n, \infty)$, tends to zero if $\lambda \to +\infty$ and becomes infinitely large when λ tends ot the end points of these intervals. Let us show that the function $f(\lambda)$ is decreasing monotonically on each of these intervals, i.e., that

$$\frac{d}{d\lambda}\left(\frac{e^+(i\lambda, x)e^-(-i\lambda, x)}{\lambda a(i\lambda)}\right) < 0 .$$

From

$$-y'' + (q(x)y = -\lambda^2 y , \quad -\left(\frac{dy}{d\lambda}\right)'' + q(x)\left(\frac{dy}{d\lambda}\right) = -2\lambda y - \lambda^2 \left(\frac{dy}{d\lambda}\right)$$

which are satisfied by the functions $e^+(i\lambda, x)$, $(d/d\lambda)(e^+(i\lambda, x))$ and $e^-(-i\lambda, x)$, $(d/d\lambda)(e^-(-i\lambda, x))$ it follows that

$$-2\lambda y^2 = -y \left(\frac{dy}{d\lambda}\right)'' + y'' \left(\frac{dy}{d\lambda}\right) = \left(y'\frac{dy}{d\lambda} - y\frac{dy'}{d\lambda}\right)' .$$

Thus,

$$-2\lambda \int_x^\infty |e^+(i\lambda, t)|^2 \, dt = -e^+(i\lambda, x)'\frac{de^+(i\lambda, x)}{d\lambda} + e^+(i\lambda, x)\left(\frac{de^+(i\lambda, x)}{d\lambda}\right)'$$

$$= \frac{d}{d\lambda}\left(\frac{e^+(i\lambda, x)'}{e^+(i\lambda, x)}\right) e^+(i\lambda, x)^2 ,$$

$$-2\lambda \int_{-\infty}^x |e^-(-i\lambda, t)|^2 \, dt = e^-(-i\lambda, x)'\frac{de^-(-i\lambda, x)}{d\lambda}$$

$$- e^-(-i\lambda, x)\left(\frac{de^-(-i\lambda, x)}{d\lambda}\right)'$$

$$= -\frac{d}{d\lambda}\left(\frac{e^-(-i\lambda, x)'}{e^-(-i\lambda, x)}\right) e^-(-i\lambda, x)^2 ,$$

whence it follows that

$$\frac{d}{d\lambda}\left[\frac{e^+(i\lambda, x)'}{e^+(i\lambda, x)} - \frac{e^-(-i\lambda, x)'}{e^-(-i\lambda, x)}\right] < 0 ,$$

and since

$$-2\lambda a(i\lambda) = \left\{ e^+(i\lambda, x), e^-(-i\lambda, x) \right\} = e^+(i\lambda, x)'e^-(-i\lambda, x)$$
$$- e^+(i\lambda, x)e^-(-i\lambda, x)' ,$$

then

$$\frac{d}{d\lambda}\left[\frac{-\lambda a(i\lambda)}{e^+(i\lambda, x)e^-(-i\lambda, x)} \right] < 0 ,$$

and therefore

$$\frac{d}{d\lambda}f(\lambda) = \frac{d}{d\lambda}\left[\frac{e^+(i\lambda, x)e^-(-i\lambda, x)}{\lambda a(i\lambda)} \right] < 0 .$$

Thus, the function $f(\lambda)$ decreases monotonically on each of these intervals $(0, \kappa_1), (\kappa_1, \kappa_2), \dots , (\kappa_{n-1}, \kappa_n), (\kappa_n, \infty)$, becomes infinitely large in the vicinity of the end points $0, \kappa_1, \kappa_2, \dots \kappa_n$ and tends to zero if $\lambda \rightarrow +\infty$. Therefore, in each interval $(0, \kappa_1), \dots (\kappa_{n-1}, \kappa_n)$ there lies one simple root of the function $f(\lambda)$ and it has no other roots on the semiaxis $(0, \infty)$. Since the roots of this function coincide with those of the function $e^+(i\lambda, x)e^-(-i\lambda, x)$ which according to Lemma 1.1 are equal to $|\lambda_k(x)|$ $(k = 1, 2, \dots , n)$, then for $x \notin K$ strict inequalities are fulfilled

$$0 < |\lambda_{k_1}(x)| < \kappa_1 < |\lambda_{k_2}(x)| < \kappa_2 < \dots < \kappa_{n-1} < |\lambda_{k_n}(x)| < \kappa_n .$$

$$(1.10')$$

Since the set K is finite, and the functions $\lambda_k(x)$ are continuous, then the nonstrict inequalities (1.10) are correct for all $x \in (-\infty, \infty)$.

Note also that from Lemma 1.1 and (1.6) there follow the identities

$$e^{-\kappa_k x} \prod_{j=1}^{n} \frac{\kappa_k - \lambda_j(x)}{\kappa_k + \kappa_j} \equiv C_k^+ e^{\kappa_k x} \prod_{j=1}^{n} \frac{\kappa_k - \lambda_j(x)}{\kappa_k + \kappa_j} \qquad (1.11)$$

that show that, if for the given value x one of the roots $\lambda_j(x)$ appears to be equal to κ_k, then there is sure to be found another one equal to $-\kappa_k$. Therefore, in accordance with Lemma 1.2, at each x all the roots are subdivided into two groups: the first one includes those for which the strict inequalities

$$\kappa_{\alpha-1}^2 < \lambda_{k_\alpha}(x) < \kappa_\alpha^2$$

hold; the second one includes the pairs differing only by their sign, and satisfying the equalities

$$-\lambda_{k_\beta}(x) = \lambda_{k_{\beta+1}}(x) = \kappa_\beta .$$

It is obvious that for $x \in K$ all the roots belong to the first group, so that the second one appears only for a finite number of values x.

Lemma 1.3. All the $\lambda_k(x)$ are monotonically decreasing, i.e., for all $x \in (-\infty, \infty)$, $k = 1, 2, \ldots, n$ the strict inequalities $-\lambda'_k(x) > 0$ hold. If $\lambda_{k_\alpha}(x)$ belongs to the first group, then

$$-\lambda'_{k_\alpha}(x) = [\kappa_\alpha^2 - \lambda_{k_\alpha}^2(x)] \prod_{j \neq \alpha} \left(\frac{\kappa_j^2 - \lambda_{k_\alpha}(x)^2}{\lambda_{k_j}(x)^2 - \lambda_{k_\alpha}(x)^2} \right),$$

and if the pair $-\lambda_{k_\beta}(x) = \lambda_{k_{\beta+1}}(x) = \kappa_\beta$ belongs to the second group, then their derivatives can be found from

$$-\left(\lambda'_{k_\beta}(x) + \lambda'_{k_{\beta+1}}(x) \right) = (\kappa_{\beta+1}^2 - \kappa_\beta^2) \prod_{j \neq \beta, \beta+1} \frac{\kappa_j^2 - \kappa_\beta^2}{\lambda_{k_j}(x)^2 - \kappa_\beta^2}$$

$$-\lambda'_{k_{\beta+1}}(x) e^{-\kappa_\beta x} \prod_{j \neq \beta, \beta+1} \left(\kappa_\beta - \lambda_{k_j}(x) \right)$$

$$= \lambda'_{k_\beta}(x) C_\beta^+ e^{\kappa_\beta x} \prod_{j \neq \beta, \beta+1} \left(\kappa_\beta + \lambda_{k_j}(x) \right).$$

Proof. For real λ the solutions $e^+(\pm\lambda, x)$ satisfy the asymptotic equalities

$$e^+(\pm\lambda, x) = e^{\pm i\lambda x} + O(1), \quad e^+(\pm\lambda, x) = \pm i\lambda e^{\pm i\lambda x} + O(1) \quad (x \to +\infty)$$

from which it follows that their Wronskian is equal to $2i\lambda$. Therefore,

$$2i\lambda = e^+(\lambda, x)' e^+(-\lambda, x) - e^+(\lambda, x) e^+(-\lambda, x)'$$

$$= e^+(\lambda, x) e^+(-\lambda, x) \frac{d}{dx} \ln \frac{e^+(\lambda, x)}{e^+(-\lambda, x)}$$

and according to Lemma 1.1

$$2i\lambda = \prod_{k=1}^n \frac{\lambda^2 + \lambda_k(x)^2}{\lambda^2 + \kappa_k^2} \left[2i\lambda + i \sum_{k=1}^n \frac{-\lambda_k(x)}{\lambda - i\lambda_k(x)} + i \sum_{k=1}^n \frac{-\lambda_k(x)}{\lambda + i\lambda_k(x)} \right]$$

i.e.,

$$1 + \sum_{k=1}^n \frac{-\lambda'_k(x)}{\lambda^2 + \lambda_k(x)^2} = \prod_{k=1}^n \frac{\lambda^2 + \kappa_k^2}{\lambda^2 + \lambda_k(x)^2} \tag{1.12}$$

whence it immediately follows that

$$-\lambda'_{k_\alpha}(x) = [\kappa_\alpha^2 - \lambda_{k_\alpha}(x)^2] \prod_{j \neq \alpha} \left(\frac{\kappa_j^2 - \lambda_{k_\alpha}(x)^2}{\lambda_{k_j}(x)^2 - \lambda_{k_\alpha}(x)^2} \right), \tag{1.13}$$

if $\kappa_{\alpha-1}^2 < \lambda_{k_\alpha}(x)^2 < \kappa_\alpha^2$ and

$$-\left(\lambda'_{k_\beta}(x)+\lambda'_{k_{\beta+1}}(x)\right)=(\kappa_{\beta+1}^2-\kappa_\beta^2)\prod_{j\neq\beta,\beta+1}\frac{\kappa_j^2-\kappa_\beta^2}{\lambda_{k_j}(x)^2-\kappa_\beta^2} \qquad (1.13')$$

if $\lambda_{k_{\beta+1}}(x)=-\lambda_{k_\beta}(x)=\kappa_\beta$. Besides, in the second case from identity (1.11) (if $k=\beta$) after differentiation with respect to x we find that

$$-\lambda'_{k_{\beta+1}}(x)e^{-\kappa_\beta x}\prod_{j\neq\beta,\beta+1}\left(\kappa_\beta-\lambda_{k_j}(x)\right)$$

$$=\lambda'_{k_{\beta+1}}(x)C_\beta^+e^{\kappa_\beta x}\prod_{j\neq\beta,\beta+1}\left(\kappa_\beta+\lambda_{k_j}(x)\right), \qquad (1.14)$$

since

$$\kappa_\beta-\lambda_{k_{\beta+1}}(x)=\kappa_\beta+\lambda_{k_\beta}(x)=0, \quad \kappa_\beta-\lambda_{k_\beta}(x)=\kappa_\beta+\lambda_{k_{\beta+1}}(x)=2\kappa_\beta.$$

If $x\notin K$, then all the roots belong to the first group and from (1.10′) it follows that the right parts of (1.13) are strictly positive. Therefore, for $x\notin K$ strict inequalities $-\lambda'_j(x)>0$ hold for all $j=1,2,\ldots,n$, and since the set K is finite and the functions $\lambda'_j(x)$ are continuous, then the nonstrict inequalities $-\lambda'_j(x)\geq 0$ are always correct. Finally, from (1.13′) and Lemma 1.2 it follows that $-(\lambda'_{k_\beta}(x)+\lambda'_{k_{\beta+1}}(x))>0$. Comparing this inequality with (1.14) we find that $-\lambda'_{k_\beta}(x)\neq 0$, $\lambda'_{k_{\beta+1}}(x)\neq 0$ and since these derivatives are nonnegative, then $-\lambda'_{k_\beta}(x)>0$, $-\lambda'_{k_{\beta+1}}(x)>0$. Thus, strict inequalities $-\lambda'_j(x)>0$ in fact hold for all $x\in(-\infty,\infty)$, $j=1,2,\ldots,n$.

Remark. From (1.11, 14) it follows that

$$C_\alpha^+=e^{-2\kappa_\alpha x}\prod_{s=1}^n\frac{\kappa_\alpha-\lambda_{k_s}(x)}{\kappa_\alpha+\lambda_{k_s}(x)}$$

if $\kappa_\alpha\neq\lambda_j(x)$ for all $j=1,2,\ldots,n$, and

$$C_\alpha^+=-\left(\frac{\lambda'_{k_{\alpha+1}}(x)}{\lambda'_{k_\alpha}(x)}\right)e^{-2\kappa_\alpha x}\prod_{s\neq\alpha,\alpha+1}\frac{\kappa_\alpha-\lambda_{k_s}(x)}{\kappa_\alpha+\lambda_{k_s}(x)}$$

if $\lambda_{k_{\alpha+1}}(x)=-\lambda_{k_\alpha}(x)=\kappa_\alpha$. Since in the reflectionless case

$$a(\lambda)=\prod_{k=1}^n\frac{\lambda-i\kappa_k}{\lambda+i\kappa_k}, \quad i\dot a(i\kappa_\alpha)=\frac{1}{2\kappa_\alpha}\prod_{s\neq\alpha}\frac{(\kappa_\alpha-\kappa_s)}{(\kappa_\alpha+\kappa_s)}$$

then, substituting the expressions obtained for C_α^+, $i\dot a(i\kappa_\alpha)$ into (1.7), we find that

$$(m_\alpha^+)^{-2}=\frac{e^{-2\kappa_\alpha x}}{2\kappa_\alpha}\frac{\kappa_\alpha-\lambda_{k_\alpha}(x)}{\kappa_\alpha+\lambda_{k_\alpha}(x)}\prod_{s\neq\alpha}\frac{(\kappa_\alpha-\lambda_{k_s}(x))(\kappa_\alpha-\kappa_s)}{(\kappa_\alpha+\lambda_{k_s}(x))(\kappa_\alpha+\kappa_s)}$$

$$=\frac{e^{-2\kappa_\alpha x}}{2\kappa_\alpha}\frac{\kappa_\alpha^2-\lambda_{k_\alpha}(x)^2}{[\kappa_\alpha+\lambda_{k_\alpha}(x)]^2}\prod_{s\neq\alpha}\left(\frac{\kappa_\alpha^2-\lambda_{k_s}(x)^2}{\kappa_\alpha^2-\kappa_s^2}\right)\left(\frac{\kappa_\alpha-\kappa_s}{\kappa_\alpha+\lambda_{k_s}(x)}\right)^2$$

if $\kappa_\alpha^2 \neq \lambda_j(x)^2$ $(j = 1, 2, \dots, n)$, and

$$
(m_\alpha^+)^{-2} = -\frac{\lambda'_{k_{\alpha+1}}(x)}{\lambda'_{k_\alpha}(x)} \frac{e^{-2\kappa_\alpha x}}{2\kappa_\alpha} \frac{\kappa_\alpha - \kappa_{\alpha+1}}{\kappa_\alpha + \kappa_{\alpha+1}}
$$

$$
\times \prod_{s \neq \alpha, \, \alpha+1} \frac{\kappa_\alpha - \lambda_{k_s}(x)}{\kappa_\alpha + \lambda_{k_s}(x)} \frac{\kappa_\alpha - \kappa_s}{\kappa_\alpha + \kappa_s}
$$

$$
= \frac{\lambda'_{k_{\alpha+1}}(x)}{\lambda'_{k_\alpha}(x)} \frac{e^{-2\kappa_\alpha x}}{2\kappa_\alpha} \frac{\kappa_{\alpha+1} - \kappa_\alpha}{\kappa_{\alpha+1} + \kappa_\alpha}
$$

$$
\times \prod_{s \neq \alpha, \, \alpha+1} \frac{\kappa_\alpha^2 - \lambda_{k_s}(x)^2}{\kappa_\alpha^2 - \kappa_s^2} \left(\frac{\kappa_\alpha - \kappa_s}{\kappa_\alpha + \lambda_{k_s}(x)} \right)^2 ,
$$

if $\lambda_{k_{\alpha+1}}(x) = -\lambda_{k_\alpha}(x) = \kappa_\alpha$. In particular, assuming $x = 0$, we find that

$$
(m_\alpha^+)^{-2} = \frac{\kappa_\alpha^2 - \lambda_{k_\alpha}^2(0)}{2\kappa_\alpha(\kappa_\alpha + \lambda_{k_\alpha}(0))^2} \prod_{s \neq \alpha} \left(\frac{\kappa_\alpha^2 - \lambda_{k_s}^2(0)}{\kappa_\alpha^2 - \kappa_s^2} \right) \left(\frac{\kappa_\alpha - \kappa_s}{\kappa_\alpha + \lambda_{k_s}(0)} \right)^2 \quad (1.15)
$$

if $\kappa^2 \neq \lambda^2(0)$ $(j = 1, 2, \dots, n)$, and

$$
(m_\alpha^+)^{-2} = \frac{\lambda'_{k_{\alpha+1}}(0)}{2\kappa_\alpha \lambda'_{k_\alpha}(0)} \left(\frac{\kappa_{\alpha+1} - \kappa_\alpha}{\kappa_{\alpha+1} + \kappa_\alpha} \right)
$$

$$
\times \prod_{s \neq \alpha, \, \alpha+1} \left(\frac{\kappa_\alpha^2 - \lambda_{k_s}(x)^2}{\kappa_\alpha^2 - \kappa_s^2} \right) \left(\frac{\kappa_\alpha - \kappa_s}{\kappa_\alpha + \lambda_{k_s}(x)} \right)^2 \quad (1.15')
$$

if $\lambda_{k_{\alpha+1}}(0) = -\lambda_{k_\alpha}(0) = \kappa_\alpha$.

Let us call the sequences

$$
\lambda'_1(0), \, \lambda_1(0); \quad \lambda'_2(0), \, \lambda_2(0); \quad \dots; \quad \lambda'_n(0), \, \lambda_n(0)
$$

the spectral data of corresponding operators (1.1) with reflectionless potentials. The spectral data also uniquely determine the reflectionless potentials. Indeed, from (1.12) it follows that the polynomial $\prod_{k=1}^n (z - \kappa_k^2)$ can be reconstructed from the spectral data by

$$
\prod_{k=1}^n (z - \kappa_k^2) = \prod_{k=1}^n (z - \lambda_k(0)^2) \left[1 - \sum_{k=1}^n \frac{-\lambda'_k(0)}{z - \lambda_k(0)^2} \right] \quad (1.16)
$$

and the normalizing coefficients m_k^+ are reconstructed from the roots of this polynomial and spectral data according to (1.15), (1.15') in which κ_α should be understood as a positive root of κ_α^2. Thus, the scattering data $\{i\kappa_k, m_k^+\}$ and, consequently, the reflectionless potential, can be uniquely reconstructed from the spectral data.

Lemma 1.4. The sequence

$$\lambda_1', \lambda_1; \ldots; \lambda_n', \lambda_n$$

of n pairs of real numbers is the spectral data of a certain operator (1.1) with the reflectionless potential $q(x)$ if and only if $\lambda_j' < 0$ $(1 \le j \le n)$; $\lambda_j \ne \lambda_k$ when $j \ne k$. In this case, $q(x) \in B(-\mu^2)$, where

$$\max_{1 \le k \le n} \{\lambda_k^2\} < \mu^2 \le \max_{1 \le k \le n} \{\lambda_k^2\} + \sum_{k=1}^{n} (-\lambda_k') .$$

Proof. The necessity of these conditions follows from Lemmas 1.1, 1.3. To show that these conditions are also sufficient, we will construct according to the given sequence the polynomials

$$\Lambda(z) = \prod_{k=1}^{n} (z - \lambda_k^2) \ ; \ \ N(z) = \Lambda(z) \left[1 - \sum_{k=1}^{n} \frac{-\lambda_k'}{z - \lambda_k^2} \right] \tag{1.16'}$$

and consider the location of the roots of the polynomial $N(z)$. Let us arrange the sequence of nonnegative numbers λ_k^2 in increasing order: $0 \le \lambda_{k_1}^2 \le \lambda_{k_2}^2 \le \ldots \le \lambda_{k_n}^2$. If $\lambda_{k_\alpha}^2 < \lambda_{k_{\alpha+1}}^2$, then in the interval $(\lambda_{k_\alpha}^2, \lambda_{k_{\alpha+1}}^2)$ the function

$$\varphi(z) = 1 - \sum_{k=1}^{n} \frac{-\lambda_k'}{z - \lambda_k^2}$$

is monotonically increasing from $-\infty$ to $+\infty$, whence it follows that one simple root κ_α^2 of the polynomial $N(z)$ lies in each such interval $(\lambda_{k_\alpha}^2 < \kappa_\alpha^2 < \lambda_{k_{\alpha+1}}^2)$. If, however, $|\lambda_{k_\beta}| = |\lambda_{k_{\beta+1}}| = \kappa_\beta$, then from the inequalities $\lambda_j \ne \lambda_k$ $(j \ne k)$ it follows that one of the numbers λ_{k_β}, $\lambda_{k_{\beta+1}}$ is equal to κ_β, and the other to $-\kappa_\beta$, and $\lambda_j^2 \ne \kappa_\beta^2$ for all $j \ne k_\beta, k_{\beta+1}$. Therefore, κ_β^2 is a double root of the polynomial $\Lambda(z)$ and a simple root of the polynomial $N(z)$. Thus, the polynomial $N(z)$ has $n - 1$ simple roots κ_α^2, satisfying the inequalities $\lambda_{k_\alpha}^2 \le \kappa_\alpha^2 \le \lambda_{k_{\alpha+1}}^2$. Since on the semiaxis $(\lambda_{k_n}^2, \infty)$ the function $\varphi(z)$ is monotonically increasing from $-\infty$ to 1, then one more root κ_n^2 of the polynomial $N(z)$ lies to the right of the point $\lambda_k^2 = \max_{1 \le j \le n} \lambda_j^2$. Supposing $\kappa_n^2 = \lambda_{k_n}^2 + x$, we find that

$$0 = \varphi\left(\kappa_n^2\right) = 1 - \sum_{s=1}^{n} \frac{-\lambda_s'}{\lambda_{k_n}^2 + x - \lambda_s^2} \ge 1 - \frac{1}{x} \sum_{s=1}^{n} (-\lambda_s')$$

whence it follows that $x \le \sum_{s=1}^{n} (-\lambda_s')$ and $\kappa_n^2 \le \lambda_{k_n}^2 + \sum_{s=1}^{n} (-\lambda_s')$. Thus, the polynomial $N(z)$ has simple roots κ_α^2 $(1 \le \alpha \le n)$ alternating with the roots of the polynomial $\Lambda(z)$:

$$0 \le \lambda_{k_1}^2 \le \kappa_1^2$$

$$\le \lambda_{k_2}^2 \le \ldots \le \lambda_{k_{n-1}}^2 \le \kappa_{n-1}^2 \le \lambda_{k_n}^2 \le \kappa_n^2 \le \lambda_{k_n}^2 + \sum_{s=1}^{n} (-\lambda_s') . \tag{1.17}$$

We note that $\kappa_1^2 > 0$. It is obvious when $\lambda_{k_1}^2 > 0$; if $\lambda_{k_1}^2 = 0$, then $\lambda_{k_2}^2 > 0$, and from the foregoing it follows that the strict inequality $0 = \lambda_{k_1}^2 < \kappa_1^2 < \lambda_{k_2}^2$ holds.

From the given sequence λ'_j, λ_j and the roots κ_α^2 of the polynomial $N(z)$ we will define the numbers $m_\alpha^+ > 0$ by

$$(m_\alpha^+)^{-2} = \frac{\kappa_\alpha^2 - \lambda_{k_\alpha}^2}{2\kappa_\alpha(\kappa_\alpha + \lambda_{k_\alpha})^2} \prod_{s \neq \alpha} \left(\frac{\kappa_\alpha^2 - \lambda_{k_s}^2}{\kappa_\alpha^2 - \kappa_s^2} \right) \left(\frac{\kappa_\alpha - \kappa_s}{\kappa_\alpha + \lambda_{k_s}} \right)^2 \tag{1.18}$$

if $\alpha = n$ or $\lambda_{k_\alpha}^2 < \kappa_\alpha^2 < \lambda_{k_{\alpha+1}}^2$ and by

$$(m_\alpha^+)^{-2} = \frac{\lambda'_{k_{\alpha+1}}}{2\kappa_\alpha \lambda'_{k_\alpha}} \left(\frac{\kappa_{\alpha+1} - \kappa_\alpha}{\kappa_{\alpha+1} + \kappa_\alpha} \right) \prod_{s \neq \alpha, \alpha+1} \frac{\kappa_\alpha^2 - \lambda_{k_s}^2}{\kappa_\alpha^2 - \kappa_s^2} \left(\frac{\kappa_\alpha - \kappa_s}{\kappa_\alpha + \lambda_{k_s}} \right)^2 \tag{1.18'}$$

if $\lambda_{k_{\alpha+1}} = -\lambda_{k_\alpha} = \kappa_\alpha$, where κ_α denotes positive square roots of κ_α^2. Since according to the condition that all the numbers λ'_j are negative, then from inequalities (1.17) it follows that the thus-defined numbers $(m_\alpha^+)^2$ are positive. Consequently, the constructed set $\{i\kappa_1, i\kappa_2, \dots, i\kappa_n; m_1^+, m_2^+, \dots, m_n^+\}$ is the right scattering data of a certain operator with the reflectionless potential $q(x) \in B(-\mu^2)$, where $\mu^2 = \kappa_n^2$, and according to (1.17)

$$\max_{1 \leq k \leq n} \lambda_k^2 < \mu^2 \leq \max_{1 \leq k \leq n} \lambda_k^2 + \sum_{k=1}^{n} (-\lambda'_k) .$$

Now let us show that the spectral data

$$\lambda'_1(0), \lambda_1(0); \lambda'_2(0), \lambda_2(0); \dots \; \lambda'_n(0), \lambda_n(0)$$

of this operator coincide with the given sequence $\lambda'_1, \lambda_1; \lambda'_2, \lambda_2; \dots; \lambda'_n, \lambda_n$. We note, first of all, that the polynomials

$$\Lambda_1(z) = \prod_{k=1}^{n} (z - \lambda_k) , \quad \Lambda_1(z,0) = \prod_{k=1}^{n} [z - \lambda_k(0)]$$

for all $\alpha = 1, 2, \dots, n$ satisfy the equalities

$$\Lambda_1(\kappa_\alpha) \frac{(m_\alpha^+)^2}{2\kappa_\alpha} \prod_{s \neq \alpha} \left(\frac{\kappa_\alpha - \kappa_s}{\kappa_\alpha + \kappa_s} \right) = \Lambda_1(-\kappa_\alpha)(-1)^n \tag{1.19}$$

$$\Lambda_1(\kappa_\alpha; 0) \frac{(m_\alpha^+)^2}{2\kappa_\alpha} \prod_{s \neq \alpha} \left(\frac{\kappa_\alpha - \kappa_s}{\kappa_\alpha + \kappa_s} \right) = \Lambda_1(-\kappa_\alpha; 0)(-1)^n . \tag{1.19'}$$

Indeed, if $\lambda_{k_\alpha}^2 < \kappa_\alpha^2 < \lambda_{k_{\alpha+1}}^2$ $[\lambda_{k_\alpha}^2(0) < \kappa_\alpha^2 < \lambda_{k_{\alpha+1}}^2(0)]$ equality (1.19) [(1.19')] is equivalent to equality (1.18) [(1.15)], if $\lambda_{k_\alpha}^2 = \kappa_\alpha^2 = \lambda_{k_{\alpha+1}}^2$ $[\lambda_{k_\alpha}^2(0) = \kappa_\alpha^2 = \lambda_{k_{\alpha+1}}^2(0)]$ then both the numbers $\pm\kappa_\alpha$ are the roots of the polynomial $\Lambda_1(z)$

$[\Lambda_1(z;0)]$ so that both parts of (1.19) [(1.19')] are equal to zero. From (1.19) (1.19') it follows that the odd polynomial $\Lambda_1(z)\Lambda_1(-z;0) - \Lambda_1(-z)\Lambda_1(z;0)$ of degree $\leq 2n - 1$ has $2n$ different roots $\pm\kappa_1, \pm\kappa_2, \dots, \pm\kappa_n$ i.e.

$$\Lambda_1(z)\Lambda_1(-z;0) \equiv \Lambda_1(-z)\Lambda_1(z;0) . \tag{1.20}$$

From the alternation of the sequence κ_α^2 with each of the sequences λ_j^2, $\lambda_j^2(0)$ it follows that it is possible to arrange them so that

$$\lambda_{k_1}^2 \leq \kappa_1^2 \leq \lambda_{k_2}^2 \leq \dots \leq \lambda_{k_n}^2 \leq \kappa_n^2 ,$$

$$\lambda_{k_1}^2(0) \leq \kappa_1^2 \leq \lambda_{k_2}^2(0) \leq \dots \leq \lambda_{k_n}^2(0) \leq \kappa_n^2 ,$$

and if

$$\lambda_{k_\alpha}^2 = \lambda_{k_{\alpha+1}}^2 \quad \left(\lambda_{k_\alpha}^2(0) = \lambda_{k_{\alpha+1}}^2(0)\right)$$

then

$$-\lambda_{k_\alpha} = \lambda_{k_{\alpha+1}} = \kappa_\alpha \quad \left(-\lambda_{k_\alpha}(0) = \lambda_{k_{\alpha+1}}(0) = \kappa_\alpha\right) . \tag{1.21}$$

The proof of the equalities $\lambda_{k_\alpha} = \lambda_{k_\alpha}(0)$ will be carried out inductively. If $\lambda_{k_1}^2 < \lambda_{k_2}^2$ [or $\lambda_{k_1}^2(0) < \lambda_{k_2}^2(0)$], then $\lambda_{k_1}[\lambda_{k_1}(0)]$ is the root of the polynomial $\Lambda_1(z)$ $[\Lambda_1(z;0)]$ but $-\lambda_{k_1}[-\lambda_{k_1}(0)]$ is not:

$$\Lambda_1\left(\lambda_{k_1}\right) = 0, \; \Lambda_1\left(-\lambda_{k_1}\right) \neq 0 \quad \left[\Lambda_1\left(\lambda_{k_1}(0);0\right) = 0, \; \Lambda_1\left(-\lambda_{k_1}(0);0\right) \neq 0\right] .$$

Hence, and also from identity (1.20) it follows that $\Lambda_1(\lambda_{k_1};0) = 0$ $[\Lambda_1(\lambda_{k_1}(0)) = 0]$ and, consequently, $\lambda_{k_1} = \lambda_{k_1}(0)$. If $\lambda_{k_1}^2 = \lambda_{k_2}^2 = \kappa_1^2$ and $\lambda_{k_1}^2(0) = \lambda_{k_2}^2(0) = \kappa_1^2$ then $-\lambda_{k_1} = \lambda_{k_2} = \kappa_1; \; -\lambda_{k_1}(0) = \lambda_{k_2}(0) = \kappa_1$. Thus, in all the cases $\lambda_{k_1} = \lambda_{k_1}(0)$. Suppose that the equalities $\lambda_{k_\alpha} = \lambda_{k_\alpha}(0)$ are true for all $\alpha = 1, 2, \dots \alpha_0$. Then

$$\lambda_{k\alpha_0}^2 = \lambda_{k\alpha_0}^2(0) \leq \kappa_{\alpha_0}^2 \leq \left\{ \begin{matrix} \lambda_{k_{\alpha_0+1}}^2 \\ \lambda_{k_{\alpha_0+1}}^2(0) \end{matrix} \right\} \leq \kappa_{\alpha_0+1}^2 ,$$

and, if $\kappa_{\alpha_0}^2 = \lambda_{k_{\alpha_0}}^2 = \lambda_{k_{\alpha_0}}^2(0)$, then, necessarily, $\lambda_{k_{\alpha_0+1}} = \lambda_{k_{\alpha_0+1}}(0) = \kappa_{\alpha_0}$. If, however, $\kappa_{\alpha_0}^2 > \lambda_{k_{\alpha_0}}^2$, then either one of the numbers $\lambda_{k_{\alpha_0+1}}^2$, $\lambda_{k_{\alpha_0+1}}^2(0)$ is strictly less than $\kappa_{\alpha_0+1}^2$, or $\lambda_{k_{\alpha_0+1}}^2 = \lambda_{k_{\alpha_0+1}}^2(0) = \kappa_{\alpha_0+1}^2$. In the former case $\lambda_{k_{\alpha_0+1}}$ [or $\lambda_{k_{\alpha_0+1}}(0)$] is the root of the polynomial $\Lambda_1(z)$ [or $(\Lambda_1(z;0)]$, while $-\lambda_{k_{\alpha_0+1}}$ $[-\lambda_{k_{\alpha_0+1}}]$ is not. Hence, and from identity (1.29) it follows that here $\lambda_{k_{\alpha_0+1}} = \lambda_{k_{\alpha_0+1}}(0)$. In the latter case, according to (1.21), $\lambda_{k_{\alpha_0+1}} = \lambda_{k_{\alpha_0+1}}(0) = -\kappa_{\alpha_0+1}$. Thus, in all the cases $\lambda_{k_{\alpha_0+1}} = \lambda_{k_{\alpha_0+1}}(0)$, whence, inductively, it follows that these equalities are true for all $\alpha = 1, 2, \dots, n$. Finally, the proved equalities $\lambda_j = \lambda_j(0)$ and (1.16) (1.16') imply the identity

$$\sum_{\alpha=1}^n \frac{-\lambda'_{k_\alpha}}{z - \lambda_{k_\alpha}^2} \equiv \sum_{\alpha=1}^n \frac{-\lambda'_{k_\alpha}(0)}{z - \lambda_{k_\alpha}^2} ,$$

from which it follows that $\lambda'_{k_\alpha} = \lambda'_{k_\alpha}(0)$, if $\lambda^2_{k_\alpha} \neq \lambda^2_j$ for all $j \neq k_\alpha$, and $\lambda'_{k_\alpha} + \lambda'_{k_{\alpha+1}} = \lambda'_{k_\alpha}(0) + \lambda'_{k_{\alpha+1}}(0)$, if $\lambda^2_{k_\alpha} = \lambda^2_{k_{\alpha+1}}$, while in the latter case (1.15'), (1.18') also imply that

$$\lambda'_{k_{\alpha+1}}/\lambda'_{k_\alpha} = \lambda'_{k_\alpha}(0)/\lambda'_{k_\alpha}(0) > 0 \ .$$

Consequently, $\lambda'_{k_\alpha} = \lambda'_{k_\alpha}(0)$ for all $\alpha = 1, 2, \dots, n$.

The scattering data are convenient spectral characteristic in the class of rapidly decreasing potentials. In this class, reflectionless potentials are singled out by the condition that the reflection coefficient is identically equal to zero. In the class of all real locally integrable potentials the Weyl functions are the universal spectral characteristic. Therefore, to single out reflectionless potentials in the class of all real locally integrable potentials, it is necessary to characterize them in terms of the Weyl functions.

Let us denote by $c(\lambda, x)$, $s(\lambda, x)$ the solutions of

$$-y'' + q(x)y = \lambda^2 y \quad (-\infty < x < \infty) \tag{1.22}$$

which satisfy the following initial data

$$c(\lambda, 0) = s'(\lambda, 0) = 1 \ , \quad c'(\lambda, 0) = s(\lambda, 0) = 0 \ .$$

According to the classical Weyl theorem, there exist holomorphic functions $m^+(z)$, $m^-(z)$, outside the real axis, such that for $\lambda^2 = z$ the solutions of (1.22)

$$\psi^\pm(z, x) = c\left(\sqrt{z}, x\right) + m^\pm(z)s\left(\sqrt{z}, x\right) \quad (\text{Im } z \neq 0)$$

are square integrable on the positive and negative semiaxes, respectively:

$$\psi^+(z, x) \in L_2(0, \infty) \ , \quad \psi^-(z, x) \in L_2(-\infty, 0) \ ,$$

while $m^\pm(\bar{z}) = \overline{m^\pm(z)}$ and

$$\int_0^\infty |\psi^+(z, x)|^2 \, dx = \frac{\text{Im } m^+(z)}{\text{Im } z} \ , \quad \int_{-\infty}^0 |\psi^-(z, x)|^2 \, dx = -\frac{\text{Im } m^-(z)}{\text{Im } z} \ . \tag{1.23}$$

The solutions $\psi^\pm(z, x)$ are called the Weyl solutions, and the functions $m^\pm(z)$ the Weyl functions of (1.22).

The Weyl functions can be replaced by one function $n(\lambda)$ defined outside the coordinate axes by

$$n(\lambda) = \begin{cases} m^+(\lambda^2) & \text{Im } \lambda > 0 \\ m^-(\lambda^2) & \text{Im } \lambda < 0 \end{cases}$$

and the Weyl solutions can be replaced by the solution

$$\psi(\lambda, x) = c(\lambda, x) + n(\lambda)s(\lambda, x) \ ,$$

which belongs to the space $L_2(0, \infty)$ [$L_2(-\infty, 0)$] when λ lies in the upper (lower) half-plane. We shall adhere to the original notations [$\psi(\lambda, x)$ for the Weyl solution, and $n(\lambda)$ for the Weyl function] because the difference can be reduced to the choice of the argument "momentum" λ instead of "energy" $z = \lambda^2$.

The functions $n(\lambda)$ is holomorphic outside the coordinate axes, assumes adjoint values at the points, symmetrical with respect to the imaginary axis, its imaginary part is positive in the right half-plane and negative in the left one.

The Weyl functions $m^\pm(z)$ are expressed in terms of the spectral functions $\varrho^\pm(\mu)$ of boundary value problems generated on the semiaxes $(0, \infty), (-\infty, 0)$ by the equation $-y'' + q(x)y = \mu y$ and the boundary condition $y'(0) = 0$:

$$\frac{1}{m^\pm(z)} = \pm \int_{-\infty}^{\infty} \frac{d\varrho^\pm(\mu)}{z - \mu} .$$

The spectral functions $\varrho^\pm(\mu)$ are nondecreasing, have finite limits $\varrho^\pm(-\infty)$ when $\mu \to -\infty$, and at $\mu \to +\infty$ satisfy the asymptotic formulae

$$\varrho^\pm(\mu) = \frac{2}{\pi}\sqrt{\mu} + \varrho^\pm(-\infty) + o(1)$$

which allows them to be represented in the form

$$\varrho^\pm(\mu) = \varrho^\pm(-\infty) + \frac{2}{\pi}\chi(\mu)\sqrt{\mu} + \delta^\pm(\mu) ,$$

where

$$\chi(\mu) = \begin{cases} 1 & \mu \geq 0 \\ 0 & \mu < 0 \end{cases} ; \quad \lim_{|\mu| \to \infty} \delta^\pm(\mu) = 0 .$$

Therefore,

$$\frac{1}{m^\pm(z)} = \pm \left(\frac{2}{\pi} \int_0^{\infty} \frac{d\sqrt{\mu}}{z - \mu} + \int_{-\infty}^{\infty} \frac{d\delta^\pm(\mu)}{z - \mu} \right)$$

and since

$$\frac{2}{\pi} \int_0^{\infty} \frac{d\sqrt{\mu}}{z - \mu} = \frac{1}{i\sqrt{z}} , \quad \int_{-\infty}^{\infty} \frac{d\delta^\pm(\mu)}{z - \mu} = -\int_{-\infty}^{\infty} \frac{d\delta^\pm(\mu)}{(z - \mu)^2}d\mu$$

then,

$$\pm\frac{1}{m^\pm(z)} = \frac{1}{i\sqrt{z}} - \int_{-\infty}^{\infty} \frac{d\delta^\pm(\mu)}{(z - \mu)^2}d\mu = \frac{1 - \varepsilon^\pm(z)}{i\sqrt{z}} \tag{1.24}$$

where

$$\varepsilon^\pm(z) = i\sqrt{z} \int_{-\infty}^{\infty} \frac{d\delta^\pm(\mu)}{(z - \mu)^2}d\mu \tag{1.24'}$$

and $\sqrt{z}$ is the branch of the radical that is holomorphic in the plane cut along the positive semiaxis, assuming positive values on the upper boundary of a cut. Let us introduce the following notation

$$\delta = \sup_{-\infty < \mu < \infty} (\max \{|\delta^+(\mu)|, |\delta^-(\mu)|\})$$

$$\delta(t) = \begin{cases} \sup_{\mu \geq t} (\max\{|\delta^+(\mu)|, |\delta^-(\mu)|\}) & t > 0 \\ \sup_{\mu \leq t} (\max\{|\delta^+(\mu)|, |\delta^-(\mu)|\}) & t < 0 \end{cases}.$$

Lemma 1.5. In the domain $|z| > 4$ the functions $\varepsilon^\pm(z)$ satisfy the inequalities

$$|\varepsilon^\pm(z)| \leq \frac{8\delta}{|z|} + \frac{\pi}{2} \left(1 + \frac{\operatorname{Re} z}{|\operatorname{Re} z|}\right) \left(\frac{\delta(-\sqrt{|z|})}{2\sqrt{|z|}} + \frac{\sqrt{|z|}\,\delta(\sqrt{|z|})}{|\operatorname{Im} z|}\right)$$
$$+ \frac{\pi}{2} \left(1 - \frac{\operatorname{Re} z}{|\operatorname{Re} z|}\right) \left(\frac{\sqrt{|z|}\,\delta(-\sqrt{|z|})}{|\operatorname{Im} z|} + \frac{\delta(\sqrt{|z|})}{2\sqrt{|z|}}\right). \tag{1.25}$$

Proof. Making use of the introduced notation we find that

$$\left| \int_{-\infty}^{\infty} \frac{\delta^\pm(\mu)}{(z-\mu)^2} d\mu \right| \leq \delta \int_{-\sqrt{|z|}}^{\sqrt{|z|}} \frac{d\mu}{|z-\mu|^2} + \delta\left(\sqrt{|z|}\right) \int_{\sqrt{|z|}}^{\infty} \frac{d\mu}{|z-\mu|^2}$$
$$+ \delta\left(-\sqrt{|z|}\right) \int_{-\infty}^{-\sqrt{|z|}} \frac{d\mu}{|z-\mu|^2},$$

and since

$$\int_{-\sqrt{|z|}}^{\sqrt{|z|}} \frac{d\mu}{|z-\mu|^2} \leq \frac{2\sqrt{|z|}}{|z|^2(1-1/\sqrt{|z|})^2} \leq 8|z|^{-3/2}$$

when $|z| > 4$, we have

$$|\varepsilon^\pm(z)| \leq 8/|z| + \sqrt{|z|}$$
$$\times \left\{\delta\left(\sqrt{|z|}\right) \int_{\sqrt{|z|}}^{\infty} \frac{d\mu}{|z-\mu|^2} + \delta\left(-\sqrt{|z|}\right) \int_{-\infty}^{-\sqrt{|z|}} \frac{d\mu}{|z-\mu|^2}\right\}.$$

Let $z = x + iy$. Then for all $x \in (-\infty, \infty)$, $y \neq 0$

$$\int_{-\infty}^{-\sqrt{|z|}} \frac{d\mu}{|z-\mu|^2} + \int_{\sqrt{|z|}}^{\infty} \frac{d\mu}{|z-\mu|^2} \leq \int_{-\infty}^{\infty} \frac{d\mu}{(x-\mu)^2 + y^2} = \frac{\pi}{|y|},$$

and for all $x \geq 0$ ($x \leq 0$)

$$\int_{-\infty}^{-\sqrt{|z|}} \frac{d\mu}{|z-\mu|^2} = \int_{-\infty}^{-\sqrt{|z|}} \frac{d\mu}{x^2+y^2+\mu^2-2\mu x} \le \int_{-\infty}^{0} \frac{d\mu}{x^2+y^2+\mu^2} = \frac{\pi}{2|z|}$$

$$\left(\int_{\sqrt{|z|}}^{\infty} \frac{d\mu}{|z-\mu|^2} = \int_{\sqrt{|z|}}^{\infty} \frac{d\mu}{x^2+y^2+\mu^2-2\mu x} \le \int_{0}^{\infty} \frac{d\mu}{x^2+y^2+\mu^2} = \frac{\pi}{2|z|} \right).$$

Hence, in all the domain $|\mathrm{Im}\, z| > 0$, these inequalities hold

$$\int_{-\infty}^{-\sqrt{|z|}} \frac{d\mu}{|z-\mu|^2} \le \frac{\pi}{2|\mathrm{Im}\, z|} \left(1 - \frac{\mathrm{Re}\, z}{|\mathrm{Re}\, z|} \right) + \frac{\pi}{4|z|} \left(1 + \frac{\mathrm{Re}\, z}{|\mathrm{Re}\, z|} \right)$$

$$\int_{\sqrt{|z|}}^{\infty} \frac{d\mu}{|z-\mu|^2} \le \frac{\pi}{2|\mathrm{Im}\, z|} \left(1 + \frac{\mathrm{Re}\, z}{|\mathrm{Re}\, z|} \right) + \frac{\pi}{4|z|} \left(1 - \frac{\mathrm{Re}\, z}{|\mathrm{Re}\, z|} \right).$$

Comparing the obtained inequalities, we confirm that the inequality (1.25) is correct.

Corollary. If $|\lambda| \to \infty$ the function $n(\lambda)$ satisfies the asymptotic formula

$$n(\lambda) = i\lambda + O(1)$$

uniformly outside arbitrary small angles, which contain the coordinate axes.

Indeed, the definition of the function $n(\lambda)$ and (1.24) imply that

$$n(\lambda) = \frac{i\lambda}{1 - \varepsilon(\lambda)}, \quad \varepsilon(\lambda) = \begin{cases} \varepsilon^+(\lambda^2) & \mathrm{Im}\, \lambda > 0 \\ \varepsilon^-(\lambda^2) & \mathrm{Im}\, \lambda < 0 \end{cases},$$

and from the inequality (1.25) ($z = \lambda^2$, $\lambda = \alpha + i\beta$) it follows that

$$|\varepsilon(\lambda)| \le \frac{8\delta}{|\lambda|^2} +$$

$$\begin{cases} \pi \left(\dfrac{\delta(-|\lambda|)}{2|\lambda|} + \dfrac{\sqrt{\alpha^2+\beta^2}}{2|\alpha||\beta|} \delta(|\lambda|) \right) & \alpha^2 \ge \beta^2 \\[2mm] \pi \left(\dfrac{\sqrt{\alpha^2+\beta^2}}{2|\alpha||\beta|} \delta(-|\lambda|) + \dfrac{\delta(|\lambda|)}{2|\lambda|} \right) & \alpha^2 \le \beta^2 \end{cases}.$$

From the latter inequality it follows that the estimate

$$|\varepsilon(\lambda)| \le \frac{8\delta}{|\lambda|^2} + \frac{\pi}{\sqrt{2}} \left(\frac{\delta(-|\lambda|)}{|\mathrm{Re}\, \lambda|} + \frac{\delta(|\lambda|)}{|\mathrm{Im}\, \lambda|} \right)$$

holds in all the domain $|\lambda| > 2$. Thus,

$$n(\lambda) - i\lambda = \frac{i\lambda \varepsilon(\lambda)}{1 - \varepsilon(\lambda)}$$

where

$$|i\lambda\varepsilon(\lambda)| \leq \frac{8\delta}{|\lambda|} + \frac{\pi}{\sqrt{2}}\left(\frac{|\lambda|}{|\operatorname{Re}\lambda|}\delta(-|\lambda|) + \frac{|\lambda|}{|\operatorname{Im}\lambda|}\delta(|\lambda|)\right)$$

and since $\lim_{|t|\to\infty}\delta(t) = 0$, then $i\lambda\varepsilon(\lambda)$ and, therefore, also $n(\lambda) - i\lambda$, tends to zero uniformly outside arbitrary small angles containing the coordinate axes when $|\lambda|\,rightarrow\,\infty$.

In the general case $[q(x)$ is an arbitrary real continuous function] equation (1.22) does not determine uniquely the corresponding Weyl functions. In order that this equation should uniquely determine the Weyl function, it is sufficient for the potential to be bounded from below $(\inf q(x) > -\infty)$. In this case any solution $\Phi(\lambda, x)$ of (1.22), belonging to the space $L_2(0, \infty)$ if $\operatorname{Im}\lambda > 0$ and to the space $L_2(-\infty, 0)$ if $\operatorname{Im}\lambda < 0$, differs from the Weyl solution $\psi(\lambda, x)$ by a constant factor, and the Weyl function $n(\lambda)$ is equal to the logarithmic derivative of this solution at the point $x = 0$:

$$\Phi(\lambda, x) = \Phi(\lambda, 0)\left[c(\lambda, x) + n(\lambda)s(\lambda, x)\right] , \quad n(\lambda) = \frac{\Phi'(\lambda, 0)}{\Phi(\lambda, 0)} .$$

In particular, if the potential $q(x)$ is reflectionless the solution $e^+(\lambda, x)$ $[e^-(\lambda, x)]$ belongs to the space $L_2(0, \infty)$ $[L_2(-\infty, 0)]$ at $\operatorname{Im}\lambda > 0$ $(\operatorname{Im}\lambda < 0)$ whence, according to Lemma 1.1, it follows that

$$n(\lambda) = i\lambda + i\sum_{k=1}^{n}\frac{-\lambda_k'(0)}{\lambda - i\lambda_k(0)} . \tag{1.26}$$

Thus, the Weyl function $n(\lambda)$ of (1.22) with a reflectionless potential is a rational fraction. Conversely, let the Weyl function $\tilde{n}(\lambda)$ of (1.22) with real locally integrable potential $\tilde{q}(x)$ be a rational fraction. From the Weyl function properties, it follows that its poles can lie only on the imaginary axis and should be simple because otherwise their imaginary part cannot be positive in the right and negative in the left half-planes. By the same reason, the residue at the poles, located on the imaginary axis, must have a positive imaginary part, and their real part must be equal to zero. And, finally, the corollary of Lemma 1.5 implies that the Weyl rational function necessarily has the following form:

$$\tilde{n}(\lambda) = i\lambda + i\sum_{k=1}^{n}\frac{-\lambda_k'}{\lambda - i\lambda_k}$$

where λ_k are distinct real numbers, and λ_k' are arbitrary negative numbers. Since the sequence of pairs $\lambda_1', \lambda_1; \lambda_2', \lambda_2; \dots; \lambda_N', \lambda_N$ satisfies the conditions of Lemma 1.4, it is the spectral data of a certain operator (1.1) with the reflectionless potential $q(x)$ and, according to (1.26), the Weyl function of (1.22) with this potential is equal to $n(\lambda)$. Since the Weyl function determines the potential uniquely, $\tilde{q}(x)$ coincides with the reflectionless potential $q(x)$. Summarizing the results obtained, we arrive at the following theorem.

Theorem 1.1. In order for the real locally integrable potential $q(x)$ to be reflectionless, it is necessary and sufficient that the Weyl function $n(\lambda)$ of (1.22) be a rational fraction. The Weyl function $n(\lambda)$ of (1.22) with the reflectionless potential is expressed in terms of the spectral data of the corresponding operator according to (1.26).

2. Closure of the Sets $B(-\mu^2)$

In this section we consider the potentials belonging to the closure $\overline{B(-\mu^2)}$ of the set of reflectionless potentials $B(-\mu^2)$ in the topology of uniform convergence of functions on each compact of the real axis.

Lemma 2.1. Let $q(x)$ be an arbitrary reflectionless potential. Then the functions $\sigma_p(x)$ defined by recurrent formulas (1.5) are connected with the numbers κ_k and functions $\lambda_k(x)$ $(k = 1, 2, \ldots, n)$ by

$$\int_x^\infty \sigma_p(t)dt = -\frac{2^p}{p} \sum_{k=1}^n \{\kappa_k^p - [-\lambda_k(x)]^p\} \tag{2.1}$$

$$\sigma_p(x) = -2^p \sum_{k=1}^n [-\lambda_k'(x)] [-\lambda_k(x)]^{p-1} . \tag{2.2}$$

Proof. According to Lemma 1.1

$$e^+(\lambda, x) = e^{i\lambda x} \prod_{k=1}^n \frac{\lambda - i\lambda_k(x)}{\lambda + i\kappa_k} = e^{i\lambda x} \prod_{k=1}^n \left[1 + \frac{\lambda_k(x)}{i\lambda}\right] \left(1 - \frac{\kappa_k}{i\lambda}\right)^{-1}$$

whence it follows that in the domain $|\lambda| > \kappa_n$ the function $\ln e^+(\lambda, x)$ can be expanded in the Laurent series:

$$\ln e^+(\lambda, x) = i\lambda x + \sum_{p=1}^\infty (i\lambda)^{-p} \left(\frac{1}{p} \sum_{k=1}^\infty \{\kappa_k^p - [-\lambda_k(x)]^p\}\right) .$$

On the other hand, from (1.3), (1.4) it follows that in the upper half-plane, $\ln e^+(\lambda, x)$ can be expanded as an asymptotic series

$$\ln e^+(\lambda, x) \cong i\lambda x + \sum_{p=1}^\infty (i\lambda)^{-p} \left(-2^p \int_x^\infty \sigma_p(t)dt\right) \qquad (|\lambda| \to \infty) .$$

Comparing these expansions we obtain (2.1), differentiating them with respect to x we obtain (2.2).

Corollary. If $q(x) \in B(-\mu^2)$, then

$$0 < -q(x) = -\sigma_1(x) = 2 \sum_{k=1}^{n} -\lambda'_k(x) \le 2\mu^2 \tag{2.3}$$

$$|\sigma_p(x)| \le 2^p \mu^{p+1} \quad (p = 1, 2, \ldots) . \tag{2.3'}$$

Indeed, assuming $p = 1$ in (1.5), (2.2), and $p = 2$ in (1.5), (2.1) we find that

$$-q(x) = -\sigma_1(x) = 2 \sum_{k=1}^{n} -\lambda'_k(x) ,$$

$$\int_x^{\infty} \sigma_2(t)dt = q(x) = -2 \sum_{k=1}^{n} \left\{ \kappa_k^2 - \lambda_k(x)^2 \right\}$$

$$= -2 \left\{ \sum_{j=1}^{n} \left[\kappa_{j-1}^2 - \lambda_{k_j}(x)^2 \right] + \kappa_n^2 \right\} \ge -2\kappa_n^2$$

and since

$$-\lambda'_k(x) > 0 , \quad \kappa_n^2 \le \mu^2 ,$$

then

$$0 < -q(x) = -\sigma_1(x) = 2 \sum_{k=1}^{n} -\lambda'_k(x) \le 2\mu^2 .$$

Moreover, since $|\lambda_k(x)| \le \kappa_n \le \mu$, then from the latter inequality and equalities (2.2) it follows that

$$|\sigma_p(x)| \le 2^p \sum_{k=1}^{n} -\lambda'_k(x) |\lambda_k(x)|^{p-1} \le 2^p \mu^{p-1} \sum_{k=1}^{n} -\lambda'_k(x) \le 2^p \mu^{p+1} .$$

Now we note that the functions $\sigma_p(x)$ are defined by (1.5) for any infinitely differentiable potential, and the following lemma is true for all infinitely differentiable potentials.

Lemma 2.2. If the functions $\sigma_p(x)$ satisfy inequalities $|\sigma_p(x)| \le 2^p \mu^{p+1}$ for all $p = 1, 2, \ldots$ and $x \in (-\infty, \infty)$, then all their derivatives satisfy inequalities

$$\left| \frac{d^k \sigma_p(x)}{dx^k} \right| \le 2^p \mu^{p+k+1} \frac{(p+1)!}{p!} . \tag{2.4}$$

Proof. According to the lemma condition these inequalities are true when $k = 0$. Let us suppose that they are true when $k = 0, 1, \ldots, N$ $(N \ge 0)$. Differentiating (1.5) N times with respect to x, we get

$$-\frac{d^{N+1} \sigma_p(x)}{dx^{N+1}} = \frac{d^N \sigma_{p+1}(x)}{dx^N} + \sum_{j=1}^{p-1} \sum_{i=0}^{N} C_N^i \frac{d^{N-i} \sigma_{p-j}(x)}{dx^{N-i}} \frac{d^i \sigma_j(x)}{dx^i}$$

whence, according to the accepted supposition, it follows that

$$
\left| \frac{d^{N+1}\sigma_p(x)}{dx^{N+1}} \right| \le 2^{p+1}\mu^{p+N+2}\frac{(p+N+1)!}{(p+1)!} + 2^p\mu^{p+N+2}\sum_{j=1}^{p-1}\sum_{i=0}^{N}
$$

$$
\times C_N^i \frac{(p-j+N-i)!}{(p-j)!}\frac{(j+i)!}{j!}
$$

$$
= 2^p\mu^{p+N+2}\left[2\frac{(p+N+1)!}{(p-1)!} + \sum_{j=1}^{p-1}\sum_{i=0}^{N} \right.
$$

$$
\left. \times C_N^i \frac{(p-j+N-i)!}{(p-j)!}\frac{(j+i)!}{j!} \right]. \tag{2.5}
$$

Differentiating the identity $z^{-(p+2)} = z^{-(p-j+1)}z^{-(j+1)}$ N times with respect to z we obtain the identity

$$
(-1)^N\frac{(p+2+N-1)!}{(p+1)!}z^{-(p+2+N)}
$$

$$
= \sum_{i=0}^{N}C_N^i\frac{(p-j+N-i)!}{(p-j)!}\frac{(j+i)!}{j!}(-1)^N z^{-(p+2+N)},
$$

from which it follows that

$$
\frac{(p+2+N-1)!}{(p+1)!} = \sum_{i=0}^{N}C_N^i\frac{(p-j+N-i)!}{(p-j)!}\frac{(j+i)!}{j!}.
$$

Therefore, the right-hand side of inequality (2.5) is equal to

$$
2^p\mu^{p+N+2}\frac{(p+N+1)!}{(p+1)!}(2+p-1) = 2^p\mu^{p+N+2}\frac{(p+N+1)!}{p!}
$$

and consequently

$$
\left| \frac{d^{N+1}\sigma_p(x)}{dx^{N+1}} \right| \le 2^p\mu^{p+N+2}\frac{(p+N+1)!}{p!}
$$

whence it follows inductively that inequalities (2.4) are true for all $k = 1, 2, \dots$.

Corollary. If the potential $q(x) \in B(-\mu^2)$, then functions $\sigma_p(x)$ satisfy inequalities (2.4); in particular,

$$
\left| \frac{d^k q(x)}{dx^k} \right| \le 2\mu^{k+2}(k+1)! . \tag{2.6}
$$

Lemma 2.3. Potentials belonging to the set $\overline{B(-\mu^2)}$ are infinitely differentiable; the functions $\sigma_p(x)$, connected with them, satisfy inequalities (2.4), and the sets

$\overline{B(-\mu^2)}$ are compact in the topology of the uniform convergence on each finite interval of functions and their derivatives of all the orders.

Proof. If the potential $q(x)$ belongs to the set $\overline{B(-\mu^2)}$, then there exists a sequence of reflectionless potentials $q_n(x)$, belonging to the set $B(-\mu^2)$ which converges to $q(x)$ uniformly on each compact of the real axis. Since all the potentials $q_n(x)$ of this sequence satisfy inequalities (2.6), then, according to the Arzela theorem, it is possible to extract from it the subsequence $q_{n'}(x)$ which converges uniformly on each compact together with the derivatives of all orders. Therefore, the potential $q(x) = \lim q_{n'}(x)$ is infinitely differentiable, and all its derivatives are the limits of the derivatives of the potentials $p_{n'}(x)$ of this subsequence, whence, according to the definition of the functions $\sigma_p(x)$, it follows that these functions and their derivatives are limits of functions $\sigma_p(x; n')$ connected with potentials $q_{n'}(x) \in B(-\mu^2)$ and their derivatives. Since, according to the corollary of Lemma 2.2, functions $\sigma_p(x; n')$ satisfy inequalities (2.4), then their limits $\sigma_p(x)$ also satisfy these inequalities.

Compactness of the set $\overline{B(-\mu^2)}$ immediately follows from inequalities (2.6) and the Arzela theorem.

Corollary. The potentials $q(x)$, belonging to the set $\overline{B(-\mu^2)}$, can be analytically continued into the strip $|\operatorname{Im} z| < \mu^{-1}$ and satisfy in it the inequality

$$|q(x + iy)| \le 2\mu^2 \left(1 - \mu|y|\right)^{-2} .$$

Indeed, the estimates (2.6) show that in the intervals $(x - \mu^{-1}, x + \mu^{-1})$ the potential can be expanded in the Taylor series

$$q(z) = \sum_{k=0}^{\infty} \frac{q^{(k)}(x)}{k!}(z - x)^k , \quad (x - \mu^{-1} < z < x + \mu^{-1})$$

which provide analytical continuation of this potential into the discs $|z-x| < \mu^{-1}$ covering the strip $|\operatorname{Im} z| < \mu^{-1}$ when x runs over all the real axis. Consequently, the potentials $q(x) \in \overline{B(-\mu^2)}$ can be analytically continued into this strip, and, besides, according to (2.6)

$$|q(x + iy)| = \left| \sum_{k=0}^{\infty} \frac{q^{(k)}(x)}{k!}(iy)^k \right| \le \sum_{k=0}^{\infty} \frac{|q^{(k)}(x)|}{k!}|y|^k$$

$$\le 2\mu^2 \sum_{k=0}^{\infty}(k + 1)|\mu y|^k = 2\mu^2 \left(1 - \mu|y|\right)^{-2} .$$

Let us find out what is the form of the Weyl functions of (1.22) with potentials from the set $\overline{B(-\mu^2)}$.

According to (1.26), Weyl functions of (1.22) for the reflectionless potentials have the form

$$n(\lambda) = i\lambda + \int \frac{d\sigma(\xi)}{\xi - i\lambda} \tag{2.7}$$

where $d\sigma(\xi)$ is the discrete measure concentrated at a finite number of points $-\lambda_k(0)$, whose measures are $-\lambda'_k(0)$. Besides, if the potential belongs to the set $B(-\mu^2)$, then, according to Lemma 1.2, the support of the measure lies in the segment $[-\mu, \mu]$ [i.e., $(-\lambda_k(0))^2 < \mu^2$], and, according to (2.3)

$$\int_{-\infty}^{\infty} d\sigma(\xi) = \sum -\lambda'_k(0) \leq \mu^2 .$$

Conversely, from Theorem 1.1 and Lemma 1.4 it follows that, if the measure $d\sigma(\xi)$ is concentrated on the finite number of points of the segment $[-a, a]$ and $\int d\sigma(\xi) = \sigma$, then (2.7) defines the Weyl function $n(\lambda)$ of some (1.22) with the reflectionless potential $q(x) \in B(-\mu^2)$, where $\mu^2 \leq a^2 + \sigma$.

Theorem 2.1. The function $\overline{n(\lambda)}$ is the Weyl function of (1.22) with the potential from a certain set $\overline{B(-\mu^2)}$ if and only if it may be represented in the form of (2.7), where $d\sigma(\xi)$ is an arbitrary measure with a finite support and, besides, if its support lies in the segment $[-a, a]$ and $\int d\sigma(\xi) = \sigma$, then $\mu^2 \leq a^2 + \sigma \leq 2\mu^2$.

The necessity: If the potential $q(x) \in \overline{B(-\mu^2)}$, then there exists a sequence of potentials $q_k(x) \in B(-\mu^2)$ converging to it uniformly in each finite interval. Let

$$n_k(\lambda) = i\lambda + \int \frac{d\sigma_k(\xi)}{\xi - i\lambda} ; \quad \psi_k(\lambda, x) = c_k(\lambda, x) + n_k(\lambda)s_k(\lambda, x) \qquad (2.8)$$

by Weyl functions and Weyl solutions of (1.22) with potentials $q_k(x)$. Then all the measures $d\sigma_k(\xi)$ are concentrated on the segment $[-\mu, \mu]$ and bounded: $\int d\sigma_k(\xi) \leq \mu^2$. Besides,

$$\int_0^{\pm\infty} |\psi_k(\lambda, x)|^2 \, dx = \frac{\operatorname{Im} n_k(\lambda)}{2 \operatorname{Im} \lambda \operatorname{Re} \lambda} \qquad (2.8')$$

and the solutions $c_k(\lambda, x)$, $s_k(\lambda, x)$ on each compact converge uniformly to solutions $c(\lambda, x)$, $s(\lambda, x)$ of (1.22) with the potential $q(x)$. This permits, using the Helly theorem, passing to the limit in formulas (2.8) on the appropriately chosen subsequence k':

$$\lim_{k' \to \infty} n_{k'}(\lambda) = i\lambda + \int \frac{d\sigma(\xi)}{\xi - i\lambda} = n(\lambda)$$

$$\lim_{k' \to \infty} \psi_{k'}(\lambda, x) = c(\lambda, x) + n(\lambda)s(\lambda, x) = \psi(\lambda, x)$$

and, besides, from (2.8') it follows that $\psi(\lambda, x)$ is the Weyl solution, and $n(\lambda)$ is the Weyl function of (1.22) with the potential $q(x)$. It remains to note that the limit of measures $\sigma(\xi) = \lim \sigma_{k'}(\xi)$ is, obviously, also concentrated on the segment $[-\mu, \mu]$ and satisfies the inequality $\int d\sigma(\xi) \leq \mu^2$.

The sufficiency: Let the measure $d\sigma(\xi)$ in (2.7) be concentrated on the segment $[-a, a]$ and $\int d\sigma(\xi) = \sigma$. Let us divide the segment $[-a, a]$ into k nonintersecting half-intervals of equal length, choose a point in each of them, and concentrate the measure of these half-intervals on chosen points. As a result we obtain the discrete measure $d\sigma_k(\xi)$ concentrated on a finite number of points in the segment $[-a, a]$, and $\int d\sigma_k(\xi) = \sigma$. Obviously, the sequence of measures $d\sigma_k(\xi)$ converges weakly to the measure $d\sigma(\xi)$. According to Theorem 1.1 and Lemma 1.4 the functions

$$n_k(\lambda) = i\lambda + \int \frac{d\sigma_k(\xi)}{\xi - i\lambda}$$

are Weyl functions of (1.22) with the potentials $q_k(x) \in B(-\mu^2)$, where $\mu^2 \le a^2 + \sigma$. Therefore, it is possible to choose a subsequence $q_{k'}(x)$ converging to some potential $q(x) \in \overline{B(-\mu^2)}$ uniformly in each finite interval. Repeating the last part of the proof of the necessity, we see that the function $n(\lambda) = \lim n_k(\lambda)$ is the Weyl function of (1.22) with the potential $q(x) \in \overline{B(-\mu^2)}$, and $\mu^2 \le a^2 = \sigma$.

We shall denote the union of all the sets $\overline{B(-\mu^2)}$ through $\tilde{B}$:

$$\tilde{B} = \cup_{\mu > 0} \overline{B(-\mu^2)}.$$

Thus, $q(x) \in \tilde{B}$ if $q(x) \in \overline{B(-\mu^2)}$ for some μ.

Corollary. The potential $q(x)$ belongs to the set $\tilde{B}$ if and only if the Weyl function $n(\lambda)$ of (1.22) is holomorphic outside a finite segment of the imaginary axis.

The necessity of this condition follows immediately from Theorem 2.1. To prove the sufficiency we note that from the holomorphness of the function $n(\lambda)$ outside the compact of the imaginary axis there follows the holomorphness of the function $n(-iz)$ outside the compact of the real axis. Since imaginary parts of the Weyl functions $n(\lambda)$ are positive in the right half-plane, then the imaginary part $n(-iz)$ is positive in the upper half-plane. Therefore, the function $n(-iz)$ can be represented in the form

$$n(-iz) = az + b + \int \frac{d\varrho(\xi)}{\xi - z}$$

where $a > 0$, $\text{Im } b = 0$ and the measure $d\varrho(\xi)$ is concentrated on the compact Δ of the real axis. Thus

$$n(\lambda) = ai\lambda + b + \int \frac{d\varrho(\xi)}{\xi - i\lambda},$$

and, besides, according to the corollary of Lemma 1.5, $a = 1$, $b = 0$, i.e. the function $n(\lambda)$ can be represented in the form of (2.7), whence, according to Theorem 2.1, it follows that the potential $q(x)$ belongs to some set $\overline{B(-\mu^2)}$.

Remark. From the definition of the function $n(\lambda)$ it follows that it is holomorphic outside the imaginary axis if and only if

$$\lim_{\varepsilon \to 0} \left[m^+(x + i\varepsilon) - m^-(x - i\varepsilon) \right] = 0$$

for all $x > 0$.

Let us denote by $R^1 + \tilde{B}$ the set of all the potentials of the form $q(x) = c + \tilde{q}(x)$, where $c \in R^1$, $\tilde{q}(x) \in \tilde{B}$. Thus, $q(x) \in \overline{(R^1 + \tilde{B})}$ if such real numbers c, μ can be found that $q(x) = c + \tilde{q}(x)$, where $\tilde{q}(x) \in \overline{B(-\mu^2)}$. We note that the solutions $\psi^{\pm}(z, x)$, $\tilde{\psi}^{\pm}(z, x)$ and the Weyl functions $m^{\pm}(z)$, $\tilde{m}^{\pm}(z)$ of equations (1.22) with any real potentials $q(x)$, $\tilde{q}(x)$ differing by a constant c, $q(x) = c + \tilde{q}(x)$, are bound by obvious equalities

$$\psi^{\pm}(z + c, x) = \tilde{\psi}^{\pm}(z, x) , \quad m^{\pm}(z + c) = \tilde{m}^{\pm}(z) .$$

For corresponding solutions $\psi(\lambda, x)$, $\tilde{\psi}(\lambda, x)$ and functions $n(\lambda)$, $\tilde{n}(\lambda)$ there follow the relations

$$\tilde{\psi}(\lambda, x) = \psi\left(\sqrt{\lambda^2 + c}, x\right) , \quad \psi(\lambda, x) = \tilde{\psi}\left(\sqrt{\lambda^2 - c}, x\right),$$
$$\tilde{n}(\lambda) = n\left(\sqrt{\lambda^2 + c}\right) , \quad n(\lambda) = \tilde{n}\left(\sqrt{\lambda^2 - c}\right) \tag{2.9}$$

where by $\sqrt{\lambda^2 + A}$ we denote the branch of the radical that is holomorphic in the plane with cut $-i\sqrt{A}$, $i\sqrt{A}$ and assumes positive values on the semiaxis $\lambda > \sqrt{|A|}$.

Lemma 2.4. The real potential $q(x)$ belongs to the set $R^1 + \tilde{B}$ if and only if the corresponding Weyl function $n(\lambda)$ is holomorphic outside some disc.

The necessity: If $\tilde{q}(x) \in \tilde{B}$ then, according to the corollary of Theorem 2.1 the function $\tilde{n}(\lambda)$ is holomorphic outside a finite segment of the imaginary axis, whence it follows that the Weyl function $n(\lambda) = \tilde{n}(\sqrt{\lambda^2 - c})$ of (1.22) with the potential $q(x) = c + \tilde{q}(x)$ is holomorphic outside some disc.

The sufficiency: If the function $n(\lambda)$ is holomorphic outside the disc $|\lambda| \leq r$, then the Weyl function $\tilde{n}(\lambda)$ of equation (1.22) with the potential $\tilde{q}(x) = q(x) - r^2$ is equal to $n(\sqrt{\lambda^2 + r^2})$ and, therefore, is holomorphic outside the segment $[-2ir, 2ir]$ of the imaginary axis, whence it follows that $\tilde{q}(x) \in \tilde{B}$ and $q(x) = r^2 + \tilde{q}(x)$ belongs to the set $R^1 + \tilde{B}$.

Lemma 2.5. If the real potential $q(x)$ is infinitely differentiable and the functions $\sigma_p(x)$, defined by (1.5), satisfy the inequalities $|\sigma_p(x)| \leq (2R)^p R$ $(R < \infty)$, then the function

$$\psi(\lambda, x) = \exp\left[i\lambda x + \int_0^x \sum_{p=1}^{\infty}(2i\lambda)^{-p}\sigma_p(t)dt\right] \tag{2.10}$$

is the Weyl solution of (1.22) with this potential in the domain $|\lambda| > R$.

Proof. From the inequalities $|\sigma(x)| \leq (2R)^p R$ and Lemma 2.2 it follows that in the domain considered the series $\sum_{p=1}^{\infty}(2i\lambda)^{-p}\sigma_p(x)$ converges, it can be differentiated term by term with respect to x and, since functions $\sigma_p(x)$ satisfy

(1.5), the right-hand side of (2.10) is the solution of (1.22), normalized by the condition $\psi(\lambda, 0) = 1$. Further we have

$$\left| \int_0^x \sum_{p=1}^\infty (2i\lambda)^{-p} \sigma_p(t)dt \right| \leq |x| R \sum_{p=1}^\infty |R\lambda^{-1}|^p$$

$$= |x|R^2 \left(|\lambda| - R \right)^{-1} \quad (|\lambda| > R)$$

and consequently,

$$|\psi(\lambda, x)| \leq \exp \left[\text{Im} \left(\lambda x \right) + |x|R^2 \left(|\lambda| - R \right)^{-1} \right].$$

whence it follows that in the domain $\text{Im} \, \lambda > 3R$ ($\text{Im} \, \lambda < -3R$) the solution $\psi(\lambda, x)$ exponentially decreases when $x \to +\infty$ ($x \to -\infty$). Therefore, $\psi(\lambda, x)$ is the Weyl solution in the domain $|\text{Im} \, \lambda| > 3R$ and, consequently, in the domain of the holomorphness of the right-hand side of (2.10), i.e. when $|\lambda| > R$.

Theorem 2.2. The real potential $q(x)$ belongs to the set $R^1 + \tilde{B}$ if and only if it is infinitely differentiable and the functions $\sigma_p(x)$, defined by (1.5), for some $R < \infty$ satisfy the inequalities $|\sigma_p(x)| < (2R)^p R$.

Proof. The necessity: Let $q(x) \in R^1 + \tilde{B}$, i.e. $q(x) = c + \tilde{q}(x)$, where $c \in R$, $q(x) \in B(-\mu^2)$. Then $|\tilde{\sigma}_p(x)| \leq \mu(2\mu)^p$ and, according to Lemma (2.5), the Weyl solution of (1.22) with the potential $\tilde{q}(x)$ is equal to

$$\tilde{\psi}(\lambda, x) = \exp \left[\int_0^x \left(i\lambda + \sum_{p=1}^\infty (2i\lambda)^{-p} \tilde{\sigma}_p(t) \right) dt \right] \quad (|\lambda| > \mu) .$$

Hence, according to (2.9), a representation for the Weyl solution of (1.22) with the potential $q(x) = c + \tilde{q}(x)$ is:

$$\psi(\lambda, x) = \tilde{\psi} \left(\sqrt{\lambda^2 + c}, x \right)$$

$$= \exp \left\{ \int_0^x \left[i\sqrt{\lambda^2 - c} + \sum_{p=1}^\infty \left(2i\sqrt{\lambda^2 - c} \right)^{-p} \tilde{\sigma}_p(t) \right] dt \right\} .$$

The function in brackets is holomorphic in the domain $|\lambda| > \mu + \sqrt{|c|}$ and can be expanded there in the Laurent series

$$\left[i\sqrt{\lambda^2 - c} + \sum_{p=1}^\infty \left(2i\sqrt{\lambda^2 - c} \right)^{-p} \tilde{\sigma}_p(t) \right] = i\lambda + \sum_{p=1}^\infty (2i\lambda)^{-p} a_p(t)$$

with the coefficients

$$a_p(t) = (2\mathrm{i})^p \frac{1}{2\pi\mathrm{i}} \oint_{|\lambda|=R_1} \left[\mathrm{i}\sqrt{\lambda^2 - c} - \mathrm{i}\lambda \right.$$

$$\left. + \sum_{k=1}^{\infty} \left(2\mathrm{i}\sqrt{\lambda^2 - c} \right)^{-k} \tilde{\sigma}_k(t) \right] \lambda^{p-1} d\lambda \tag{2.11}$$

where R_1 is any number greater than $\mu + \sqrt{|c|}$. Therefore,

$$\psi(\lambda, x) = \exp\left\{ \int_0^x \left[\mathrm{i}\lambda + \sum_{p=1}^{\infty} (2\mathrm{i}\lambda)^{-p} a_p(t) \right] dt \right\} \qquad \left(|\lambda| > \mu + \sqrt{|c|} \right)$$

and, since $\psi(\lambda, x)$ is the solution of (1.22), functions $a_p(x)$ and the potential $q(x)$ are connected by the same recurrent formulas as $\sigma_p(x)$, i.e. $a_p(x) = \sigma_p(x)$. Hence, according to (2.11), it follows that $|\sigma_p(x)| \le 2^p M(R_1) R_1^p$, where $R_1 > \mu + \sqrt{|c|}$ and

$$M(R_1) = \max_{|\lambda|=R_1} \left| \mathrm{i}\sqrt{\lambda^2 - c} - \mathrm{i}\lambda + \sum_{k=1}^{\infty} \left(2\mathrm{i}\sqrt{\lambda^2 - c} \right)^{-k} \tilde{\sigma}_k(x) \right| .$$

Since $|\tilde{\sigma}_k(x)| \le \mu(2\mu)^k$ and on the circle $|\lambda| = R_1 > \mu + \sqrt{|c|}$ the inequalities

$$\left| \sqrt{\lambda^2 - c} \right| \ge R_1 - \sqrt{|c|}, \quad \left| \sqrt{\lambda^2 - c} - \lambda \right| \le \sqrt{|c|}$$

are true, then

$$M(R_1) \le \sqrt{|c|} + \mu \sum_{k=1}^{\infty} \left(\frac{\mu}{R_1 - \sqrt{|c|}} \right)^k = \sqrt{|c|} + \mu^2 \left(R_1 - \sqrt{|c|} - \mu \right)^{-1}$$

whence, for $R_1 = 2\mu + \sqrt{|c|}$, it follows that $M(R_1) \le \sqrt{|c|} + \mu < R_1$ and, therefore, $|\sigma_p(x)| \le R_1 (2R_1)^p$.

The sufficiency: If $|\sigma_p(x)| \le (2R)^p R$ then, according to Lemma 2.5, the Weyl solution of (1.22) is equal to

$$\psi(\lambda, x) = \exp\left\{ \mathrm{i}\lambda x + \int_0^x \left[\sum_{p=1}^{\infty} (2\mathrm{i}\lambda)^{-p} \sigma_p(t) \right] dt \right\} \qquad (|\lambda| > R)$$

and the Weyl function that equals

$$n(\lambda) = \frac{\psi'_x(\lambda, 0)}{\psi(\lambda, 0)} = \mathrm{i}\lambda + \sum_{p=1}^{\infty} (2\mathrm{i}\lambda)^{-p} \sigma_p(0) ,$$

is holomorphic in the domain $|\lambda| > R$, whence, according to Lemma 2.4, it follows that the potential $q(x)$ belongs to the set $R^1 + \tilde{B}$.

Corollary. The Weyl solution $\psi(\lambda, x) = c(\lambda, x) + n(\lambda)s(\lambda, x)$ of (1.22) with the potential $q(x)$ from the set $R^1 + \tilde{B}$ is holomorphic with respect to the variable λ outside some disc and when $|\lambda| \to \infty$ it satisfies the asymptotic equality

$$e^{-i\lambda x}\psi(\lambda, x) \equiv e^{-i\lambda x}[c(\lambda, x) + n(\lambda)s(\lambda, x)]$$

$$= 1 + \frac{1}{2i\lambda}\int_0^x q(t)dt + O\left(\lambda^{-2}\right) . \tag{2.12}$$

Let us give some examples of potentials belonging to the set $R^1 + \tilde{B}$.

1) Decreasing potentials. Formulas (1.2), (1.2′) are true for (1.22) with the potentials satisfying the condition

$$\int_{-\infty}^{\infty} (1 + |x|) \, |q(x)| dx < \infty , \tag{2.13}$$

and, since $e^+(\lambda, x) \in L_2(0, \infty)$, if $\operatorname{Im} \lambda > 0$, and $e^-(\lambda, x) \in L_2(-\infty, 0)$, if $\operatorname{Im} \lambda < 0$, we have in this case

$$n(\lambda) = \begin{cases} e^+(\lambda, 0)^{-1}e^+(\lambda, 0)' & \operatorname{Im} \lambda > 0 \\ e^-(\lambda, 0)^{-1}e^-(\lambda, 0)' & \operatorname{Im} \lambda < 0 \end{cases} .$$

Therefore, for the Weyl function $n(\lambda)$ to be holomorphic outside some disc $|\lambda| < R$, it is necessary and sufficient that

$$n(\lambda + i0) - n(\lambda - i0) = \frac{e^-(\lambda, 0)e^+(\lambda, 0)' - e^+(\lambda, 0)e^-(\lambda, 0)'}{e^+(\lambda, 0)e^-(\lambda, 0)} = 0$$

for $\lambda \in (R, \infty)$. Hence, according to the definition of the reflection coefficient, it follows that the real potential $q(x)$ satisfying inequality (2.13) belongs to the set $R^1 + \tilde{B}$ if and only if the corresponding reflection coefficient is equal to zero outside some compact.

2) The finite-zone potentials. The potential $q(x)$ is called a finite-zone one if the Weyl functions $m^+(z)$, $m^-(z)$ coincide with the same function $m(z)$ meromorphic on the Riemann surface of the radical

$$\sqrt{(z - E_N^-) \prod_{j=1}^{N-1} (z - E_j^-)(z - E_j^+)} ;$$

$$(E_1^- < E_1^+ < E_2^- < E_2^+ < \ldots < E_{N-1}^+ < E_{N-1}^- < E_N^-)$$

i.e. $m(z) = m^+(z)$ when z lies on the upper, and $m(z) = m^-(z)$ when z lies on the lower band of this surface. In this case $m^+(x + i0) = m^-(x - i0)$ if $x \in (E_N^-, \infty)$ hence it follows that the function $n(\lambda)$ is holomorphic outside the disc $|\lambda| \le R$, where $R^2 = \max_{1 \le k \le N} |E_k^-|$. Thus, the finite-zone potentials belong to the set $R^1 + \tilde{B}$, moreover, they belong to the set $\tilde{B}$, if $E_N^- = 0$.

Theorem 2.3. The Cauchy problem for the KdV equation

$$u_t = 6uu_x - u_{xxx} , \quad u(x,0) = q(x) \tag{2.14}$$

with the initial function $q(x)$ from the set $R^1 + \tilde{B}$ has a bounded solution.

Proof. Problem (2.14) is solvable if $q(x) \in B(-\mu^2)$ and, besides, its solution at each fixed t also belongs to the set $B(-\mu^2)$. Therefore, estimates (2.6) hold for the solutions $u(x,t)$ of the KdV equation with the initial data from the set $B(-\mu^2)$, and from them, according to the KdV equation itself, follow the inequalities

$$\left| \frac{\partial^{k+l} u(x,t)}{\partial x^k \partial t^l} \right| \leq C(k,l,\mu)$$

with constants $C(k,l,\mu)$ independent of the initial data $q(x) \in B(-\mu^2)$. For example

$$C(k,0,\mu) = 2\mu^{k+2}(k+1)! , \quad C(k,1,\mu) = 5\mu^{k+5}(k+4)! , \quad \text{etc.}$$

Therefore, the set of all solutions of the KdV equation with the initial data from $B(-\mu^2)$ is compact with respect to the uniform convergence of functions with all their derivatives on each bounded part of the (x,t) plane. We see that the closure of this set consists of solutions of the KdV equation and contains its solutions with any initial data from the set $\overline{B(-\mu^2)}$.

Thus, the Cauchy problem (2.14) is solvable if $q(x) \in \overline{B(-\mu^2)}$. Now let $q(x) = c + \tilde{q}(x)$ where $c \in R$ and $\tilde{q}(x) \in \overline{B(-\mu^2)}$. According to the previous discussion, the solution of the KdV equation with the initial data $\tilde{q}(x)$ exists. Let us denote it by $\tilde{u}(x,t)$ and let

$$u(x,t) = \tilde{u}(x + 6ct, t) + c . \tag{2.15}$$

A simple verification shows that the function $u(x,t)$ is a bounded solution of the Cauchy problem (2.14) with the initial function $q(x) = c + \tilde{q}(x)$. Therefore, for all $q(x) \in R^1 + \tilde{B}$ the Cauchy problem (2.14) has a bounded solution.

3. The Inverse Problem

As is known, the Weyl functions $m^+(z)$, $m^-(z)$ [or one function $n(\lambda)$] uniquely determine the potential. It is possible to reconstruct the potential from the given Weyl functions, for instance, in the following way: first, to find the spectral functions $\varrho^\pm(\mu)$ by the known formulas, then to use the Gelfand–Levitan equation. Unfortunately, this algorithm cannot be adjusted for solving non-linear equations. In this section another method for solving the inverse problem is given which also makes it possible to solve the Cauchy problem for the KdV equation with the initial data from the set $R^1 + \tilde{B}$. Transformation (2.15) shows that it is possible to confine oneself to the discussion of the set $\tilde{B}$.

According to Theorem 2.1, the Weyl functions $n(\lambda)$ of (1.22) with the potentials from the set $\tilde{B}$ have the following form:

$$n(\lambda) = i + \int_{-\infty}^{\infty} \frac{d\sigma(\xi)}{\xi - i\lambda} ,$$

where $d\sigma(\xi)$ is the positive Borel measure with the compact support Ω. We remind the reader that the support of the positive measure $d\varrho(\xi)$ is the set of all the points ξ, such that the ϱ-measure of any interval containing the point ξ is positive. The support is always closed, and if it is bounded, as in our case, it is compact.

We want to reconstruct the Weyl solution

$$\psi(\lambda, x) = c(\lambda, x) + n(\lambda)s(\lambda, x)$$

from the given function $n(\lambda)$ without using the approximation of reflectionless potentials. In order to deal with the usual classes of analytical functions, we make the substitution $\lambda = -iz$ and get

$$n_1(z) = n(-iz) = z + \int_{-\infty}^{+\infty} \frac{d\sigma(\xi)}{\xi - z} \tag{3.1}$$

$$\psi_1(z, x) = \psi(-iz, x) = c(iz, x) + n_1(z)s(iz, x) . \tag{3.2}$$

The functions $n_1(z)$ and $-n_1(-z)$ have a positive imaginary part in the upper half plane, i.e. they are the Nevanlinna functions. The important role in the spectral theory of one-dimensional Schrödinger operators is played also by the functions

$$M_{11}(z) = -\frac{n_1(z)n_1(-z)}{n_1(z) - n_1(-z)} , \quad M_{22}(z) = -\frac{1}{n_1(z) - n_1(-z)} ;$$

$$M_{21}(z) = M_{12}(z) = \frac{1}{2} \frac{n_1(z) + n_1(-z)}{n_1(z) - n_1(-z)}$$

$$= \frac{n_1(z)}{n_1(z) - n_1(-z)} - \frac{1}{2} = \frac{n_1(-z)}{n_1(z) - n_1(-z)} + \frac{1}{2} . \tag{3.3}$$

The Nevanlinna representations for these functions have the following form:

$$M_{11}(z) = \frac{1}{2}z + \int_{-\infty}^{+\infty} \frac{d\varrho_{11}(\xi)}{\xi - z} , \quad M_{22}(z) = \int_{-\infty}^{+\infty} \frac{d\varrho_{22}(\xi)}{\xi - z} ,$$

$$M_{12}(z) = M_{21}(z) = \int_{-\infty}^{+\infty} \frac{d\varrho_{12}(\xi)}{\xi - z} ; \tag{3.4}$$

where

$$\int_a^b d\varrho_{ij}(t) = \lim_{\varepsilon \downarrow 0} \frac{1}{\pi} \int_a^b \operatorname{Im} M_{ij}(t + i\varepsilon)dt .$$

The measures $d\varrho_{11}(\xi)$, $d\varrho_{22}(\xi)$ are positive and all three measures are absolutely continuous with respect to the measure

$$d\varrho(\xi) = d\varrho_{11}(\xi) + d\varrho_{22}(\xi) ,$$

i.e. $d\varrho_{ij}(t) = \delta_{ij}(t)d\varrho(t)$ where

$$0 \le \delta_{ij}(t) \le 1 , \quad \delta_{11}(t) + \delta_{22}(t) = 1 , \quad \delta_{12}^2 \le \delta_{11}(t)\delta_{22}(t) .$$

Lemma 3.1. The function $(2z)^{-1}[n_1(z) - n_1(-z)]$ can be represented in the form

$$\frac{n_1(z) - n_1(-z)}{2z} = \exp \frac{1}{\pi} \int_0^\infty \left(\frac{1}{t - z} + \frac{1}{t + z} \right) \eta_1(t)dt$$

where

$$\eta_1(t) = \lim_{\varepsilon \downarrow 0} \arg [n_1(t + i\varepsilon) - n_1(-t - i\varepsilon)] . \tag{3.5}$$

Proof. From (3.1) it follows that

$$n_1(z) - n_1(-z) = 2z + \int_{-\infty}^{+\infty} \left(\frac{1}{\xi - z} - \frac{1}{\xi + z} \right) d\sigma(\xi)$$

$$= 2z \left[1 + \int_{-\infty}^{+\infty} \frac{d\sigma(\xi)}{\xi^2 - z^2} \right]$$

and

$$\frac{n_1(z) - n_1(-z)}{2z} = F(z^2) \tag{3.6}$$

where

$$F(w) = 1 + \int \frac{d\sigma(\xi)}{\xi^2 - w} .$$

Since the imaginary part of the function $F(w)$ is positive in the upper half-plane, then $0 < \arg F(w) < \pi$ (Im $w > 0$), the function $\ln F(w) = \ln |F(w)| + i \arg F(w)$ is Nevanlinna's and almost everywhere there exists a limit

$$\varphi(t) = \lim_{\varepsilon \downarrow 0} \arg F(t + i\varepsilon) = \lim_{\varepsilon \downarrow 0} \operatorname{Im} \ln F(t + i\varepsilon)$$

and according to the Lebesgue theorem

$$\lim_{\varepsilon \downarrow 0} \frac{1}{\pi} \int_a^b \operatorname{Im} \ln F(t + i\varepsilon)dt = \frac{1}{\pi} \int_a^b \varphi(t)dt .$$

On the real axis the function $F(t)$ is positive outside the finite segment of the positive semiaxis. Therefore, outside this segment $\varphi(t) \equiv 0$, and, since

$\lim_{|w| \to \infty} \ln F(w) = 0$, the Nevanlinna's formula for the function $\ln F(w)$ assumes the following form:

$$\ln F(w) = \frac{1}{\pi} \int_0^\infty \frac{\varphi(t)}{t - w} dt ,$$

where the integration is, in fact, carried out along the finite segment of the positive semiaxis.

Therefore

$$\frac{n_1(z) - n_1(-z)}{2z} = F(z^2) = \exp\left[\frac{1}{\pi} \int_0^\infty \frac{\varphi(t)}{t - z^2} dt\right] = \exp\left[\frac{1}{\pi} \int_0^\infty \frac{2\tau\varphi(\tau^2)}{\tau^2 - z^2} \tau\right]$$

$$= \exp\left[\frac{1}{\pi} \int_0^\infty \left(\frac{1}{\tau - z} + \frac{1}{\tau + z}\right) \varphi(\tau^2) d\tau\right]$$

and to conclude the proof, it suffices to remark that, according to (3.6), for $t > 0$

$$\eta_1(t) = \lim_{\varepsilon \downarrow 0} \arg\left[n_1(t + i\varepsilon) - n_1(-t - i\varepsilon)\right]$$

$$= \lim_{\varepsilon \downarrow 0} \arg F\left(t^2 - \varepsilon^2 + 2it\varepsilon\right) = \varphi(t^2) .$$

Let us decompose the measure $d\sigma(\xi)$ into the absolutely continuous $d\sigma_a(\xi) = \sigma'(\xi)d\xi$ and singular $d\sigma_s(\xi)$ parts, the supports of which are denoted by Ω_a and Ω_s respectively:

$$d\sigma(\xi) = d\sigma_a(\xi) + d\sigma_s(\xi) , \quad \Omega = \Omega_a \cup \Omega_s .$$

We agree also to denote by $f(\mathcal{U})$ the image of the set $\mathcal{U}$ for the mapping $x \to f(x)$. For example, $-\mathcal{U}$, $|\mathcal{U}|$, $\mathcal{U}^2$ are the images of the set $\mathcal{U}$ for mappings $x \to -x$, $x \to |x|$, $x \to x^2$.

The set

$$\Omega(2) = \Omega \cap (-\Omega) = \{\xi : \xi \in \Omega, -\xi \in \Omega\}$$

is called the symmetrical part of the support Ω and its complement

$$\Omega(1) = \Omega/\Omega(2) = \{\xi : \xi \in \Omega, -\xi \in \Omega\}$$

is called the asymmetrical part. It is clear that

$$\Omega_a(2) = \Omega_a \cap \Omega(2) , \quad \Omega_s(2) = \Omega_s \cap \Omega(2) , \quad \Omega(2) = \Omega_a(2) \cup \Omega_s(2) ,$$

$$\Omega = \Omega(1) \cup \Omega_a(2) \cup \Omega_s(2) .$$

The sets Ω, $\Omega(2)$, $\Omega_a(2)$, $\Omega_s(2)$ are compact, and the set $\Omega(1)$ is compact if and only if dist$(\Omega(1), \Omega(2)) > 0$; moreover, in this case the distance between the sets $\Omega(1)$ and $-\Omega(1)$ is also positive.

Later on it is supposed that the support Ω of the measure $d\sigma(\xi)$ does not contain the origin, the set $\Omega_a(2)$ consists of a finite number of nonintersecting segments $\pm\Delta_i(a)$ and all three sets $\Omega(1)$, $\Omega_a(2)$, $\Omega_s(2)$ are placed at a positive distance from one another. In this case the sets $\Omega(1)$, $-\Omega(1)$, $\Omega_a(2)$, $\Omega_s(2)$ are compact and do not intersect, which makes it possible to cover the sets $\Omega(1)$ and $\Omega_s(2)$ by a finite number of segments $\Delta_j(1)$ and $\pm\Delta_l(s)$ so that all the segments $|\Delta_i(a)|$, $|\Delta_j(1)|$, $|\Delta_l(s)|$ are strictly positive and do not have common points.

Thus, the imposed restriction is equivalent to the following condition.

A) There exists a finite system of segments $\Delta_i(a)$, $\Delta_l(s)$, $\Delta_j(1)$ such that the segments $|\Delta_i(a)|$, $|\Delta_l(s)|$, $|\Delta_j(1)|$ are strictly positive, do not have common points and

$$\Delta(1) = \bigcup_j \Delta_j(1) \supset \Omega(1) , \quad \Delta_a(2) = \bigcup_l \{\Delta_l(s) \cup -\Delta_l(s)\} \supset \Omega_s(2)$$

$$\Delta_a(2) = \bigcup_i \{\Delta_i(a) \cup -\Delta_i(a)\} = \Omega_a(2) .$$

We will introduce, for the sake of brevity, the notations $\Delta(2) = \Delta_a(2) \cup \Delta_s(2)$, $\Delta = \Delta(1) \cup \Delta_a(2) \cup \Delta_s(2)$. The complement of the set $|\Delta|$ up to the positive semiaxis $(0, \infty)$ consists of the finite number of intervals which we denote by $(|b_{i-1}|, |a_i|)$:

$$0 = |b_0| < |a_1| < |b_1| < \ldots < |a_N| < |b_N| < |a_{N+1}| = \infty .$$

We shall need special factorization of the function $2z[n_1(z) - n_1(-z)]^{-1}$ which takes into account the position of the sets $\Omega(1)$, $\Omega_a(2)$, $\Omega_s(2)$.

From the equality

$$n_1(z) - n_1(-z) = 2z + \int_{-\infty}^{\infty} \frac{d\sigma(\xi)}{\xi - z} - \int_{-\infty}^{\infty} \frac{d\sigma(\xi)}{\xi + z}$$

it can be seen that in the right half-plane the function $n(z)-n(-z)$ is holomorphic outside the set $|\Omega| \subset |\Delta|$. In particular, it is holomorphic and real in the intervals $(|b_{i-1}|, |a_i|)$, and, since its derivative

$$2 + \int_{-\infty}^{\infty} \frac{d\sigma(\xi)}{(\xi - z)^2} + \int_{-\infty}^{\infty} \frac{d\sigma(\xi)}{(\xi + z)^2}$$

is positive therein, then each interval $(|b_{i-1}|, |a_i|)$ contains not more than one root of this function denoted by $|c_{i-1}|$ assuming $|c_{i-1}| = |b_{i-1}|$ $(|c_{i-1}| = |a_i|)$ if $n_1(t) - n_1(-t) > 0$ (< 0) in all the interval $(|b_{i-1}|, |a_i|)$. We note that $|c_0| = 0$ since the function $n_1(z) - n_1(-z)$ is odd. Consequently, on the positive semiaxis function (3.5) can be different from zero only for $t \in |\Delta|$ and in the intervals $(|b_i|, |c_i|)$ $(i \geq 1)$ where it is equal to π. Hence, according to Lemma 3.1 it follows that

$$\frac{n_1(z) - n_1(-z)}{2z} = \exp\frac{1}{\pi}\left\{\sum_{i=1}^{N}\int_{|b_i|}^{|c_i|}\left(\frac{1}{t-z}+\frac{1}{t+z}\right)\pi dt\right.$$

$$\left.+\int_{|\Delta|}\left(\frac{1}{t-z}+\frac{1}{t+z}\right)\eta(t)dt\right\}$$

$$=\prod_{i=1}^{N}\frac{c_i^2-z^2}{b_i^2-z^2}\exp\left[\frac{1}{\pi}\int_{|\Delta|}\left(\frac{1}{t-z}+\frac{1}{t+z}\right)\eta(t)dt\right].$$

Further, since $|\Delta| = |\Delta(1)| \cup |\Delta(2)|$ and the identity

$$\int_{|\Delta_j|}\left(\frac{1}{t-z}+\frac{1}{t+z}\right)\eta(t)dt = \int_{\Delta_j}\left(\frac{1}{t-z}+\frac{1}{t+z}\right)\frac{t}{|t|}\eta\left(|t|\right)dt$$

is true both for positive and negative segments Δ_j, then

$$\int_{|\Delta|}\left(\frac{1}{t-z}+\frac{1}{t+z}\right)\eta(t)dt = \int_{\Delta(1)\cup\Delta(2)|}\left(\frac{1}{t-z}+\frac{1}{t+z}\right)\frac{t}{|t|}\eta\left(|t|\right)dt$$

and

$$\frac{2z}{n_1(z) - n_1(-z)}$$

$$=\prod_{i=1}^{N}\frac{b_i^2-z^2}{c_i^2-z^2}\exp\left[-\frac{1}{\pi}\int_{\Delta(1)\cup\Delta(2)|}\left(\frac{1}{t-z}+\frac{1}{t+z}\right)\frac{t}{|t|}\eta\left(|t|\right)dt\right].$$

Assuming

$$r(z) = \prod_{i=1}^{N}\frac{b_i - z}{c_i - z}\exp\left[-\frac{1}{\pi}\int_{\Delta(1)\cup\Delta(2)|}\frac{t}{|t|}\eta\left(|t|\right)\frac{1}{t-z}dt\right] \qquad (3.7)$$

we get the desired factorization

$$\frac{2z}{n_1(z) - n_1(-z)} = r(z)r(-z).$$

Let us explain the choice of signs of the numbers b_i, c_i. If $|b_i|$ is the right endpoint of segment $|\Delta_j(1)|$ and $-|\Delta_j(1)| \in \Delta(1)$, then we assume $-b_i = |b_i|$. In all other cases $b_i = |b_i|$. If $|c_i|$ is the left endpoint of segment $|\Delta_p(1)|$ and $-|\Delta_p(1)| \in \Delta(1)$, then $-c_i = |c_i|$. In all the other cases $c_i = |c_i|$. The set of all numbers c_i whose moduli lie strictly in the corresponding intervals ($|b_i| < |c_i| < |a_{i+1}|$) we shall denote by C.

Formula (3.7) shows that for such a choice of signs the functions $r(z)$, $r(-z)$ have the following properties:

1) The function $r(z)$ is holomorphic outside the set $\Delta(1) \cup |\Delta(2)| \cup C$ and the function $r(z)^{-1}$ is holomorphic outside the set $\Delta(1) \cup |\Delta(2)|$.

2) The domain of holomorphy of the functions $r(-z)$, $r(-z)^{-1}$ contains the set $\Delta(1) \cup |\Delta(2)| \cup C$.

3) Almost everywhere (with respect to the Lebesgue measure) there exist the limits $\lim_{\varepsilon \downarrow 0} r(t \pm i\varepsilon) = r^{\pm}(t)$, and $r^{+}(t) = \overline{r^{-}(t)}$,

$$r^{+}(t) = |r(t)| \exp \left[-\frac{i}{2} \left(\frac{t}{|t|} + 1 \right) m\left(|t| \right) \right] \qquad [t \in \Delta_a(2)] . \qquad (3.8)$$

Remark. From the identity

$$\int_{-\infty}^{\infty} \frac{\alpha(|t|)}{t - z} dt = - \int_{-\infty}^{\infty} \frac{\alpha(|t|)}{t + z} dt$$

it follows that the function

$$a(z) = \exp \left[\int_{-\infty}^{\infty} \frac{\alpha(|t|)}{t - z} dt \right]$$

satisfies the identity $a(z)a(-z) = 1$. Therefore, together with $r(z)$, the functions $r(z)a(z)$ also realize the factorization of the function $2z[n_1(z) - n_1(-z)]^{-1}$. The choice of the function $r(z)$ is motivated by the following two conditions: first, the function $r(z)$ must be meromorphic outside the support of the measure $d\sigma(\xi)$ [i.e. in the domain of holomorphy of the function $n_1(z)$] and, second, in the resulting integral equation, the function $p(s)$ must be the Muckenhoupt weight.

When solving the inverse problem, the main role belongs to the function of the complex variable z

$$g(z) = g(z, x) = e^{-zx} \psi_1(z, x) r(z) = e^{-zx} [r(z)c(iz, x) + r(z)n_1(z)s(iz, x)]$$

for which a convenient integral equation will be obtained. Let us enumerate the main properties of this function considering the real parameter x arbitrarily fixed.

Since $c(iz, x)$, $s(iz, x)$ are even entire functions and the function $n_1(z)$ is holomorphic outside the support Ω of the measure $d\sigma(\xi)$, then the function $e^{-zx} \psi_1(z, x)$ is also holomorphic outside the set Ω and, according to (2.12) in the neighborhood of the infinitely remote point

$$e^{-zx} \psi_1(z, x) = 1 + (2z)^{-1} \int_0^x q(t) dt + O(z^{-2}) .$$

The function $r(z)$ is holomorphic outside the set $\Delta(1) \cup |\Delta_s(2)| \cup \Delta_a(2) \cup C$ and $r(z) = 1 + z^{-1} r_0 + O(z^{-2})$ if $|z| \rightarrow \infty$. Therefore, the function $y(z) = e^{-zx} \psi_1(z, x) r(z)$ is holomorphic outside the set $\Delta \cup C$ and in the neighborhood of the infinitely remote point it expands in a series

$$g(z) = 1 + z^{-1} \left(\frac{1}{2} \int_0^x q(t) dt + r_0 \right) + O(z^{-2})$$

hence, by the way, a convenient formula for the potential follows:

$$q(x) = 2\frac{d}{dx} \lim_{z \to \infty} z\left(g(z) - 1\right).$$ (3.9)

For almost all the values $t \in (-\infty, +\infty)$ (according to the Lebesgue measure) the function $g(z)$ has the limiting values

$$g^{\pm}(t) = \lim_{\varepsilon \downarrow 0} g(t \pm \varepsilon) = e^{-tz}\left[r^{\pm}(t)c(it, x) + r^{\pm}(t)n_1^{\pm}(t)s(it, x)\right]$$ (3.10)

satisfying the equality $g^+(t) = \overline{g^-(t)}$.

The following lemma connects the jump of the function $g(z)$ at the point t with its limiting values at the point $-t$.

Lemma 3.2. Almost everywhere, with respect to the Lebesgue measure, on the set $\Delta_a(2) = \Omega_a(2)$ the equality

$$g^+(t) - g^-(t) = e^{-2tz}\left[a(t)g^+(-t) + b(t)g^-(-t)\right]$$ (3.11)

holds, where

$$a(t) = -\overline{b(t)} = -\frac{|r(t)|}{2t|r(-t)|}$$
$$\times \left[\frac{v(t)(|t| - t) + v(-t)(|t| + t)}{v(-t)}\right] e^{i\eta_1(|t|)}$$

and $v(t) = \operatorname{Im} n^+(t) = \pi\sigma'(t)$.

Proof. From (3.10) it follows that (3.11) holds if and only if the functions $a(t)$, $b(t)$ satisfy the equations

$$r^+(t) - r^-(t) = a(t)r^+(-t) + b(t)r^-(-t)$$

$$r^+(t)n_1^+(t) - r^-(t)n_1^-(t) = a(t)r^+(-t)n_1^+(-t) + b(t)r^-(-t)n_1^-(-t).$$

Solving this system we find that

$$a(t) = \frac{r^+(t)[n_1^+(t) - n_1^-(-t)] - r^-(t)[n_1^-(t) - n_1^-(-t)]}{r^+(-t)[n_1^+(t) - n_1^-(-t)]}$$
$$b(t) = \frac{r^-(t)[n_1^-(t) - n_1^+(-t)] - r^+(t)[n_1^+(t) - n_1^+(-t)]}{r^-(-t)[n_1^+(-t) - n_1^-(-t)]}.$$ (3.12)

And since the equality $f^+(t) = \overline{f^-(-t)}$ is true for all the functions in the right-hand parts of these equalities then $a(t) = \overline{b(t)}$. Further, from the identity

$$n_1^+(t) - n_1^-(-t) = -\overline{[n_1^+(t) - n_1^-(-t)]}$$

and from the definition of the argument of $\eta_1(t)$ it follows that

$$n_1^+(t) - n_1^-(-t) = \frac{t}{|t|}\left|n_1^+(t) - n_1^-(-t)\right| \exp\left[\frac{it}{|t|}\eta_1(|t|)\right],$$

and from formulas (3.8) it follows that

$$\frac{r^+(t)}{r^+(-t)} = \frac{|r(t)|}{|r(-t)|} \exp\left[-\frac{it}{|t|}\eta_1(|t|)\right] ,$$

$$\frac{r^-(t)}{r^+(-t)} = \frac{|r(t)|}{|r(-t)|} \exp\left[i\eta_1(|t|)\right] .$$

Inserting these expressions into the right-hand parts of (3.12) we find that

$$a(t) = \frac{|r(t)|}{2iv(-t)|r(-t)|}\left\{\frac{t}{|t|}\left|n_1^+(t) - n_1^-(-t)\right|\right.$$

$$\left. - e^{i\eta_1(|t|)}\left[n_1^+(t) - n_1^-(-t) + n_1^-(t) - n_1^+(t)\right]\right\}$$

$$= \frac{|r(t)|e^{i\eta_1(|t|)}}{2iv(-t)|r(-t)|}$$

$$\times \left\{2iv(t) + \frac{t}{|t|}\left|n_1^+(t) - n_1^-(-t)\right|\left[e^{i\eta_1(|t|)} - e^{i(t/|t|)\eta_1(|t|)}\right]\right\} ,$$

where $v(t) = \operatorname{Im} n_1^+(t)$. Finally, using the equalities

$$\exp\left[-i\eta_1(|t|)\right] - \exp\left[i\frac{t}{|t|}\eta_1(|t|)\right] = -i\left(1 + \frac{t}{|t|}\right)\sin \eta_1(|t|) ,$$

$$v(t) + v(-t) = \operatorname{Im}\left[n_1^+(t) - n_1^-(-t)\right] = \left|n_1^+(t) - n_1^-(-t)\right|\sin \eta_1(|t|)$$

by simple transformations we get the final expression for $a(t)$:

$$a(t) = -\frac{|r(t)|}{|r(-t)|}\left\{\frac{v(t)(|t| - t) + v(-t)(|t| + t)}{2tv(-t)}\right\}e^{i\eta_1(|t|)} .$$

Corollary. Almost everywhere, with respect to Lebesgue measure, on the set $\Delta_a(2) = \Omega_a(2)$ the equality

$$\frac{g^+(-t) + g^-(-t)}{2} = \frac{e^{2tx}[g^+(t) - g^-(t)]}{a(t) + b(t)} - \frac{a(t) - b(t)}{2[a(t) + b(t)]}[g^+(-t) - g^-(-t)]$$

$$\tag{3.13}$$

holds, where

$$a(t) + b(t) = -i\frac{|r(t)|}{|r(-t)|}\left[\frac{v(t)(|t| - t) + v(-t)(|t| + t)}{tv(-t)}\right]\sin \eta_1(|t|)$$

$$a(t) - b(t) = -\frac{|r(t)|}{|r(-t)|}\left[\frac{v(t)(|t| - t) + v(-t)(|t| + t)}{tv(-t)}\right]\cos \eta_1(|t|) .$$

These formulae follow from Lemma 3.2 in the obvious way if (3.11) is transformed into the form

$$g^+(t) - g^-(t) = e^{-2tz} \left\{ \frac{a(t) + b(t)}{2} [g^+(-t) + g^-(-t)] \right.$$
$$\left. + \frac{a(t) - b(t)}{2} [g^+(-t) - g^-(-t)] \right\} .$$

Let us denote by Γ a family of closed nonintersecting contours running anticlockwise, which surround the segments $\Delta_j(1)$, $|\Delta_i(a)|$, $-|\Delta_i(a)|$, $|\Delta_l(s)|$, $-|\Delta_l(s)|$ and the points $c_p \in C$. In the domain outside the contours, the function $g(z)$ is holomorphic and tends to 1 as $|z| \to \infty$, which, with the help of the Cauchy integral formula, permits us to represent it in the form

$$g(z) = 1 - \frac{1}{2\pi i} \int_\Gamma \frac{g(\xi)}{\xi - z} d\xi \qquad (3.14)$$

where z is an arbitrary point of the domain outside the contours. In order to calculate the integral in the right-hand part of this equality, it is convenient to split the family Γ into four parts containing the contours which surround, respectively, the set $\Delta(1) \cup C$ [part $\Gamma(1)$], the sets $|\Delta_s(2)|$, $-|\Delta_s(2)|$, [parts $\Gamma_s^+(2)$, $\Gamma_s^-(2)$] and the set $\Delta_a(2)$ [part $\Gamma_a(2)$]. From properties 1 and 2 of the functions $r(z)$, $r(-z)$ it follows that all the contours can be pulled so closely to the sets which they surround, that the function $r(z)$ will be holomorphic inside the contours from the family $\Gamma_s^-(2)$ and the functions $r(-z)$, $r(-z)^{-1}$ will be holomorphic inside the contours belonging to the families $\Gamma(1)$ and $\Gamma_s^+(2)$. It should also be noted that, since the set $\Delta(1) \cup C$ lies in the domain of the holomorphy of the function $n_1(-z)$, it will also be holomorphic inside the contours of the family $\Gamma(1)$.

We now demonstrate the calculation of the integral along the contours of the family $\Gamma(1)$.

Because the functions $c(i\xi, x)$, $s(i\xi, x)$ are even, it follows that

$$g(\xi) = e^{-2\xi z} \frac{r(\xi)}{r(-\xi)} e^{\xi x} \{r(-\xi)c(-i\xi, x) + r(-\xi)n_1(-\xi)s(-i\xi, x)$$
$$+ r(-\xi)[n_1(\xi) - n_1(-\xi)]s(i\xi, x)\}$$
$$= \frac{e^{-2\xi x} g(-\xi)}{r(-\xi)^2} r(\xi)r(-\xi) + \frac{e^{-\xi x} s(i\xi, x)}{r(-\xi)} r(\xi)r(-\xi)[n_1(\xi) - n_1(-\xi)]$$

and since

$$r(\xi)r(-\xi)[n_1(\xi) - n_1(-\xi)] = 2\xi ,$$

then

$$g(\xi) = \frac{2\xi e^{-2\xi x} g(-\xi)}{r(-\xi)^2} \frac{1}{n_1(\xi) - n_1(-\xi)} + \frac{2\xi e^{-\xi x} s(i\xi, x)}{r(-\xi)} .$$

From the previous considerations it can be seen that the function

$$-\frac{1}{n_1(\xi) - n_1(-\xi)} = M_{22}(\xi) = \int_{-\infty}^\infty \frac{d\varrho_{22}(t)}{t - \xi}$$

is real in the intervals complementary to the set $\Delta \cup |\Delta| = \Delta \cup (-\Delta)$ and has poles at the points $\pm c_p$ ($c_p \in C$) and also at the point $c_0 = 0$. Therefore, the support of the measure $d\varrho_{22}(t)$ lies in the set $\Delta \cup (-\Delta) \cup C \cup (-C) \cup \{0\}$ and

$$-\frac{1}{n_1(\xi) - n_1(-\xi)} = \int_{\Delta(1) \cup C} \frac{d\varrho_{22}(t)}{t - \xi} + f(\xi)$$

where $f(\xi)$ is a holomorphic function on the $\Delta(1) \cup C$ and in its neighborhood. The functions $r(-\xi)^{-1}$, $n_1(-\xi)$ and, consequently, $g(-\xi)$ are also holomorphic in this domain. Therefore,

$$\frac{g(\xi)}{\xi - z} = \frac{2\xi[e^{-\xi x} s(i\xi, x) r(-\xi) - e^{-2\xi x} g(-\xi) f(\xi)]}{r^2(-\xi)(\xi - z)}$$

$$- \frac{2\xi e^{-2\xi x} g(-\xi)}{r^2(-\xi)(\xi - z)} \int_{\Delta(1) \cup C} \frac{d\varrho_{22}(t)}{t - \xi} ,$$

and the first term in the right-hand side and the coefficient of the integral are holomorphic in the domain which is surrounded by the contours $\Gamma(1)$. Therefore,

$$\frac{1}{2\pi i} \int_{\Gamma(1)} \frac{g(\xi)}{\xi - z} d\xi = -\frac{1}{2\pi i} \int_{\Gamma(1)} \frac{2\xi e^{-2\xi x} g(-\xi)}{r(-\xi)^2(\xi - z)} \int_{\Delta(1) \cup C} \frac{d\varrho_{22}(t)}{t - \xi} d\xi$$

$$= \int_{\Delta(1) \cup C} \left\{ \frac{1}{2\pi i} \int_{\Gamma(1)} \frac{2\xi e^{-2\xi x} g(-\xi)}{r(-\xi)^2(\xi - z)(\xi - t)} d\xi \right\} d\varrho_{22}(t)$$

and from the residue theorem

$$\frac{1}{2\pi i} \int_{\Gamma(1)} \frac{g(\xi)}{\xi - z} d\xi = \int_{\Delta(1) \cup C} \frac{2te^{-2tz} g(-t)}{r(-t)^2(t - z)} d\varrho_{22}(t) . \tag{3.15}$$

We now show the calculation of the integral along the contours of the family $\Gamma_s^+(2)$. Using the equalities

$$g(\xi) = \frac{e^{-\xi x}}{r(-\xi)} [r(\xi) r(-\xi) c (i\xi, x) + r(\xi) r(-\xi) n_1(\xi) s(i\xi, x)]$$

$$= \frac{2\xi e^{-\xi x}}{r(-\xi)} \left[\frac{c(i\xi, x)}{n_1(\xi) - n_1(-\xi)} + \frac{n_1(\xi) s(i\xi, x)}{n_1(\xi) - n_1(-\xi)} \right]$$

$$\frac{1}{n_1(\xi) - n_1(-\xi)} = -\int_{-\infty}^{\infty} \frac{d\varrho_{22}(t)}{t - \xi} , \quad \frac{n_1(\xi)}{n_1(\xi) - n_1(-\xi)} = \frac{1}{2} \int_{\infty}^{\infty} \frac{d\varrho_{12}(t)}{t - \xi}$$

and the holomorphy of the function $r(-\xi)^{-1}$ inside the contours of the family $\Gamma_s^+(2)$, we find by analogy with the foregoing that

$$\frac{1}{2\pi i} \int_{\Gamma_s^+(2)} \frac{g(\xi)}{\xi - z} d\xi$$

$$= \int_{|\Delta_s(2)|} \frac{2te^{-tz}}{r(-t)(t - z)} [c(it, x) d\varrho_{22}(t) - s(it, x) d\varrho_{12}(t)]$$

$$= \int_{|\Delta_s(2)|} \frac{2te^{-tz}}{r(-t)(t - z)} [c(it, x) \delta_{22}(t) - s(it, x) \delta_{12}(t)] d\varrho(t) \tag{3.16}$$

where

$$\delta_{ij}(t) = \frac{d\varrho_{ij}(t)}{d\varrho(t)} \ , \quad d\varrho(t) = d\left(\varrho_{11}(t) + \varrho_{22}(t)\right) \ .$$

Now we note that on the set $\Delta_s(2)$ the measure $d\sigma(\xi)$ is purely singular. This makes it possible to use the Katz theorem [2] which states that almost everywhere according to the measure $d\varrho(t)$ on the set $\Delta_s(2)$ either $\delta_{22}(t) = \delta_{12}(t) = 0$ or there exists a finite real limit $-n_1^+(-t) = \lim_{\varepsilon \downarrow 0} -n_1(-t + i\varepsilon)$ and

$$\delta_{12}(t) = \delta_{22}(t)\left[-n_1^+(-t)\right] = -n_1^+(-t)\delta_{22}(t) \ .$$

Consequently, almost everywhere according to the measure $d\varrho(t)$ on the set $|\Delta_s(2)|$ the equality

$$c(it, x)\delta_{22}(t) - s(it, x)\delta_{12}(t)$$
$$= \frac{e^{-tx}}{r(-t)} \left\{e^{tx}\left[c(it, x)r(-t) + r(-t)n_1^+(-t)s(it, x)\right]\right\} \delta_{22}(t)$$
$$= \frac{e^{-tx}}{r(-t)} g(-t)\delta_{22}(t)$$

holds, from which, according to (3.16), it follows that

$$\frac{1}{2\pi i} \int_{\Gamma_s^+(2)} \frac{g(\xi)}{\xi - z} d\xi = \int_{|\Delta_s(2)|} \frac{2te^{-tx}}{r^2(-t)(t - z)} g(-t) d\varrho_{22}(t) \qquad (3.17)$$

where

$$g(-t) = \tfrac{1}{2} \lim_{\varepsilon \downarrow 0} \left[g(-t + i\varepsilon) + g(-t - i\varepsilon)\right] \qquad (3.18)$$

and, besides, this limit exists almost for all $t \in |\Delta_s(2)|$ and according to the measure $d\varrho_{22}(t)$.

We now demonstrate the calculation of the integral along the contours of the family $\Gamma_s^-(2)$.

From (3.1) it follows that

$$n_1(\xi) = \int_{-|\Delta_s(2)|} \frac{d\sigma(t)}{t - \xi} + f(\xi)$$

where $f(\xi)$ is the function holomorphic on the set $-|\Delta_s(2)|$ and in its neighborhood. Therefore,

$$\frac{g(\xi)}{\xi - z} = \frac{e^{-\xi x}r(\xi)[c(i\xi, x) + f(\xi)s(i\xi, x)]}{\xi - x} + \frac{e^{-\xi x}r(\xi)s(i\xi, x)}{\xi - z} \int_{-|\Delta_s(2)|} \frac{d\sigma(t)}{t - \xi} \ .$$

The first term in the right-hand side and the factor multiplying the integral are holomorphic in the domain which is surrounded by the contours of the family $\Gamma_s^-(2)$. Hence, by analogy with the foregoing, we find that

$$\frac{1}{2\pi i}\int_{\Gamma_\varepsilon^-(2)}\frac{g(\xi)}{\xi-z}d\xi = \int_{-|\Delta_s(2)|}\frac{e^{-tz}r(t)s(it,x)}{t-z}d\sigma(t) . \tag{3.19}$$

Now we shall find the value of the limit (3.18) for almost all points of the set $-|\Delta_s(2)|$ with respect to the measure $d\sigma(t)$. Since the measure $d\sigma(t)$ is singular on this set, so is σ, almost everywhere that $d\sigma(t)/dt = +\infty$ and, consequently, also σ-almost everywhere $\lim_{\varepsilon\downarrow 0}\mathrm{Im}\, n_1(t+i\varepsilon) = +\infty$. Therefore, it is sufficient to find the limit (3.18) on the subset $N \subset |\Delta_s(2)|$ of those points t where

$$\lim_{\varepsilon\downarrow 0}\mathrm{Im}\, n_1(t+i\varepsilon) = +\infty .$$

From the inequality $\mathrm{Im}\,\{n_1(t+i\varepsilon) - n_1(-t-i\varepsilon)\} > \mathrm{Im}\, n_1(t+i\varepsilon)$, the identity

$$r(-t-i\varepsilon) = \frac{2(t+i\varepsilon)}{r(t+i\varepsilon)[n_1(t+i\varepsilon) - n_1(-t-i\varepsilon)]}$$

and the holomorphy of the function $r(z)^{-1}$ on the set $-|\Delta_s(2)|$ it follows that for $t \in N$, $\lim_{\varepsilon\downarrow 0} r(-t-i\varepsilon) = 0$ and consequently

$$\lim_{\varepsilon\downarrow 0} g(-t-i\varepsilon) = \frac{2te^{tz}s(it,x)}{r(t)}\lim_{\varepsilon\downarrow 0}\frac{n_1(-t-i\varepsilon)}{n_1(t+i\varepsilon) - n_1(-t-i\varepsilon)} \tag{3.20}$$

if the limit on the right-hand side exists.

Let us introduce an auxiliary measure $d\tilde\sigma(t) = d\sigma(t) + d(-\sigma(-t))$. The measure $d\sigma(t)$ is, obviously, absolutely continuous with respect to $d\tilde\sigma(t)$ so that

$$d\sigma(t) = \delta(t)d\tilde\sigma(t) , \quad \delta(t) + \delta(-t) = 1 , \quad 0 \leq \delta(t) \leq 1 .$$

We shall assume the following condition to be true:

B) On the set $\Delta(2)$ the function $\delta(t)$ is different from zero and satisfies the Lipshitz condition: $|\delta(t+h) - \delta(t)| \leq c|h|$.

We note that due to the compactness of the set $\Delta(2)$

$$0 < \inf_{t\in\Delta(2)} \delta(t) \leq \sup_{t\in\Delta(2)} \delta(t) < 1 .$$

Besides, the Lipshitz condition obviously guarantees the boundedness (for $\varepsilon\downarrow 0$) of the integrals

$$\int\frac{\delta(\xi) - \delta(t)}{\xi - t - i\varepsilon}d\tilde\sigma(\xi) , \quad \int\frac{\delta(\xi) - \delta(-t)}{\xi + t + i\varepsilon}d\tilde\sigma(\xi) . \tag{3.21}$$

Further, we have

$$n_1(t+i\varepsilon) = t + i\varepsilon + \int\frac{\delta(\xi)}{\xi - t - i\varepsilon}d\tilde\sigma(\xi)$$

$$= \delta(t)\int\frac{d\tilde\sigma(\xi)}{\xi - t - i\varepsilon} + \left[t + i\varepsilon + \int\frac{\delta(\xi) - \delta(t)}{\xi - t - i\varepsilon}d\tilde\sigma(\xi)\right]$$

$$n_1(-t - i\varepsilon) = \delta(-t) \int \frac{d\tilde{\sigma}(\xi)}{\xi + t + i\varepsilon} + \left[-t - i\varepsilon + \int \frac{\delta(\xi) - \delta(-t)}{\xi + t + i\varepsilon} d\tilde{\sigma}(\xi) \right]$$

and, since the measure $d\tilde{\sigma}(\xi)$ is even

$$\int \frac{d\tilde{\sigma}(\xi)}{\xi + t + i\varepsilon} = -\int \frac{d\tilde{\sigma}(\xi)}{-\xi + t + i\varepsilon} = -\int \frac{d\tilde{\sigma}(\xi)}{\xi - t - i\varepsilon}$$

and integrals (3.21) are bounded,

$$n_1(t + i\varepsilon)\delta(-t) + n_1(-t - i\varepsilon)\delta(t) = O(1) \qquad (\varepsilon \downarrow 0) .$$

Hence, it follows that for $t \in N$ there exists a limit

$$\lim_{\varepsilon \downarrow 0} \frac{n_1(-t - i\varepsilon)}{n_1(t + i\varepsilon)} = -\frac{\delta(-t)}{\delta(t)} .$$

Therefore, for $t \in N$ the limit (3.18) also exists and

$$\lim_{\varepsilon \downarrow 0} g(-t - i\varepsilon) = -\frac{2te^{tx} s(it, x)}{r(t)} \delta(-t) ,$$

so that

$$s(it, x) = -\frac{r(t)e^{-tx} g(-t)}{2t\delta(-t)}$$

almost everywhere on the set $-|\Delta_s(2)|$ with respect to the measure $d\sigma(t)$. Substituting this expression into the right-hand side of (3.19) we finally get

$$\frac{1}{2\pi i} \int_{\Gamma_s^-(2)} \frac{g(\xi)}{\xi - z} d\xi = \int_{-|\Delta_s(2)|} \frac{e^{-2tx} r^2(t)g(-t)}{(t - z)2t\delta(-t)} d\sigma(t) . \tag{3.22}$$

Here we show the calculation of the integral along the contours of the family $\Gamma_s(2)$.

The proof of (3.16), (3.19) used only the holomorphy of the functions $r(\xi)$ and $r(-\xi)^{-1}$ on the sets $-|\Delta_s(2)|$ and $|\Delta_s(2)|$, respectively. Since these properties of the functions $r(\xi)$, $r(-\xi)^{-1}$ are true also on the sets $-|\Delta_a(2)|$, $|\Delta_a(2)|$, so (3.16), (3.9) (with the substitution of $|\Delta_s(2)|$ for $|\Delta_a(2)|$) also hold for contours surrounding the sets $-|\Delta_a(2)|$, $|\Delta_a(2)|$ i.e.

$$\frac{1}{2\pi i} \int_{\Gamma_a^-(2)} \frac{g(\xi)}{\xi - z} d\xi = -\int_{-|\Delta_a(2)|} \frac{e^{-tx} r(t)s(it, x)}{t - z} d\sigma(t)$$

$$+ \int_{|\Delta_a(2)|} \frac{2te^{-tx}}{(t - z)r(-t)} [c(it, x)d\varrho_{22}(t) - s(it, x)d\varrho_{12}(t)] . \tag{3.23}$$

In addition to the restrictions obtained above on the measure $d\sigma(\xi)$ we shall assume one more condition to be true.

C) Functions $\varrho_{22}(t)$, $\varrho_{12}(t)$ are absolutely continuous on the set $\Delta_a(2) = \Omega_a(2)$.

We shall not discuss here the character of the imposed restrictions A, B, C and limit ourselves only to the following remark: in the space of measures with strong topology, dist $(d\sigma_1(\xi), d\sigma_2(\xi)) = \text{var}\,[\sigma_1(\xi) - \sigma_2(\xi)]$, measures satisfying the imposed restrictions form a dense set. For us it is relevant that in this case the contours $\Gamma_a(2)$ can be pulled to the twice repeated segments from the set $\Delta_a(2)$. Indeed, if condition C is fulfilled, then for $t \in \Delta_a(2)$

$$d\sigma(t) = \sigma'(t)dt \; ; \quad d\varrho_{22}(t) = \varrho'_{22}(t)dt \; ; \quad d\varrho_{12}(t) = \varrho'_{12}(t)dt \; .$$

On the other hand, from the holomorphy of the function $r(\xi)$ on the set $-|\Delta_a(2)|$ it follows that for $t \in |\Delta_a(2)|$

$$g^+(t) - g^-(t) = e^{-tz} r(t) s(it, x) \left[n_1^+(t) - n_1^-(t)\right] = 2\pi i e^{-tz} r(t) s(it, x) \sigma'(t)$$

and from the equalities

$$r(\xi) = 2\xi r(-\xi)^{-1} \left[n_1(\xi) - n_1(-\xi)\right]^{-1} = -2\xi r(-\xi)^{-1} M_{22}(\xi)$$

$$r(\xi)n_1(\xi) = -2\xi r(-\xi)^{-1} M_{22}(\xi) n_1(\xi) = 2\xi r(-\xi)^{-1} \left[0.5 + M_{12}(\xi)\right]$$

and the holomorphy of the function $r(-\xi)^{-1}$ on the set $|\Delta_a(2)|$ it follows that for $t \in |\Delta_a(2)|$

$$g^+(t) - g^-(t)$$
$$= -2tr(-t)^{-1}e^{-tz} \left\{c(it, x)\left[M_{22}^+(t) - M_{22}^-(t)\right]\right.$$
$$\left. - s(it, x)\left[M_{12}^+(t) - M_{12}^-(t)\right]\right\}$$
$$= -2\pi i 2tr(-t)^{-1}e^{-tz} \left[c(it, x)\varrho'_{22}(t) - s(it, x)\varrho'_{12}(t)\right] \; .$$

Hence, according to (3.23)

$$\frac{1}{2\pi i} \int_{\Gamma_a(2)} \frac{g(\xi)}{\xi - z} d\xi = -\frac{1}{2\pi i} \int_{\Delta_a(2)} \frac{g^+(t) - g^-(t)}{t - z} dt \; . \tag{3.24}$$

We shall introduce a new function and a measure assuming

$$y(t) = \begin{cases} \dfrac{1}{2}\left[g^+(-t) + g^-(-t)\right]e^{-2tz} & t \in R^1/\Omega_a(2) \\[2mm] \dfrac{1}{t}\left[g^+(t) - g^-(t)\right] & t \in \Omega_a(2) \end{cases}$$

$$d\nu(t) = \begin{cases} \dfrac{2}{r^2(-t)}d\varrho_{22}(t), & t \in \Delta(1) \cup |\Delta_s(2)| \cup C \\[2mm] \dfrac{r^2(t)}{2t^2\delta(-t)}d\sigma(t), & t \in -|\Delta_s(2)| \\[2mm] \dfrac{1}{2\pi}dt, & t \in \Omega_a(2) \; . \end{cases}$$

From (3.15), (3.17), (3.22), (3.24) and (3.14) it follows that

$$g(z) = 1 - \int \frac{y(\xi)\xi}{\xi - z} d\nu(\xi) \tag{3.25}$$

and the limit

$$\tfrac{1}{2} \left[g^+(-t) + g^-(-t) \right] = \lim_{\varepsilon \downarrow 0} \tfrac{1}{2} \left[g^+(-t + i\varepsilon) + g^-(-t - i\varepsilon) \right]$$

exists almost everywhere with respect to the measure $d\nu(t)$. Therefore, almost everywhere there exists a limit

$$\lim_{\varepsilon \downarrow 0} \int \left(\frac{1}{\xi + t - i\varepsilon} + \frac{1}{\xi + t + i\varepsilon} \right) y(\xi)\xi \, d\xi = 1 - \int \frac{y(\xi)\xi}{\xi + t} d\nu(\xi)$$

and the equality

$$\frac{1}{2} \left[g^+(-t) + g^-(-t) \right] = 1 - \int \frac{y(\xi)\xi}{\xi + t} d\nu(\xi) \tag{3.26}$$

holds. Further, from the definition of the function $y(t)$ and (3.13) it follows that

$$\tfrac{1}{2} \left[g^+(-t) + g^-(-t) \right] = \begin{cases} e^{2tx} y(t) & t \notin \Omega_a(2) \\ \dfrac{te^{2tx} y(t)}{i[a(t) + b(t)]} + \dfrac{t[a(t) - b(t)]}{2i[a(t) + b(t)]} y(-t) & t \in \Omega_a(2) \end{cases} .$$

Therefore (3.26) generates the following integral equation for the function $y(\xi)$:

$$y(\xi) + e^{-2\xi x} p(\xi) \left[m(\xi)y(-\xi) + \int \frac{y(\xi')}{\xi' + \xi} \xi' d\nu(\xi') - 1 \right] = 0 \tag{3.27}$$

where

$$m(t) = \chi_a(t) \frac{t[a(t) - b(t)]}{2i[a(t) + b(t)]} = -\frac{t}{2} \chi_a(t) \operatorname{ctg} \eta_1 \left(|t| \right) ,$$

$$p(t) = [1 - \chi_a(t)] + \chi_a(t) \frac{i[a(t) + b(t)]}{t}$$

$$= [1 - \chi_a(t)] + \chi_a(t) \frac{|r(t)|}{|r(-t)|} \frac{[v(t)(|t| - t) + v(-t)(|t| + t)]}{t^2 v(-t)} \sin \eta_1 \left(|t| \right)$$

and $\chi_a(t)$ is a characteristic function of the set $\Omega_a(2)$. Thus, to solve the inverse problem one should find functions $p(t)$, $m(t)$ and the measure $d\nu(t)$ from the given Weyl function $n(\lambda)$ and solve integral equations (3.27). According to (3.9, 25), the potential is restored by

$$q(x) = 2 \frac{d}{dx} \int y(\xi)\xi \, d\nu(\xi) .$$

The integral equation (3.27) differs from (0.1), to which the finding of solutions of the KdV equation was reduced in [4], only by the coefficient: in (0.1), $\exp[-2\xi(x - 4\xi^2 t)]$ is present instead of $\exp(-2\xi x)$. Therefore, to solve the Cauchy problem (2.14), one should:

1) Starting from the initial data $q(x) = u(x,0)$ find the corresponding Weyl function $n(z)$;
2) using the above described method carry out the factorization which allows one to find the functions $p(\xi)$, $m(\xi)$ and the measure $d\nu(\xi)$;
3) solve the integral equations

$$z(\xi) + e^{-2\xi(x-4\xi^2 t)} p(\xi) \left[m(\xi)z(-\xi) + \int \frac{z(\xi')}{\xi' + \xi} \xi' d\nu(\xi') - 1 \right] = 0 .$$

$$(3.28)$$

The solution $u(x,t)$ of the Cauchy problem is expressed in terms of the solutions $z(\xi; x, t)$ of (3.28) in the following way:

$$u(x,t) = 2\frac{d}{dx} \int z(\xi; x, t)\xi d\nu(\xi) .$$

The unique solvability of (3.28) in the space $L_2(d\nu)$ for all real values x, t is established in [4].

Remark. Let us explain the role of the factor $r(z)$. If we omit it, then the function

$$g_1(z) = g_1(z, x) = e^{-zx}\psi_1(z, x) = e^{-zx} [c(iz, x) + n_1(z)s(iz, x)] \qquad (3.29)$$

as is easily seen, will also satisfy (3.9, 14). Since

$$\frac{g_1(\xi)}{\xi - z} = \frac{e^{-\xi x}[c(i\xi, x) + \xi s(i\xi, x)]}{\xi - z} + \frac{e^{-\xi x} s(i\xi, x)}{\xi - x} \int \frac{d\sigma(t)}{t - \xi} ,$$

and all the functions in the right-hand side of this equality, except the function $\int d\sigma(t)/(t - \xi)$, are holomorphic inside the contour Γ, then, according to the residue theorem,

$$g_1(z) = 1 - \frac{1}{2\pi i} \int_\Gamma \frac{g_1(\xi)}{\xi - z} d\xi = 1 - \frac{1}{2\pi i} \int_\Gamma \frac{e^{-\xi x} s(i\xi, x)}{\xi - z} \int \frac{d\sigma(t)}{t - \xi} d\xi$$

$$= 1 + \int \left[\frac{1}{2\pi i} \int_\Gamma \frac{e^{-\xi x} s(i\xi, x)}{(\xi - z)(\xi - t)} d\xi \right] d\sigma(t) = 1 + \int \frac{e^{-tz} s(it, x)}{t - z} d\sigma(t) .$$

Hence, according to (3.29), (3.9) it follows that

$$c(iz, x) + zs(iz, x) = e^{zx} \left[1 + \int \frac{e^{-tz} s(it, x) - e^{-zx} s(iz, x)}{t - z} d\sigma(t) \right] ,$$

$$q(x) = -2\frac{d}{dx} \int e^{-tx} s(it, x)d\sigma(t) .$$

The evenness of functions $c(it, x)$, $s(it, x)$ allows us to eliminate $c(it, x)$. As a result of the elimination, we get the following equation for the function $y(z) = e^{-zx} s(iz, x)$:

$$e^{zx} \left[a(z)y(z) - \frac{1}{2z} \int \frac{y(t) - y(z)}{t - z} d\sigma(t) \right]$$

$$- e^{-zx} \left\{ [a(z) - 1] y(-z) - \frac{1}{2z} \int \frac{y(t) - y(-z)}{t + z} d\sigma(t) \right\}$$

$$= (2z)^{-1} \left(e^{zx} - e^{-zx} \right) ,$$

where $a(z)$ is an arbitrary function.

We again obtain an equation which differs from the general equations in [4] only by a factor. Therefore, if the equation

$$e^{z(x - 4z^2 t)} \left\{ a(z)y(z) - \frac{1}{2z} \left[\int \frac{y(z') - y(z)}{z' - z} d\sigma(z') - 1 \right] \right\}$$

$$= e^{-z(x - 4z^2 t)} \left\{ [a(z) - 1] y(-z) - \frac{1}{2z} \left[\int \frac{y(z') - y(-z)}{z' + z} d\sigma(z') - 1 \right] \right\}$$

$$\tag{3.30}$$

is uniquely solvable with respect to $y(z)$, then the function

$$u(x, t) = -2 \frac{d}{dx} \int y(z) d\sigma(z)$$

satisfies the KdV equation and the initial data $u(x, 0) = q(x)$. Thus, the solution of the Cauchy problem (2.14) can be obtained by solving (3.30) with the function $a(z)$, that ensures its unique solvability in the space. But it is not known whether it is possible to obtain it for a wide enough class of measures $d\sigma(\xi)$.

The proof of the unique solvability of the corresponding integral equations is the most important part of inverse problems. The factor $r(z)$ was introduced to obtain well-defined and uniquely solvable integral equations.

References

1   V. Bargman: "On the connection between phase shifts and scattering potential," Rev. Mod. Phys., 30–45 (1949)

2   I.S. Katz: "The spectrum of the differential operator," Izv. Akad. Nauk. SSSR Ser. Mat., 27, 1081–1112 (1963)

3   D.S. Lundina: "Compactness of the set of reflectionless potentials," Teoriya funktsyi, funktsyonal'nyi analiz i ikh prilozheniya, 44, 57–66 (1985)

4   V.A. Marchenko: Nonlinear Equations and Operator Algebras (D. Reidel, Dordrecht 1987)

Subject Index

B.N. Zakhariev, A.A. Suzko

Direct and Inverse Problems

Potentials in Quantum Scattering

1990. XIV, 223 pp. 42 figs. Softcover DM 48,– ISBN 3-540-52484-3

This textbook can almost be viewed as a "how-to" manual for solving quantum inverse problems, that is, for deriving the potential from spectra and/or scattering data. The formal exposition of inverse methods is paralleled by a discussion of the direct problem.

In part differential and finite-difference equations are presented side by side. A variety of solution methods is presented. Their common features and (dis)advantages are analyzed. To foster a better understanding, the physical meaning of the mathematical quantities are discussed in detail. Wave confinement in continuum bound states, resonance and collective tunneling, and the spectral and phase equivalence of various interactions are some of the physical problems covered.

A.G. Sitenko

Scattering Theory

1990. Approx. 300 pp. 32 figs. (Springer Series in Nuclear and Particle Physics) Hardcover DM 88,– ISBN 3-540-51953-X

This mathematically rigorous introduction to nonrelativistic scattering theory addresses upper level undergraduates in physics. The relationship between the scattering matrix and physical observables is discussed in detail. Among the emphasized topics are the stationary formulation of the scattering problem, the inverse scattering problem, dispersion relations, three-particle bound states and their scattering, collisions of particles with spin and polarization phenomena. The analytical properties of the scattering matrix are discussed. Problems are included to help the reader to gain some experience and more expertise in scattering theory.

Springer-Verlag

Berlin
Heidelberg
New York
London
Paris
Tokyo
Hong Kong
Barcelona

D. Park, Williams College, Williamstown, MA

Classical Dynamics and Its Quantum Analogues

2nd enl. and updated ed. 1990. IX, 333 pp. 101 figs. Hardcover DM 78,–
ISBN 3-540-51398-1

The primary purpose of this textbook is to introduce students to the principles of classical dynamics of particles, rigid bodies, and continuous systems while showing their relevance to subjects of contemporary interest. Two of these subjects are quantum mechanics and general relativity. The book shows in many examples the relations between quantum and classical mechanics and uses classical methods to derive most of the observational tests of general relativity. A third area of current interest is in nonlinear systems, and there are discussions of instability and of the geometrical methods used to study chaotic behaviour. In the belief that it is most important at this stage of a student's education to develop clear conceptual understanding, the mathematics is for the most part kept rather simple and traditional.

This book devotes some space to important transitions in dynamics: the development of analytical methods in the 18th century and the invention of quantum mechanics.

A. Hasegawa, AT & T Bell Laboratories, Murray Hill, NJ

Optical Solitons in Fibers

2nd enl. ed. 1990. XII, 79 pp. 25 figs. Softcover DM 48,– ISBN 3-540-51747-2

Already after six months high demand made a new edition of this textbook necessary. The most recent developments associated with two topical and very important theoretical and practical subjects are combined: **Solitons** as analytical solutions of nonlinear partial differential equations and as lossless signals in dielectric **fibers.** The practical implications point towards technological advances allowing for an economic and undistorted propagation of signals revolutionizing telecommunications. Starting from an elementary level readily accessible to undergraduates, this pioneer in the field provides a clear and up-to-date exposition of the prominent aspects of the theoretical background and most recent experimental results in this new and rapidly evolving branch of science. This well-written book makes not just easy reading for the researcher but also for the interested physicist, mathematician, and engineer. It is well suited for undergraduate or graduate lecture courses.

Springer-Verlag

Berlin
Heidelberg
New York
London
Paris
Tokyo
Hong Kong
Barcelona